NAYFEH—Perturbation Methods
NAYFEH and MOOK—Nonlinear Oscillations
ODEN and REDDY—An Introduction to the Mathematical Theory of Finite Elements
PAGE—Topological Uniform Structures
PASSMAN—The Algebraic Structure of Group Rings
PRENTER—Splines and Variational Methods
RIBENBOIM—Algebraic Numbers
RICHTMYER and MORTON—Difference Methods for Initial-Value Problems, 2nd Edition
RIVLIN—The Chebyshev Polynomials
RUDIN—Fourier Analysis on Groups
SAMELSON—An Introduction to Linear Algebra
SIEGEL—Topics in Complex Function Theory
Volume I—Elliptic Functions and Uniformization Theory
Volume 2—Automorphic Functions and Abelian Integrals
Volume 3—Abelian Functions and Modular Functions of Several Variables
STAKGOLD—Green's Functions and Boundary Value Problems
STOKER—Differential Geometry
STOKER—Nonlinear Vibrations in Mechanical and Electrical Systems
STOKER—Water Waves
WHITHAM—Linear and Nonlinear Waves
WOUK—A Course of Applied Functional Analysis

GEOMETRY AND CONVEXITY

GEOMETRY AND CONVEXITY

A Study in Mathematical Methods

PAUL J. KELLY
University of California, Santa Barbara

MAX L. WEISS
University of California, Santa Barbara

A WILEY-INTERSCIENCE PUBLICATION

JOHN WILEY & SONS New York · Chichester · Brisbane · Toronto

Library of Congress Cataloging in Publication Data

Kelly, Paul Joseph.
Geometry and convexity.

(Pure and applied mathematics)
"A Wiley-Interscience publication."
Includes index.
1. Geometry, Modern. 2. Convex bodies.
3. Convex surfaces. I. Weiss, Max L., 1933 –
joint author. II. Title.

QA473.K44 516′.04 78-21919
ISBN 0-471-04637-X

Printed in the United States of America

10 9 8 7 6 5 4 3 2 1

PREFACE

This book was written because we believe that the subject matter of basic convex body theory is ideally suited to give pregraduate students an unusual and valuable mathematical experience. The geometric theme is apparently a simple one, namely, to identify the convex bodies and surfaces of euclidean n-space and to establish some of their basic properties. This program, from the outset, involves the student with viewpoints and methods that both generalize and unify much of his previous background. It is this eclectic aspect of the material that we feel is especially valuable. Thus, although the setting and the central questions of the book are geometric, what is stressed throughout—as the subtitle indicates—is the natural way in which the development involves the interplay of concepts and methods from topology, analysis, and linear and affine algebra.

To sketch the logic of the development, the subject matter of the first two chapters was dictated largely by our central theme and judgments about the student's background. To make the definition of n-dimensional convex bodies and surfaces intelligible requires topological ideas. We do not assume that the student has had formal work in topology, and Chapter 1 establishes the basic metric topology that is needed. Here, as in many other parts of the book, we were conscious of the "half-way house" character of the material. That is, what appears to the student at this stage as a new level of abstraction will later be seen as a rather concrete illustration of a much more general point of view. Wherever possible, we sought formulations that would aid the later transition.

Chapter 2 is concerned with the n-dimensional setting of the text. The vector character of euclidean n-space is reviewed briefly to recall some facts familiar from linear algebra, and these sections could be simply assigned reading. The rest of the chapter develops the affine character of the space and the analogy between linear and affine concepts. The affine point of view is new to most students and widens their perspective of both Euclidean geometry and linear algebra.

In Chapters 3 and 4, the background of the first two chapters is applied to the central theme of the book. The n-dimensional convex bodies and surfaces of euclidean n-space are identified and a number of their basic properties are established. Many of the results are so intuitively obvious that their derivation has little appeal to the experienced geometer. It has been our experience, however, that students find satisfaction, even some excitement, in using analytic methods to formulate n-dimensional analogs of familiar geometric concepts and to verify anticipated geometric relations.

The last two chapters are intended to illustrate the application of the background developed to a number of nonintuitive results and to give the student a better appreciation of the fertility of ideas and the wide range of problems in convex body theory. The opportunities are so extensive that our choices were necessarily somewhat arbitrary. The selections in Chapter 5 were guided by two main considerations, the accessibility of the results to the methods of the earlier chapters, and the general rather than geometric importance of the results. In Chapter 6, we want to give the student the intriguing generalization of a space in which convex bodies themselves appear as "points" and to show the application of this concept to some open research questions.

To conclude, some comment should be made about omissions that we regretfully found necessary. Interesting problems about arc length, area, and volume are not considered, since consistent rigor would have involved too much background preparation. We would have liked to do more with the notion of transformation groups. We also wish that the book had more of the charm that is characteristic of ingenious, ad hoc solutions to geometric problems. Nonetheless, we believe that the material represents a valuable bridge between a student's lower division past and his potential graduate future. We also hope that it makes some contribution toward a higher regard for the value of geometry, not in isolation, but in combination with mathematics of contemporary interest.

PAUL KELLY
MAX WEISS

Santa Barbara, California
November 1978

CONTENTS

NOTATIONS

A, B, C: points (n-tuples in $\mathscr{E}^n$).
a, b, c: vectors (n-tuples in $\mathscr{E}^n$).
$\varnothing$: the null set.
$\mathrm{Cp}(\mathscr{S})$: complement of set $\mathscr{S}$.
$\mathrm{B}(A, \delta)$: the ball with center A and radius δ.
$\mathrm{S}(A, \delta)$: the sphere with center A and radius δ.
$\mathrm{N}(A, \delta)$: the δ neighborhood of point A.
$\mathrm{N}_o(A, \delta)$: the deleted δ-neighborhood of point A.
$\mathrm{In}(\mathscr{S})$: the interior of set $\mathscr{S}$.
$\mathrm{Ex}(\mathscr{S})$: the exterior to set $\mathscr{S}$.
$\mathrm{Bd}(\mathscr{S})$: the boundary to set $\mathscr{S}$.
$\mathrm{Is}(\mathscr{S})$: the isolated points of set $\mathscr{S}$.
$\mathrm{Lp}(\mathscr{S})$: the set of limit points to set $\mathscr{S}$.
$\mathrm{Cl}(\mathscr{S})$: the closure of set $\mathscr{S}$.
$\mathrm{d}(P, \mathscr{S})$: the distance from point P to set $\mathscr{S}$.
$\mathrm{Dm}(\mathscr{S})$: the diameter of set $\mathscr{S}$.
$\mathrm{n}(\mathscr{R}, \mathscr{S})$: the nearness of sets $\mathscr{R}$ and $\mathscr{S}$.
AB: the vector $\mathrm{b} - \mathrm{a}$.
$\mathrm{Ln}(AB)$: the line through the distinct points A and B.
$\mathrm{Sg}[AB]$, $\mathrm{Sg}(AB)$: the closed and open segments joining A and B.
$\mathrm{Sg}[A, B)$, $\mathrm{Sg}(A, B]$: the half-open segments joining A and B.
$\mathrm{Ry}[A, B)$, $\mathrm{Ry}(A, B]$: the closed and open rays from A through B.
$\mathrm{Pl}(ABC)$: the plane of the noncollinear points A, B, C.
$\measuredangle(p, q)$, $\measuredangle(p, q)^\circ$: angle of vectors p, q and the degree measure of the angle.
$\measuredangle BAC$, $\measuredangle BAC^\circ$: the angle of the rays $\mathrm{Ry}[A, B)$, $\mathrm{Ry}[A, C)$ and the degree measure of the angle.
$\mathrm{LS}(a_1, a_2, \ldots, a_m)$: the linear span of the vectors $a_1, a_2, \ldots, a_m$.
$\mathrm{AS}(A_1, A_2, \ldots, A_m)$: the affine span of the points $A_1, A_2, \ldots, A_m$.
$\mathrm{CS}(A_1, A_2, \ldots, A_m)$: the convex span of the points $A_1, A_2, \ldots, A_m$.
$\mathrm{CS}^+(A_1, A_2, \ldots, A_m)$: the positive convex span of the points $A_1, A_2, \ldots, A_m$.
$\mathrm{d}(t_1, t_2)$: the distance between parallel lines t_1, t_2.
$\mathrm{d}(\mathscr{H}_1, \mathscr{H}_2)$: the distance between parallel hyperplanes $\mathscr{H}_1, \mathscr{H}_2$.

CH($\mathscr{S}$): the convex hull of set $\mathscr{S}$.
Cn($\mathscr{S}$; P): the cone of set $\mathscr{S}$ and point P.
E($\mathscr{S}$): the set of extreme points of set $\mathscr{S}$.
Ep($\mathscr{S}$): the set of exposed points of set $\mathscr{S}$.
$\alpha\mathscr{R} + \beta\mathscr{S}$: the linear combination of sets $\mathscr{R}$ and $\mathscr{S}$.
D($\mathscr{R}$, $\mathscr{S}$): the Hausdorff distance between sets $\mathscr{R}$ and $\mathscr{S}$.

1

INTRODUCTORY METRIC TOPOLOGY AND EUCLIDEAN *n*-SPACE

INTRODUCTION

This chapter emphasizes the development of the elementary topology necessary to make precise and general those properties of euclidean sets which are geometric figures. At the same time, we use euclidean space to provide concrete illustrations of concepts important to abstract metric spaces.

The material on metric spaces is independent of any previous background in topology. However, we do suppose that the reader is familiar with the ideas of linear algebra and euclidean n-space. Thus the material on the spaces $\mathscr{V}^n$ and $\mathscr{E}^n$, with which we begin, is a review to establish common ground with the reader.

SECTION 1.1. THE SPACES $\mathscr{V}^n$ AND $\mathscr{E}^n$

We denote the set of all real numbers by $\mathscr{V}^1$, the set of all ordered pairs of real numbers by $\mathscr{V}^2$, and in general the set of all ordered n-tuples by $\mathscr{V}^n$. We refer to the ordered n-tuple $(x_1, x_2, \ldots, x_n)$ in $\mathscr{V}^n$ as the *point* X and also as the *vector* x. Although having two names for the n-tuple does not correspond to any mathematical necessity, we later relate the names to different points of view about the same object.

The set $\mathscr{V}^n$ is a vector space with respect to the following definitions.

1. (*Equality*) $x = y$ if and only if $x_i = y_i$, $i = 1, 2, \ldots, n$.
2. (*Addition*) $x + y = (x_1 + y_1, x_2 + y_2, \ldots, x_n + y_n)$.
3. (*Scalar multiplication*) The product of a *scalar* k, that is, a real number k, and a vector x is the vector $kx = (kx_1, kx_2, \ldots, kx_n)$.

Basic Properties

1. (*Associativity*) $x + (y + z) = (x + y) + z$.
2. (*Commutativity*) $x + y = y + x$.
3. (*Zero vector*) The zero or *null* vector is $o = (0, 0, \ldots, 0)$ and $x + o = x$. The point O of the null vector is called the *origin* of $\mathscr{V}^n$.
4. (*Additive inverse*) Corresponding to each x in $\mathscr{V}^n$, there is a vector $-x = (-x_1, -x_2, \ldots, -x_n)$ such that $x + (-x) = o$. Each of the vectors x, $-x$ is the *additive inverse* or negative of the other.
5. (*Distributivity*) For all real numbers h, k and all vectors x, y in $\mathscr{V}^n$, $h(x + y) = hx + hy$ and $(h + k)x = hx + kx$.
6. (*Associativity*) For all real numbers h, k and all vectors x in $\mathscr{V}^n$, $(hk)x = h(kx)$.
7. (*Unit*) For each x in $\mathscr{V}^n$, $1x = x$.

CONVENTIONS. The vector kx is called the k *multiple* of x and is a *positive* or *negative multiple* according as k is positive or negative. Nonzero vectors x and y are *like directed* if one is a positive multiple of the other and *opposite directed* if one is a negative multiple of the other. The zero vector is regarded as being both like and opposite directed to every vector.

The properties just enumerated give $\mathscr{V}^n$ its algebraic structure. Shortly, we define a distance for pairs of points in $\mathscr{V}^n$ that specializes $\mathscr{V}^n$ to the euclidean space $\mathscr{E}^n$, but first we introduce general distance functions, or metrics.

DEFINITION (Metric). If $\mathscr{E}$ is a set of elements of any kind, and if to each ordered pair (A, B) for A, B in $\mathscr{E}$, there is associated a real number $d(A, B)$—so d is a real valued function defined on the ordered pairs of $\mathscr{E}$—then d is a *metric* for $\mathscr{E}$ if:

(i) $d(A, B) \geq 0$, with equality if and only if $A = B$.
(ii) $d(A, B) = d(B, A)$.
(iii) $d(A, B) + d(B, C) \geq d(A, C)$ for all A, B, C in $\mathscr{E}$.

The set $\mathscr{E}$ together with d is a *metric space*.

We call attention to the abstract properties of a metric partly because $\mathscr{V}^n$ can be metrized in many different ways and partly because many of the topological properties that we consider later depend only on the fact that our space has a metric and not on the fact that the metric is euclidean.

The familiar distance formulas used in analytic geometry for the spaces $\mathscr{E}^2$ and $\mathscr{E}^3$ are metrizations for $\mathscr{V}^2$ and $\mathscr{V}^3$, respectively, and hence give a clear indication of what the euclidean metric for $\mathscr{V}^n$ should be. However, we reach that definition by stages that have analogs in more advanced mathematics and in contexts in which the metric is not obvious. We begin with the fact that there is a natural inner product for vectors in $\mathscr{V}^n$, which we now define, whose importance is hard to overstate. As our work progresses, this inner product will be seen to be one of the major links between the algebra, geometry, and analysis in our program.

DEFINITION (Inner product). The *inner product* (dot product, scalar product) of vectors x and y is the number

$$x \cdot y = \sum_1^n x_i y_i.$$

Our first theorem gives the definitive properties of an inner product.

THEOREM 1. For all x, y, z in $\mathscr{V}^n$ and all real numbers k, the inner product has the following properties:

(i) $x \cdot x \geq 0$, with equality if and only if $x = o$.
(ii) $x \cdot y = y \cdot x$.
(iii) $(kx) \cdot y = y \cdot (kx) = k(x \cdot y)$.
(iv) $x \cdot (y + z) = x \cdot y + x \cdot z$. (E)*

Next, we use the inner product to define a norm for the vectors of $\mathscr{V}^n$.

DEFINITION (Norm of a vector). The *norm*, or *length*, of the vector x is the number

$$|x| = \sqrt{x \cdot x} = \sqrt{\sum_1^n x_i^2}.\dagger$$

* The symbol (E) denotes that a proof is left as an exercise.
† Some texts use $\| x \|$ to denote the norm of x.

It is easily seen that $|kx| = |k||x|$, and that $|x| \geq 0$, with equality if and only if $x = o$. We make use of these facts in the proof of the next theorem.

THEOREM 2. For all x, y in $\mathscr{V}^n$, the norm and inner product satisfy the relations:

(i) $2(x \cdot y) = |x + y|^2 - |x|^2 - |y|^2$.

(ii) $|x \cdot y| \leq |x||y|$, with equality if and only if x and y are like or opposite directed.

PROOF. Property (i) follows by direct computation, using the distributive property (iv) of Theorem 1.

To establish (ii), first consider the case in which x and y are like or opposite directed. Then one of them is a multiple of the other, say $y = kx$ (where k may be zero). Then

$$|x \cdot y| = |x \cdot (kx)| = |k(x \cdot x)| = |k||x|^2 \tag{1}$$

and

$$|x||y| = |x||kx| = |x||k||x| = |k||x|^2 \tag{2}$$

show that equality holds in (ii).

Next, suppose that x and y are neither like nor opposite directed. Then neither x or y is null, and $x - ky$ is not null for any real number k. Therefore, for all k,

$$\begin{aligned} 0 < |x - ky|^2 &= (x - ky) \cdot (x - ky) \\ &= |x|^2 - 2k(x \cdot y) + k^2|y|^2. \end{aligned} \tag{3}$$

Corresponding to $k_1 = |x|/|y|$, (3) becomes

$$0 < |x|^2 - 2\frac{|x|}{|y|}(x \cdot y) + \frac{|x|^2}{|y|^2}|y|^2,$$

or

$$2\frac{|x|}{|y|}(x \cdot y) < 2|x|^2,$$

or

$$x \cdot y < |x||y|. \tag{4}$$

Corresponding to $k_2 = -|x|/|y|$, a similar calculation shows that (3) reduces to

$$-(x \cdot y) < |x||y|. \tag{5}$$

Together, (4) and (5) imply that when neither x nor y is a multiple of the other then

$$|x \cdot y| < |x||y|. \tag{6}$$

From the two parts of the argument, it follows that the weak inequality in (ii) is always valid and that equality occurs if and only if x and y are like or opposite directed. □

The proof for Theorem 2 implies another special equality.

COROLLARY 2.1. $x \cdot y = |x||y|$ if and only if x and y are like directed. (E)

Property (ii) in Theorem 2 is also called the Cauchy–Schwartz inequality and is often stated in the following equivalent form.

COROLLARY 2.2. If $x_1, x_2, \ldots, x_n$ and $y_1, y_2, \ldots, y_n$ are real numbers, then

$$\left(\sum_1^n x_i y_i\right)^2 \leq \left(\sum_1^n x_i^2\right)\left(\sum_1^n y_i^2\right)$$

with equality if and only if one of the n-tuples is a multiple of the other.

The next theorem establishes that $|x|$ has three properties that are characteristic of a norm.

THEOREM 3. For all x, y in $\mathscr{V}^n$ and all real numbers k, the norm has the following properties:

(i) $|x| \geq 0$, with equality if and only if $x = o$.
(ii) $|kx| = |k||x|$.
(iii) $|x + y| \leq |x| + |y|$, with equality and if and only if x and y are like directed.

PROOF. Properties (i) and (ii) follow from simple computations (we mentioned them earlier and used them in proving Theorem 2). To establish (iii), the two properties of Theorem 2 imply that

$$|x + y|^2 - |x|^2 - |y^2| = 2(x \cdot y) \leq 2|x||y|, \tag{1}$$

or

$$|x + y|^2 \leq |x|^2 + 2|x||y| + |y|^2, \tag{2}$$

or

$$|x + y|^2 \leq (|x| + |y|)^2. \tag{3}$$

Since both sides of (3) are nonegative, we can extract square roots to obtain

$$|x + y| \leq |x| + |y|. \tag{4}$$

Equality in (4) holds if and only if $x \cdot y = |x||y|$ in (1), and hence, by Corollary 2.1, if and only if x and y are like directed. □

We now make use of the (euclidean) norm to define the euclidean metric.

DEFINITION (Euclidean metric). The euclidean distance between points X and Y in $\mathscr{V}^n$ is the number $d(X, Y)$ defined by

$$d(X, Y) = |x - y| = \sqrt{\sum_1^n (x_i - y_i)^2}.*$$

We must, of course, show that $d(X, Y)$ as defined here is actually a metric, namely, that it has the three properties of an abstract metric listed earlier.

THEOREM 4. For all X, Y, Z in $\mathscr{V}^n$,

(i) $d(X, Y) \geq 0$, with equality if and only if $X = Y$.
(ii) $d(X, Y) = d(Y, X)$ (symmetry).
(iii) $d(X, Y) + d(Y, Z) \geq d(X, Z)$ (triangle inequality).

* Strictly speaking, each metric should have its own name, and $e(X, Y)$ would be a more logical notation. However, since we do not use many different metrics here, we use d, which is commonly associated with distance.

PROOF. Properties (i) and (ii) are correct because $|x - y| \geq 0$, with equality if and only if $x - y = 0$, and because $|x - y| = |y - x|$.

To establish (iii), we first observe that, by definition,

$$d(X, Y) + d(Y, Z) = |x - y| + |y - z|. \tag{1}$$

Next, by (iii) of Theorem 3,

$$|x - y| + |y - z| \geq |(x - y) + (y - z)| = |x - z|. \tag{2}$$

Combining (1) and (2), we obtain

$$d(X, Y) + d(Y, Z) \geq |x - z| = d(X, Z). \quad \square$$

From now on, "euclidean n-space," denoted by $\mathscr{E}^n$, refers to the space $\mathscr{V}^n$ with the metric $d(X, Y) = |x - y|$.

Exercises – Section 1.1

1. Prove Theorem 1.
2. Prove that $|kx| = |k||x|$.
3. Prove Corollary 2.1.
4. Prove that $|x| \geq 0$ and that $|x| = 0$ implies $x = o$.
5. Verify in $\mathscr{V}^2$ that if k_1 and k_2 are scalars, then

$$\begin{aligned} |k_1 a + k_2 b|^2 &= (k_1 a + k_2 b) \cdot (k_1 a + k_2 b) \\ &= k_1^2(a \cdot a) + 2k_1 k_2(a \cdot b) + k_2^2(b \cdot b). \end{aligned}$$

6. Show that if $d(X, Y)$ is a metric for $\mathscr{V}^n$, not necessarily the euclidean metric, then $[d(X, Y)]^{1/2}$ is also a metric for $\mathscr{V}^n$. Show also that if X, Y, Z are distinct points, then in terms of the new metric the distance from X to Z is not ever the sum of the distances from X to Y and Y to Z.
7. When does the norm $|x - y|$ have exactly the same meaning as the absolute value of $x - y$?

SECTION 1.2. LOCAL METRIC CONCEPTS

We suppose that we are dealing with some non-empty universal set $\mathscr{E}$ whose elements, called *points*, are denoted by capital letters such as A, B, C, and so on. The set $\mathscr{E}$ is a *space* in the sense that the *complement* of $\mathscr{E}$, denoted by Cp($\mathscr{E}$),* is the *empty* or *null set* $\varnothing$. Thus for any set $\mathscr{S}$, Cp($\mathscr{S}$) consists of

* We make common use of notations based on name abbreviations, since these facilitate the reading of symbolic statements. Many other notations are in common usage.

those points of $\mathscr{E}$ not in $\mathscr{S}$. We also suppose that $\mathscr{E}$ has a metric d, with the properties given in Sec. 1.1, and hence is a metric space. Because the euclidean spaces $\mathscr{E}^1$, $\mathscr{E}^2$, and $\mathscr{E}^3$ are familiar and important, they are a natural source of examples to illustrate the theory. However, the theory itself does not depend on any particular choice of the metric.

When we visualize a simple region in the plane, or simple solid object in space, certain topological features appear very clearly. There are points that are inside or interior to the object, points that are outside or exterior to the object, and the remaining points belong to the surface or common boundary to the inside and outside. In any metric space, we can define these sets precisely in terms of the following concepts.

DEFINITION (Ball, sphere, neighborhood, deleted neighborhood). Corresponding to a point A and a positive real number δ, the set

$$\mathrm{B}(A, \delta) = \{X : d(A, X) \leq \delta\}$$

is the *ball* with *center* A and *radius* δ. The set

$$\mathrm{S}(A, \delta) = \{X : d(A, X) = \delta\}$$

is the *sphere* with center A and radius δ. The set

$$\mathrm{N}(A, \delta) = \{X : d(A, X) < \delta\}$$

is the *neighborhood* of A with center A and radius δ. The set

$$\mathrm{N}_o(A, \delta) = \{X : 0 < d(A, X) < \delta\}$$

is the *deleted neighborhood* of A with center A and radius δ.

Comment.* Since $d(A, A) = 0$, A belongs to all its neighborhoods.

CONVENTIONS. One ball, sphere, or neighborhood is said to be *larger* or *smaller* than a second according as the radius of the first is larger or smaller than that of the second.

DEFINITION (Concentric balls, spheres, neighborhoods). Two or more balls, spheres, or neighborhoods are *concentric* if they have the same center.

* We occasionally highlight the importance of some simple fact by stating it as a "comment."

The next four theorems state simple but fundamental facts about neighborhoods.

THEOREM 1. The union of a finite number of concentric neighborhoods is the largest of the neighborhoods, and the intersection of the neighborhoods is the smallest of the neighborhoods. (E)

THEOREM 2. If $P \in \mathrm{N}(A, \delta)$ then there exists a neighborhood of P contained in $\mathrm{N}(A, \delta)$, and if $P \neq A$, then there is a neighborhood of P contained in $\mathrm{N}_o(A, \delta)$.

PROOF. If $P = A$, then $\mathrm{N}(P, \delta) = \mathrm{N}(A, \delta) \subset \mathrm{N}(A, \delta)$. If $P \neq A$, then $0 < d(A, P) < \delta$, hence $\delta_1 = \min\{d(A, P), \delta - d(A, P)\} > 0$, so $\mathrm{N}(P, \delta_1)$ exists. If $X \in \mathrm{N}(P, \delta_1)$, then

$$\begin{aligned} d(A, X) &\leq d(A, P) + d(P, X) < d(A, P) + \delta_1 \\ &\leq d(A, P) + \delta - d(A, P) = \delta \end{aligned}$$

shows that $X \in \mathrm{N}(A, \delta)$. Since $d(P, X) < \delta_1 \leq d(P, A)$, $X \neq A$, therefore $X \in \mathrm{N}_o(A, \delta)$. Thus $\mathrm{N}(P, \delta_1) \subset \mathrm{N}_o(A, \delta)$. □

THEOREM 3. If a point P belongs to the intersection set of a finite number of neighborhoods, then some neighborhood of P is also contained in the intersection of the neighborhoods.

PROOF. Let $\mathrm{N}(A_1, \delta_1)$, $\mathrm{N}(A_2, \delta_2), \ldots, \mathrm{N}(A_n, \delta_n)$ denote a finite set of neighborhoods such that P belongs to their intersection. By Theorem 2, $P \in \mathrm{N}(A_i, \delta_i)$ implies the existence of $\mathrm{N}(P, \delta_i') \subset \mathrm{N}(A_i, \delta_i)$, $i = 1, 2, \ldots, n$. By Theorem 1, the intersection of the n concentric neighborhoods of P is just the smallest neighborhood, say $\mathrm{N}(P, \delta_k')$. Then $\mathrm{N}(P, \delta_k') \subset \mathrm{N}(P, \delta_i') \subset \mathrm{N}(A_i, \delta_i)$, $i = 1, 2, \ldots, n$ shows that $\mathrm{N}(P, \delta_k')$ is contained in the intersection of all the A_i neighborhoods. □

THEOREM 4. If $A \neq B$, there exist neighborhoods of A and B that do not intersect. (E)

We now make use of the local notion of neighborhoods to define the sets we described informally at the beginning of this section.

DEFINITION (Interior, exterior, and boundary of a set). A point P is an *interior point* of a set $\mathscr{S}$ if there exists a neighborhood of P that is contained in $\mathscr{S}$. A point P is an *exterior point* to set $\mathscr{S}$ if there exists a neighborhood of P that is contained in the complement of $\mathscr{S}$. A point P is a *boundary point* to both $\mathscr{S}$ and $\mathrm{Cp}(\mathscr{S})$ if every neighborhood of P intersects both $\mathscr{S}$ and $\mathrm{Cp}(\mathscr{S})$.

The *interior* of $\mathscr{S}$, denoted by $\text{In}(\mathscr{S})$ is the set of all points interior to $\mathscr{S}$. The *exterior* of $\mathscr{S}$, denoted by $\text{Ex}(\mathscr{S})$ is the set of all points exterior to $\mathscr{S}$. The *boundary* of $\mathscr{S}$ and of $\text{Cp}(\mathscr{S})$, denoted by $\text{Bd}(\mathscr{S})$, is the set of all boundary points to $\mathscr{S}$ and $\text{Cp}(\mathscr{S})$.

Comment. The interior of $\mathscr{S}$ is contained in $\mathscr{S}$, and the exterior of $\mathscr{S}$ is contained in the complement of $\mathscr{S}$.

With respect to an arbitrary point P, consider the following exhaustion of possibilities:

1. Some neighborhood of P is contained in $\mathscr{S}$.
2. Some neighborhood of P is contained in $\text{Cp}(\mathscr{S})$.
3. No neighborhood of P is contained in either $\mathscr{S}$ or $\text{Cp}(\mathscr{S})$.

The first two conditions obviously exclude the third and conversely. But since P belongs to its neighborhoods and cannot be in both $\mathscr{S}$ and $\text{Cp}(\mathscr{S})$, the first two conditions are also exclusive of each other. Thus one and only one of the three conditions must be valid, so we have the following theorem.

THEOREM 5. Corresponding to any set $\mathscr{S}$ in the space $\mathscr{E}$,

$$\mathscr{E} = \text{In}(\mathscr{S}) \cup \text{Bd}(\mathscr{S}) \cup \text{Ex}(\mathscr{S}),$$

and the sets in this union are pairwise disjoint.

COROLLARY 5.1. Every point of $\mathscr{S}$ is either an interior point of $\mathscr{S}$ or a boundary point of $\mathscr{S}$, that is,

$$\mathscr{S} \subset \text{In}(\mathscr{S}) \cup \text{Bd}(\mathscr{S}).$$

PROOF. If $P \in \mathscr{S}$ then $P \notin \text{Ex}(\mathscr{S})$, hence, by Theorem 5, $P \in \text{In}(\mathscr{S}) \cup \text{Bd}(\mathscr{S})$. □

COROLLARY 5.2. $\mathscr{S} \cup \text{Bd}(\mathscr{S}) = \text{In}(\mathscr{S}) \cup \text{Bd}(\mathscr{S})$.

PROOF. Since $\text{In}(\mathscr{S}) \subset \mathscr{S}$, then $\text{In}(\mathscr{S}) \cup \text{Bd}(\mathscr{S}) \subset \mathscr{S} \cup \text{Bd}(\mathscr{S})$. From Corollary 5.1, $\mathscr{S} \cup \text{Bd}(\mathscr{S}) \subset \text{In}(\mathscr{S}) \cup \text{Bd}(\mathscr{S})$. The opposite containments imply the equality. □

The symmetric way in which $\mathscr{S}$ and $\text{Cp}(\mathscr{S})$ were used in defining interior, boundary, and exterior imply the following relations.

THEOREM 6. For any set $\mathscr{S}$, $\text{In}(\mathscr{S}) = \text{Ex}[\text{Cp}(\mathscr{S})]$, $\text{Bd}(\mathscr{S}) = \text{Bd}[\text{Cp}(\mathscr{S})]$, and $\text{Ex}(\mathscr{S}) = \text{In}[\text{Cp}(\mathscr{S})]$. (E)

Depending on the set $\mathscr{S}$, one or more but not all of the sets $\text{In}(\mathscr{S})$, $\text{Bd}(\mathscr{S})$, $\text{Ex}(\mathscr{S})$ may be empty. It is always true that $\mathscr{S}$ contains $\text{In}(\mathscr{S})$ and that $\text{Cp}(\mathscr{S})$ contains $\text{Ex}(\mathscr{S})$. But, depending on $\mathscr{S}$, $\text{Bd}(\mathscr{S})$ may be contained in $\mathscr{S}$, be contained in $\text{Cp}(\mathscr{S})$, or may intersect both $\mathscr{S}$ and $\text{Cp}(\mathscr{S})$.

We now introduce some concepts that are useful in describing alternate characterizations of the set $\text{In}(\mathscr{S}) \cup \text{Bd}(\mathscr{S})$.

DEFINITION (Limit point, derived set). A point P is a *limit point* to a set $\mathscr{S}$ if every deleted neighborhood of P intersects $\mathscr{S}$. The set of all limit points to $\mathscr{S}$ is the *limit point set* of $\mathscr{S}$, or the *derived* set of $\mathscr{S}$, denoted by $\text{Lp}(\mathscr{S})$.

DEFINITION (Isolated points of a set). A point P is an *isolated point* of set $\mathscr{S}$ if it belongs to $\mathscr{S}$ but is not a limit point to $\mathscr{S}$. The set of all isolated points of $\mathscr{S}$ is denoted by $\text{Is}(\mathscr{S})$.

The next theorem is an analog of Corollary 5.1.

THEOREM 7. Every point of $\mathscr{S}$ is either an isolated point of $\mathscr{S}$ or a limit point to $\mathscr{S}$, that is,

$$\mathscr{S} \subset \text{Is}(\mathscr{S}) \cup \text{Lp}(\mathscr{S}).$$

PROOF. Consider $P \in \mathscr{S}$. If $P \notin \text{Lp}(\mathscr{S})$, then by definition $\text{P} \in \text{Is}(\mathscr{S}) \subset \text{Is}(\mathscr{S}) \cup \text{Lp}(\mathscr{S})$. On the other hand, if $P \in \text{Lp}(\mathscr{S})$, then clearly $P \in \text{Is}(\mathscr{S}) \cup \text{Lp}(\mathscr{S})$. $\square$

COROLLARY 7. $\mathscr{S} \cup \text{Lp}(\mathscr{S}) = \text{Is}(\mathscr{S}) \cup \text{Lp}(\mathscr{S})$. (E)

Corollaries 5.2 and 7 deal with different aspects of the same set. The next theorem, which unifies and summarizes some previous relations, will be extremely useful in future work.

THEOREM 8. For any set $\mathscr{S}$,

$$\text{In}(\mathscr{S}) \cap \text{Bd}(\mathscr{S}) = \varnothing, \qquad \text{Is}(\mathscr{S}) \cap \text{Lp}(\mathscr{S}) = \varnothing,$$

and

$$\mathscr{S} \cup \text{Bd}(\mathscr{S}) = \text{In}(\mathscr{S}) \cup \text{Bd}(\mathscr{S}) = \text{Is}(\mathscr{S}) \cup \text{Lp}(\mathscr{S}) = \mathscr{S} \cup \text{Lp}(\mathscr{S}). \quad \text{(E)}$$

The set described in different ways in Theorem 8 bears an important relation to $\mathscr{S}$ and is given a separate name.

DEFINITION (Closure of a set). The *closure* of a set $\mathscr{S}$, denoted by $\text{Cl}(\mathscr{S})$, is the union of $\mathscr{S}$ with its boundary, that is, $\text{Cl}(\mathscr{S}) = \mathscr{S} \cup \text{Bd}(\mathscr{S})$.

In Theorem 8, then, we have different representations for the closure of a set.

Exercises – Section 1.2

1. Describe the sets $\text{B}(A, \delta)$, $\text{N}(A, \delta)$, and $\text{S}(A, \delta)$ in the euclidean space $\mathscr{E}^1$.
2. Prove Theorem 1.
3. Is Theorem 1 still valid if "neighborhood" is replaced throughout by "deleted neighborhood"?
4. Prove Theorem 4.
5. Prove that $\mathscr{R} \subset \mathscr{S} \Rightarrow \text{In}(\mathscr{R}) \subset \text{In}(\mathscr{S})$.
6. Give an example to show that $\text{In}(\mathscr{R}) \subset \text{In}(\mathscr{S})$ does not imply $\mathscr{R} \subset \mathscr{S}$.
7. Prove Theorem 6.
8. Prove Corollary 7.
9. Prove Theorem 8.
10. For each of the following sets $\mathscr{S}$, define the sets $\text{In}(\mathscr{S})$, $\text{Bd}(\mathscr{S})$, $\text{Lp}(\mathscr{S})$, $\text{Cl}(\mathscr{S})$, $\text{Is}(\mathscr{S})$:

    a. in $\mathscr{E}^1$, $\mathscr{S} = \{x : x \text{ is a rational number}\}$
    b. in $\mathscr{E}^1$, $\mathscr{S} = \{x : a < x \le b\}$
    c. in $\mathscr{E}^1$, $\mathscr{S} = \{x : x > a\}$
    d. in $\mathscr{E}^1$, $\mathscr{S} = \{-1\} \cup \{x : x > 0\}$
    e. in $\mathscr{E}^2$, $\mathscr{S} = \{(x_1, x_2) : x_2^2 < x_1\}$
    f. in $\mathscr{E}^2$, $\mathscr{S} = \{(x_1, x_2) : x_1^2 + x_2^2 \le 1, x_2 > 0\}$

11. Prove the theorem: "Point P is a limit point to set $\mathscr{S}$ if and only if every neighborhood of P contains infinitely many points of $\mathscr{S}$."
12. Give an example of a set $\mathscr{S}$ such that $\text{Lp}[\text{Lp}(\mathscr{S})] \ne \text{Lp}(\mathscr{S})$.
13. Prove that $\text{Lp}[\text{Lp}(\mathscr{S})] \subset \text{Lp}(\mathscr{S})$.
14. If $\mathscr{S}$ is a finite set, what are the sets $\text{In}(\mathscr{S})$, $\text{Lp}(\mathscr{S})$, $\text{Bd}(\mathscr{S})$, $\text{Cl}(\mathscr{S})$?
15. If all points of a set are isolated points, that is, $\mathscr{S} = \text{Is}(\mathscr{S})$, the set is said to be "scattered." Why is every finite set scattered? In $\mathscr{E}^1$, give an example of a scattered set that has infinitely many points and is contained in the interval $-1 \le x \le 1$.

SECTION 1.3. BASIC CATEGORIES AND RELATIONS OF METRIC SETS; RELATIVE TOPOLOGY

The two paramount concepts of topology are those of open and closed sets. From a formal point of view, open sets are traditionally taken as a starting point in axiomatic treatments of topology and closed sets appear in a dual role. From the concrete point of view of our program, it will be seen that the two concepts have a natural relation to the geometric objects we propose to study.

DEFINITION (Open set, closed set). A set $\mathscr{S}$ is an *open set* if it consists entirely of interior points, hence if $\mathscr{S} = \mathrm{In}(\mathscr{S})$. A set $\mathscr{S}$ is a *closed set* if it contains its boundary, hence if $\mathscr{S} \supset \mathrm{Bd}(\mathscr{S})$.

From the definition of an open set and an interior point, it is clear that a set $\mathscr{S}$ is open if and only if for each $P \in \mathscr{S}$ there exists $\mathrm{N}(P, \delta) \subset \mathscr{S}$. Thus, as a corollary of Theorem 2, Sec. 1.2, we have the important fact.

THEOREM 1. Neighborhoods are open sets.

We now want to consider some operations with sets that produce open or closed sets, and this involves families of sets. In dealing with these, some specific agreements are useful.

CONVENTIONS. We ordinarily use the particular capital script letters $\mathscr{F}$ and $\mathscr{G}$ for collections or families of sets, that is, for sets whose elements are themselves point sets. To denote an arbitrary collection of sets, $\mathscr{F}$, we use the notation

$$\mathscr{F} = \{\mathscr{S}_\alpha : \alpha \in \Lambda\},$$

where Λ, the capital Greek letter lambda, denotes an indexing set. If Λ is the set of positive integers, then $\mathscr{F}$ is the collection $\{\mathscr{S}_1, \mathscr{S}_2, \ldots, \mathscr{S}_n, \ldots\}$. In general, however, the notation simply indicates that $\mathscr{F}$ is a collection whose sets have been put into one to one correspondence with some appropriate set Λ.

The next theorem gives the characteristic properties of open sets.

THEOREM 2

(i) The space $\mathscr{E}$ and the empty set $\varnothing$ are open sets.
(ii) The union of any number of open sets is an open set.
(iii) The intersection of a finite number of open sets is an open set.

PROOF. (i) Since $\mathrm{N}(P, \delta) \subset \mathscr{E}$ for all $P \in \mathscr{E}$ and all $\delta > 0$, the space $\mathscr{E}$ is open. Next, in the statement "$P \in \varnothing$ implies $P \in \mathrm{In}(\varnothing)$," the hypothesis "$P \in \varnothing$" is false, hence the statement is valid and therefore $\varnothing$ is an open set.

(ii) Let $\mathscr{F} = \{\mathscr{S}_\alpha : \alpha \in \Lambda\}$ denote an arbitrary collection of open sets, and let $\mathscr{S}$ denote the union of the sets in $\mathscr{F}$. If $P \in \mathscr{S}$, then, by the definition of a union, there is some set in $\mathscr{F}$, say $\mathscr{S}_\beta$, such that $P \in \mathscr{S}_\beta$. Since $\mathscr{S}_\beta$ is open, there is a neighborhood of P, $\mathrm{N}(P, \delta)$, contained in $\mathscr{S}_\beta$. Because $\mathscr{S}_\beta \subset \mathscr{S}$, then $\mathrm{N}(P, \delta) \subset \mathscr{S}$, hence P is interior to $\mathscr{S}$. Since the points of $\mathscr{S}$ are interior points, $\mathscr{S}$ is an open set.

(iii) Let $\mathscr{F} = \{\mathscr{S}_i : i = 1, 2, \ldots, n\}$ be a finite collection of open sets, let $\mathscr{S}$ denote the intersection of the sets in $\mathscr{F}$, and consider $P \in \mathscr{S}$. Then P belongs to each set $\mathscr{S}_i$, $i = 1, 2, \ldots, n$. Because each $\mathscr{S}_i$ is open, there is a neighborhood $\mathrm{N}(P, \delta_i) \subset \mathscr{S}_i$, $i = 1, 2, \ldots, n$. The smallest of these n concentric neighborhoods of P is also a neighborhood of P, say $\mathrm{N}(P, \delta)$. Then from $\mathrm{N}(P, \delta) \subset \mathrm{N}(P, \delta_i) \subset \mathscr{S}_i$, $i = 1, 2, \ldots, n$, it follows that $\mathrm{N}(P, \delta) \subset \mathscr{S}$. Thus P is interior to $\mathscr{S}$, and $\mathscr{S}$ is an open set. □

Turning now to closed sets, these can be characterized in a variety of ways. To establish the equivalence of the different conditions listed in the next theorem, we show that each statement implies the succeeding one and finally that the last implies the first.

THEOREM 3. The following statements are equivalent:

(i) $\mathscr{S}$ is a closed set, that is, $\mathscr{S} \supset \mathrm{Bd}(\mathscr{S})$.
(ii) The limit points to $\mathscr{S}$ belong to $\mathscr{S}$, that is, $\mathscr{S} \supset \mathrm{Lp}(\mathscr{S})$.
(iii) If $\mathrm{N}(P, \delta) \cap \mathscr{S} \neq \varnothing$, for all $\delta > 0$, then $P \in \mathscr{S}$.
(iv) The complement of $\mathscr{S}$ is an open set.
(v) $\mathscr{S}$ is its own closure, that is, $\mathscr{S} = \mathrm{Cl}(\mathscr{S})$.

PROOF. Assume (i), namely, that $\mathscr{S} \supset \mathrm{Bd}(\mathscr{S})$. Since it is also true that $\mathscr{S} \supset \mathrm{In}(\mathscr{S})$, then $\mathscr{S} \supset \mathrm{In}(\mathscr{S}) \cup \mathrm{Bd}(\mathscr{S})$. Now, from Theorem 8, Sec. 1.2 we have

$$\mathscr{S} \supset \mathrm{In}(\mathscr{S}) \cup \mathrm{Bd}(\mathscr{S}) = \mathscr{S} \cup \mathrm{Lp}(\mathscr{S}) \supset \mathrm{Lp}(\mathscr{S}).$$

Thus $\mathscr{S} \supset \mathrm{Lp}(\mathscr{S})$, hence (i) $\Rightarrow$ (ii).

Next, assume (ii), namely, that $\mathscr{S} \supset \mathrm{Lp}(\mathscr{S})$. Consider a point P such that $\mathrm{N}(P, \delta) \cap \mathscr{S} \neq \varnothing$ for all $\delta > 0$. If P is not in $\mathscr{S}$, then $\mathrm{N}_o(P, \delta) \cap \mathscr{S} \neq \varnothing$ for all $\delta > 0$, hence $P \in \mathrm{Lp}(\mathscr{S})$. But if P is not in $\mathscr{S}$, then by (ii), $P \notin \mathrm{Lp}(\mathscr{S})$. Thus $P \notin \mathscr{S}$ contradicts (ii), hence P must be in $\mathscr{S}$. Therefore (ii) $\Rightarrow$ (iii).

Now assume (iii). Consider any point P in $\mathrm{Cp}(\mathscr{S})$. Then $P \notin \mathscr{S}$. Then, from (iii), it cannot be true that every neighborhood of P intersects $\mathscr{S}$. Thus there is some $\delta > 0$ for which $\mathrm{N}(P, \delta) \cap \mathscr{S} = \varnothing$, hence $\mathrm{N}(P, \delta) \subset \mathrm{Cp}(\mathscr{S})$. Therefore P is interior to $\mathrm{Cp}(\mathscr{S})$, and $\mathrm{Cp}(\mathscr{S})$ is open. Thus (iii) $\Rightarrow$ (iv).

Next, assume (iv), namely, that $\mathrm{Cp}(\mathscr{S})$ is an open set. Then $\mathrm{Cp}(\mathscr{S}) = \mathrm{In}[\mathrm{Cp}(\mathscr{S})] = \mathrm{Ex}(\mathscr{S})$. Now, by Theorems 5 and 8, Sec. 1.2, $\mathscr{S} = \mathrm{In}(\mathscr{S}) \cup \mathrm{Bd}(\mathscr{S}) = \mathrm{Cl}(\mathscr{S})$. Thus (iv) $\Rightarrow$ (v).

Finally, assume (v), that is, that $\mathscr{S} = \mathrm{Cl}(\mathscr{S})$. Then $\mathscr{S} = \mathrm{In}(\mathscr{S}) \cup \mathrm{Bd}(\mathscr{S}) \supset \mathrm{Bd}(\mathscr{S})$, so $\mathscr{S}$ is closed. Thus (v) $\Rightarrow$ (i). □

Although openess and closedness are not opposites, they are related, and (i) and (iv) of Theorem 3 show the relation precisely.

COROLLARY 3. A set is closed if and only if its complement is open, and is open if and only if its complement is closed. (E)

From elementary set theory, we know that the complement of a union is the intersection of the complements, and the complement of an intersection is the union of the complements. Using these facts and the "duality" of openess and closedness in Corollary 3, our next theorem becomes an easy consequence of Theorem 2.

THEOREM 4

(i) The space $\mathscr{E}$ and the empty set $\varnothing$ are closed sets.
(ii) The intersection of any number of closed sets is closed.
(iii) The union of a finite number of open sets is open.

PROOF. (i) Since $\varnothing$ and $\mathscr{E}$ are open, then $\mathrm{Cp}(\varnothing) = \mathscr{E}$ and $\mathrm{Cp}(\mathscr{E}) = \varnothing$ are closed.

(ii) Let $\mathscr{F} = \{\mathscr{S}_\alpha : \alpha \in \Lambda\}$ be an arbitrary collection of closed sets, and let $\mathscr{S}$ be the intersection of the sets in $\mathscr{F}$, that is,

$$\mathscr{S} = \bigcap_{\alpha \in \Lambda} \mathscr{S}_\alpha. \tag{1}$$

Then

$$\mathrm{Cp}(\mathscr{S}) = \bigcup_{\alpha \in \Lambda} \mathrm{Cp}(\mathscr{S}_\alpha). \tag{2}$$

By Corollary 3, the sets $\mathrm{Cp}(\mathscr{S}_\alpha)$ are open, and by Theorem 2 the union of open sets is open. Thus (2) shows that $\mathrm{Cp}(\mathscr{S})$ is open, hence $\mathscr{S}$ is closed.

(iii) (E) □

The next theorem gives some confirmation that the preceding abstractions provide an acceptable model for our intuitive ideas. The theory would certainly seem unsatisfactory if it did not imply the following relations.

THEOREM 5. For any set $\mathcal{S}$

(i) The interior $\text{In}(\mathcal{S})$ and the exterior $\text{Ex}(\mathcal{S})$ are open, hence $\text{In}[\text{In}(\mathcal{S})] = \text{In}(\mathcal{S})$.
(ii) The closure $\text{Cl}(\mathcal{S})$ is a closed set, hence $\text{Cl}[\text{Cl}(\mathcal{S})] = \text{Cl}(\mathcal{S})$.
(iii) The boundary $\text{Bd}(\mathcal{S})$ is a closed set, hence $\text{Bd}[\text{Bd}(\mathcal{S})] \subset \text{Bd}(\mathcal{S})$.
(iv) The derived set $\text{Lp}(\mathcal{S})$ is a closed set, hence $\text{Lp}[\text{Lp}(\mathcal{S})] \subset \text{Lp}(\mathcal{S})$.

PROOF. (i) If $A \in \text{In}(\mathcal{S})$, then there is a neighborhood $\text{N}(A, \delta) \subset \mathcal{S}$. For each X in $\text{N}(A, \delta)$, Theorem 2, Sec. 1.2 implies the existence of $\text{N}(X, \delta_X) \subset \text{N}(A, \delta) \subset \mathcal{S}$. That is, $X \in \text{N}(A, \delta)$ implies $X \in \text{In}(\mathcal{S})$. Therefore $\text{N}(A, \delta) \subset \text{In}(\mathcal{S})$. Hence $\text{In}(\mathcal{S})$ is open, so $\text{In}[\text{In}(\mathcal{S})] = \text{In}(\mathcal{S})$. Since we have shown that the interior of any set is open, then $\text{In}[\text{Cp}(\mathcal{S})] = \text{Ex}(\mathcal{S})$ is open.

(ii) By the definition of $\text{Cl}(\mathcal{S})$ and Theorem 8, Sec. 1.2, we have $\text{Cl}(\mathcal{S}) = \text{In}(\mathcal{S}) \cup \text{Bd}(\mathcal{S})$. This, with Theorem 5, Sec. 1.2, implies that $\text{Cl}(\mathcal{S}) = \text{Cp}[\text{Ex}(\mathcal{S})]$. By the proof in (i), $\text{Ex}(\mathcal{S})$ is open. Thus by property (iv) in Theorem 3, $\text{Cp}[\text{Ex}(\mathcal{S})] = \text{Cl}(\mathcal{S})$ is closed. Then, by property (v) in Theorem 3, $\text{Cl}(\mathcal{S}) = \text{Cl}[\text{Cl}(\mathcal{S})]$.

(iii) From Theorem 5, Sec. 1.2, $\text{Cp}[\text{Bd}(\mathcal{S})] = \text{In}(\mathcal{S}) \cup \text{Ex}(\mathcal{S})$. Since we have shown that $\text{In}(\mathcal{S})$ and $\text{Ex}(\mathcal{S})$ are open, and since the union of open sets is open, then $\text{Cp}[\text{Bd}(\mathcal{S})]$ is open, hence $\text{Bd}(\mathcal{S})$ is closed. A closed set, by definition, contains its boundary, hence $\text{Bd}(\mathcal{S}) \supset \text{Bd}[\text{Bd}(\mathcal{S})]$.

(iv) Consider $P \in \text{Lp}[\text{Lp}(\mathcal{S})]$, and let $\text{N}(P, \delta)$ denote an arbitrary neighborhood of P. Since P is a limit point to $\text{Lp}(\mathcal{S})$, there exists $Q \in \text{N}_o(P, \delta) \cap \text{Lp}(\mathcal{S})$. Then, by Theorem 2, Sec. 1.2, there exists $\text{N}(Q, \delta_1) \subset \text{N}_o(P, \delta)$, and so $\text{N}_o(Q, \delta_1) \subset \text{N}_o(P, \delta)$. Because $Q \in \text{Lp}(\mathcal{S})$, then $\text{N}_o(Q, \delta_1) \cap \mathcal{S} \neq \varnothing$, and therefore $\text{N}_o(P, \delta) \cap \mathcal{S} \neq \varnothing$. Thus $P \in \text{Lp}(\mathcal{S})$. Since $\text{Lp}[\text{Lp}(\mathcal{S})] \subset \text{Lp}(\mathcal{S})$, it follows from Theorem 3 that $\text{Lp}[\text{Lp}(\mathcal{S})]$ is a closed set. □

THEOREM 6. A point P belongs to the closure of a set $\mathcal{S}$ if and only if $\text{N}(P, \delta) \cap \mathcal{S} \neq \varnothing$ for all $\delta > 0$.

PROOF. Since $\mathcal{S} \subset \text{Cl}(\mathcal{S})$, if $\text{N}(P, \delta) \cap \mathcal{S} \neq \varnothing$ for all $\delta > 0$, then $\text{N}(P, \delta) \cap \text{Cl}(\mathcal{S}) \neq \varnothing$ for all $\delta > 0$. But $\text{Cl}(\mathcal{S})$ is closed, by Theorem 5, so from (ii) of Theorem 3, it follows that $P \in \text{Cl}(\mathcal{S})$.

Next, if $P \in \text{Cl}(\mathcal{S})$, then by Theorem 8, Sec. 1.2, $P \in \mathcal{S} \cup \text{Lp}(\mathcal{S})$. If $P \in \text{Lp}(\mathcal{S})$, then $\text{N}_o(P, \delta) \cap \mathcal{S} \neq \varnothing$ for all $\delta > 0$, which clearly implies that $\text{N}(P, \delta) \cap \mathcal{S} \neq \varnothing$ for all $\delta > 0$. Finally, if $P \in \mathcal{S}$ then P itself is in $\text{N}(P, \delta) \cap \mathcal{S}$ for every $\delta > 0$. □

All the concepts expressed in this section and the previous one can be traced back to two fundamental ideas. One is that we have a metric d for our set $\mathscr{E}$. The second is that $\mathscr{E}$ is a "space," by which we mean that we only consider points in $\mathscr{E}$, and we imposed the condition $\text{Cp}(\mathscr{E}) = \varnothing$. Suppose that we had not made $\mathrm{Cp}(\mathscr{E}) = \varnothing$ a necessary condition, but had simply agreed to restrict our attention to points of $\mathscr{E}$, whether or not $\mathrm{Cp}(\mathscr{E})$ was empty. Then, for example, the definition we gave for $\mathrm{N}(P, \delta)$, that is,

$$\mathrm{N}(P, \delta) = \{X : d(P, X) < \delta\}$$

would have been stated in the form

$$\mathrm{N}(P, \delta) = \{X : d(P, X) < \delta, P \in \mathscr{E}, X \in \mathscr{E}\}.$$

Previously $\mathrm{N}(P, \delta) \subset \mathscr{E}$ followed from the fact that all points were in $\mathscr{E}$. Now $\mathrm{N}(P, \delta) \subset \mathscr{E}$ would follow from the restriction in its definition. With such restrictions, all the theory of Secs. 1.2 and 1.3 would go through as before.

We can apply the observations just made in a sort of reverse form. Let $\mathscr{E}$ be the space, with metric d, that we have been dealing with in this section and Sec. 2, with $\mathrm{Cp}(\mathscr{E}) = \varnothing$, and let $\mathscr{E}'$ be a subset of $\mathscr{E}$. To show that d is also a metric for the set $\mathscr{E}'$ we must show that if X, Y, Z are points of $\mathscr{E}'$ then $d(X, Y) \geq 0$ and $d(X, Y) = 0$ if and only if $X = Y$, $d(X, Y) = d(Y, X)$, and $d(X, Z) + d(Z, Y) \geq d(X, Y)$. But these statements hold for all points X, Y, Z in $\mathscr{E}$, so they certainly hold for $X, Y, Z \in \mathscr{E}' \subset \mathscr{E}$. Thus d is a metric for $\mathscr{E}'$, and we can regard the metric set $\mathscr{E}'$ as a "relative metric space" by restricting our attention to the points of $\mathscr{E}'$. With such a restriction, each of the concepts discussed for $\mathscr{E}$ has a relative counterpart for $\mathscr{E}'$.

DEFINITION (Relative metric space, relative topology). If $\mathscr{E}$ is a metric space, with metric d, and $\mathscr{E}'$ is a subset of $\mathscr{E}$, then $\mathscr{E}'$, with metric d, is a *relative metric space.* If $P \in \mathscr{E}'$, and $\delta > 0$, the *relative δ neighborhood* of P, in $\mathscr{E}'$, is the set

$$\mathrm{N}(P, \delta)' = \{X ; d(P, X) < \delta, X \in \mathscr{E}'\}.$$

The topology defined on $\mathscr{E}'$ in terms of the $\mathscr{E}'$ neighborhoods is the *relative topology* of $\mathscr{E}'$.

It is a straightforward matter to express the concepts of the relative topology. For example, point P is a *relative interior* point of set $\mathscr{S} \subset \mathscr{E}'$ if there exists a relative neighborhood of P that is contained in $\mathscr{S}$. The *relative interior* of $\mathscr{S}$, $\mathrm{In}(\mathscr{S})'$, is the set of its relative interior points. A set $\mathscr{S} \subset \mathscr{E}'$ is a *relative open set*, or is *open in* $\mathscr{E}'$, if $\mathscr{S} = \mathrm{In}(\mathscr{S})'$. If $\mathscr{S} \subset \mathscr{E}'$, the *relative complement* of $\mathscr{S}$, $\mathrm{Cp}(\mathscr{S})'$, is the set of points in $\mathscr{E}'$ that are not in $\mathscr{S}$, that is, $\mathrm{Cp}(\mathscr{S})' = \mathscr{E}' \cap \mathrm{Cp}(\mathscr{S})$. If $\mathscr{S} \subset \mathscr{E}'$, the *relative boundary* of $\mathscr{S}$, $\mathrm{Bd}(\mathscr{S})'$, is the set of points that are not in the relative interior of either $\mathscr{S}$ or $\mathrm{Cp}(\mathscr{S})'$. A set $\mathscr{S} \subset \mathscr{E}'$ is a *relative closed set*, or is *closed in* $\mathscr{E}'$, if it contains its relative boundary, and so on.

The next theorem gives equivalent ways of expressing some of the relative concepts just described.

THEOREM 7. Let $\mathscr{E}$ be a metric space, with metric d, and let $\mathscr{E}' \subset \mathscr{E}$ be a relative metric space.

(i) A neighborhood (ball, sphere, deleted neighborhood) in $\mathscr{E}'$ is the intersection of $\mathscr{E}'$ with the neighborhood (ball, sphere, deleted neighborhood) of $\mathscr{E}$ that has the same center and radius.

(ii) A subset of $\mathscr{E}'$ is open in $\mathscr{E}'$ if and only if it is the intersection of $\mathscr{E}'$ with an open subset of $\mathscr{E}$.

(iii) A subset of $\mathscr{E}'$ is closed in $\mathscr{E}'$ if and only if it is the intersection of $\mathscr{E}'$ with a closed subset of $\mathscr{E}$.

(iv) If $\mathscr{S}$ is a subset of $\mathscr{E}'$, then the closure of $\mathscr{S}$ in $\mathscr{E}'$ is the intersection of $\mathscr{E}'$ with the closure of $\mathscr{S}$ in $\mathscr{E}$.

PROOF. (i) Let $\mathrm{N}(A, \delta)'$ be a neighborhood in $\mathscr{E}'$. By definition, $A \in \mathscr{E}'$ and

$$\mathrm{N}(A, \delta)' = \{X ; X \in \mathscr{E}' \text{ and } d(A, X) < \delta\}.$$

Then

$$\mathrm{N}(A, \delta)' = \{X : X \in \mathscr{E}'\} \cap \{X : d(A, X) < \delta\} = \mathscr{E}' \cap \mathrm{N}(A, \delta).$$

By entirely similar arguments, $\mathrm{B}(A, \delta)' = \mathscr{E}' \cap \mathrm{B}(A, \delta)$, $\mathrm{S}(A, \delta)' = \mathscr{E}' \cap \mathrm{S}(A, \delta)$, and $\mathrm{N}_o(A, \delta)' = \mathscr{E}' \cap \mathrm{N}_o(A, \delta)$.

(ii) Let $\mathscr{S}$ be a set that is open in $\mathscr{E}'$. Then for each $A \in \mathscr{S}$ there exists $\mathrm{N}(A, \delta_A)' \subset \mathscr{S}$. From (i), $\mathrm{N}(A, \delta_A)' = \mathscr{E}' \cap \mathrm{N}(A, \delta_A)$. Define set $\mathscr{H}$ by

$$\mathscr{H} = \bigcup_{A \in \mathscr{S}} \mathrm{N}(A, \delta_A).$$

By Theorem 1, each of the neighborhoods $\mathrm{N}(A, \delta_A)$ is open in $\mathscr{E}$ hence by (ii) of Theorem 2, $\mathscr{H}$ is open in $\mathscr{E}$. We now show that $\mathscr{S} = \mathscr{E}' \cap \mathscr{H}$. First, it is clear that $\mathscr{S} \subset \mathscr{H}$, and, by hypothesis, $\mathscr{S} \subset \mathscr{E}'$, so $\mathscr{S} \subset \mathscr{E}' \cap \mathscr{H}$. On the other hand,

$$\mathscr{E}' \cap \mathscr{H} = \mathscr{E}' \cap \bigcup_{A \in \mathscr{S}} \mathrm{N}(A, \delta_A) = \bigcup_{A \in \mathscr{S}} [\mathscr{E}' \cap \mathrm{N}(A, \delta_A)] = \bigcup_{A \in \mathscr{S}} \mathrm{N}(A, \delta_A)'.$$

Since $\mathrm{N}(A, \delta_A)' \subset \mathscr{S}$, for each $A \in \mathscr{S}$, it follows that $\mathscr{E}' \cap \mathscr{H} \subset \mathscr{S}$. This, with the opposite inclusion above, implies that $\mathscr{S} = \mathscr{E}' \cap \mathscr{H}$, so $\mathscr{S}$ is the intersection of $\mathscr{E}'$ with an open subset of $\mathscr{E}$.

Conversely, suppose that $\mathscr{H}$ is some open subset of $\mathscr{E}$ and that $\mathscr{S}$ is defined by $\mathscr{S} = \mathscr{E}' \cap \mathscr{H}$. To show that $\mathscr{S}$ is open in $\mathscr{E}'$, consider any point $A \in \mathscr{S}$.

Then $A \in \mathscr{E}' \cap \mathscr{H}$ implies $A \in \mathscr{H}$, and, since $\mathscr{H}$ is open, there exists $N(A, \delta) \subset \mathscr{H}$. But A is also in $\mathscr{E}'$, so $\mathscr{E}' \cap N(A, \delta)$ is the relative neighborhood $N(A, \delta)'$. Since $N(A, \delta)' \subset \mathscr{H}$ and $N(A, \delta)' \subset \mathscr{E}'$, then $N(A, \delta)' \subset \mathscr{S}$, so A is in the relative interior of $\mathscr{S}$. Because all of its points are in its relative interior, $\mathscr{S}$ is open in $\mathscr{E}'$.

(iii) By Theorem 3 and Corollary 3, a set $\mathscr{S} \subset \mathscr{E}'$ is closed in $\mathscr{E}'$ if and only if $\text{Cp}(\mathscr{S})'$ is open in $\mathscr{E}'$. In turn, by (ii), $\text{Cp}(\mathscr{S})'$ is open in $\mathscr{E}'$ if and only if there is a set $\mathscr{H}$, open in $\mathscr{E}$, and such that $\text{Cp}(\mathscr{S})' = \mathscr{E}' \cap \mathscr{H}$. Since any point of $\mathscr{S}$ is in $\mathscr{E}'$ and is not in $\text{Cp}(\mathscr{S})' = \mathscr{E}' \cap \mathscr{H}$, it is also in $\text{Cp}(\mathscr{H})$. Thus $\mathscr{S} \subset \mathscr{E}' \cap \text{Cp}(\mathscr{H})$. On the other hand, any point in $\mathscr{E}' \cap \text{Cp}(\mathscr{H})$ is not in $\text{Cp}(\mathscr{S})'$, since $\text{Cp}(\mathscr{S})' \subset \mathscr{H}$, and since it is in $\mathscr{E}'$ it must be in $\mathscr{S}$. Thus $\mathscr{E}' \cap \text{Cp}(\mathscr{H}) \subset \mathscr{S}$. The two opposite inclusions imply that $\mathscr{S} = \mathscr{E}' \cap \text{Cp}(\mathscr{H})$. Because $\mathscr{H}$ is open in $\mathscr{E}$, $\text{Cp}(\mathscr{H})$ is closed in $\mathscr{E}$. Thus $\mathscr{S}$ is closed in $\mathscr{E}'$ if and only if it is the intersection of $\mathscr{E}'$ with a set that is closed in $\mathscr{E}$.

(iv) Let $\mathscr{S}$ be a subset of $\mathscr{E}'$, with $\text{Cl}(\mathscr{S})'$ denoting its closure in $\mathscr{E}'$. By definition, $\text{Cl}(\mathscr{S})' \subset \mathscr{E}'$. To show that $\text{Cl}(\mathscr{S})' = \mathscr{E}' \cap \text{Cl}(\mathscr{S})$, we show that each set contains the other.

First, consider a point $P \in \text{Cl}(\mathscr{S})'$, and let $N(P, \delta)$ be any one of its neighborhoods, with $N(P, \delta)'$ the corresponding relative neighborhood. By Theorem 6, $N(P, \delta)' \cap \mathscr{S} \neq \varnothing$. Since $N(P, \delta)' \subset N(P, \delta)$, it follows that $N(P, \delta) \cap \mathscr{S} \neq \varnothing$. Thus, by Theorem 6, $P \in \text{Cl}(\mathscr{S})$. Also, $P \in \mathscr{E}'$, so $P \in \mathscr{E}' \cap \text{Cl}(\mathscr{S})$. Therefore, $\text{Cl}(\mathscr{S})' \subset \mathscr{E}' \cap \text{Cl}(\mathscr{S})$.

Next, suppose that $P \in \mathscr{E}' \cap \text{Cl}(\mathscr{S})$. Consider any relative neighborhood of P, $N(P, \delta)'$. Since $P \in \text{Cl}(\mathscr{S})$, then, by Theorem 6, $N(P, \delta) \cap \mathscr{S} \neq \varnothing$, so there exists $Q \in N(P, \delta) \cap \mathscr{S}$. Because $\mathscr{S} \subset \mathscr{E}'$, then $Q \in N(P, \delta) \cap \mathscr{E}' = N(P, \delta)'$. This, with $Q \in \mathscr{S}$, implies that $Q \in N(P, \delta)' \cap \mathscr{S}$. Because $N(P, \delta)' \cap \mathscr{S} \neq \varnothing$, for any $\delta > 0$, then, by Theorem 6, $P \in \text{Cl}(\mathscr{S})'$. Thus $\mathscr{E}' \cap \text{Cl}(\mathscr{S}) \subset \text{Cl}(\mathscr{S})'$. This, with the former opposite inclusion, shows that $\text{Cl}(\mathscr{S})' = \mathscr{E}' \cap \text{Cl}(\mathscr{S})$. □

In dealing with an object set in euclidean space, it is often natural that one's interest is concentrated on the set itself rather than the rest of space. This, in effect, is to think of the object as a space by itself. The notion of a relative metric space simply makes explicit this intuitive practice. It is worth note also that the relative subspaces of a given metric space provide a host of different metric spaces.

Exercises – Section 1.3

1. Which of the sets in Exercise 10 of Sec. 1.2 are open? Closed? Neither open nor closed?
2. Why does $P \in \varnothing \Rightarrow N(P, \delta) \subset \varnothing$? Why does $\mathscr{S} \subset \text{In}(\mathscr{S}) \Rightarrow \mathscr{S}$ is open?
3. Prove Corollary 3.
4. For each of the spaces $\mathscr{E}^1, \mathscr{E}^2, \mathscr{E}^3$ give an example of a collection of ope· sets whose intersection is not open.

5. Prove (iii) of Theorem 4 using a complementation argument.
6. If $\mathscr{S}_1$ and $\mathscr{S}_2$ are closed, prove that $\mathscr{S}_1 \cup \mathscr{S}_2$ is closed by showing that $P \in \mathrm{Lp}(\mathscr{S}_1 \cup \mathscr{S}_2)$ and $P \notin \mathscr{S}_1 \cup \mathscr{S}_2$ are contradictory.
7. For each of the spaces $\mathscr{E}^1$, $\mathscr{E}^2$, $\mathscr{E}^3$ give an example of a collection of closed sets whose union is not a closed set.
8. In what sense are Exercises 4 and 7 really the same problem?
9. Give an example of a set in $\mathscr{E}^1$ whose boundary is empty.
10. Prove that the interior of a set $\mathscr{S}$ is the union of all the open sets that are contained in $\mathscr{S}$.
11. Prove that the closure of a set $\mathscr{S}$ is the intersection of all the closed sets that contain $\mathscr{S}$.
12. Prove that every finite set is closed.
13. Prove that $\mathrm{Cl}[\mathrm{In}(\mathscr{S})] \subset \mathrm{Cl}(\mathscr{S})$.
14. Prove that $\mathrm{In}(\mathscr{S}) \subset \mathrm{In}[\mathrm{Cl}(\mathscr{S})]$.
15. Prove that $\mathrm{Bd}[\mathrm{Cl}(\mathscr{S})] \subset \mathrm{Bd}(\mathscr{S})$.
16. Give examples to show that the opposite inclusions in Exercises 13, 14, and 15 need not be true.
17. Let $\mathscr{E} = \mathscr{E}^1$ be the euclidean 1-space and $\mathscr{E}' = \{x:0 \le x < 1\}$ be a relative metric space. Clearly, $\mathscr{E}'$ is not an open subset of $\mathscr{E}$. Why is $\mathscr{E}'$ open in $\mathscr{E}'$? Why is the set $\mathscr{S} = \{x: 0 \le x < \frac{1}{2}\}$ open in $\mathscr{E}'$? What is the closure in $\mathscr{E}'$ of the set $\{x: \frac{1}{2} \le x < 1\}$? Does it make sense to ask whether or not the set $\{x: -1 < x < 1\}$ is open in $\mathscr{E}'$?
18. In $\mathscr{E} = \mathscr{E}^1$, let $\mathscr{E}' = \{x: 0 \le x < 1\} \cup \{x: 2 \le x \le 3\}$ and consider $\mathscr{E}'$ a relative metric space. What are the neighborhoods, balls, and spheres in $\mathscr{E}'$ that are centered at 0 and have

    a. radius 1
    b. radius $\frac{3}{2}$
    c. radius 2
    d. radius 3

19. Let $\mathscr{E}$ be a metric space and let $\mathscr{E}' \subset \mathscr{E}$ be a relative metric space. If $\mathscr{S} \subset \mathscr{E}'$, is it true that $\mathrm{In}(\mathscr{S})' = \mathscr{E}' \cap \mathrm{In}(\mathscr{S})$? Is it true that $\mathrm{Bd}(\mathscr{S})' = \mathscr{E}' \cap \mathrm{Bd}(\mathscr{S})$?

SECTION 1.4. BOUNDEDNESS, CONNECTEDNESS, COMPACTNESS

A large class of geometric figures are object sets, that is, they are mathematical idealizations of physical, solid objects. Such a set is closed, is all in one piece, has a non-empty interior and boundary, and is limited in extent.

In this section we formulate three concepts—boundedness, connectedness, and compactness—that are essential features of these basic geometric figures.

We begin with the notion of a set being limited in extent, or bounded.

DEFINITION (Bounded set). A set $\mathscr{S}$ is a *bounded set*, or simply *bounded*, if its distance set of real numbers,

$$\{d(X, Y): X, Y \in \mathscr{S}\}$$

is bounded above, that is, there exists a real number k such that $d(X, Y) \leq k$ for all X, Y in $\mathscr{S}$.

In geometric terms, boundedness is equivalent to the following condition.

THEOREM 1. A set is bounded if and only if it is contained in some ball. (E)

COROLLARY 1. A ball, a sphere, and a neighborhood are bounded sets.

THEOREM 2. A subset of a bounded set is bounded, and the union of a finite number of bounded sets is bounded. (E)

Connectedness is related to our intuitive recognition of *an* object as distinct from two or more objects. A single object, such as a block of wood, is connected. If we cut through it, we separate it into two objects, and we also disconnect it. This suggests that perhaps we can define connectedness as the absence or negation of some form of separateness. That is, we can ask, "Under what separateness of sets $\mathscr{R}$ and $\mathscr{S}$ is it natural to regard $\mathscr{R} \cup \mathscr{S}$ as a disconnected set?"

It is clear that distinctness of $\mathscr{R}$ and $\mathscr{S}$, that is, $\mathscr{R} \neq \mathscr{S}$, is not sufficient to imply the disconnectedness of $\mathscr{R} \cup \mathscr{S}$. For if $\mathscr{R}$ is any set that we accept as connected, and $\mathscr{S}$ is any proper subset of $\mathscr{R}$, then $\mathscr{R} \neq \mathscr{S}$ but $\mathscr{R} \cup \mathscr{S} = \mathscr{R}$. A stronger form of distinctness is disjointness, that is, $\mathscr{R} \cap \mathscr{S} = \varnothing$. But this distinctness is still not strong enough. Consider the example in $\mathscr{E}^3$ where $\mathscr{R} = \mathrm{B}(P, \frac{1}{2}\delta)$ and $\mathscr{S} = \{X : \frac{1}{2}\delta < d(P, X) \leq \delta\}$. Then $\mathscr{R}$ and $\mathscr{S}$ are disjoint, but $\mathscr{R} \cup \mathscr{S} = \mathrm{B}(P, \delta)$, and by any reasonable definition a euclidean ball is a connected set.

In the example just given, the points of the sphere $\mathrm{S}(P, \frac{1}{2}\delta)$ belong to $\mathscr{R}$ and also to the boundary of $\mathscr{S}$. We can exclude this possibility and obtain a stronger form of distinctness than disjointness by the condition that neither $\mathscr{R}$ nor $\mathscr{S}$ should intersect the closure of the other. We give this form of distinctness a special name.

DEFINITION (Separated sets). Sets $\mathscr{R}$ and $\mathscr{S}$ are *separated* sets if neither is empty and neither intersects the closure of the other.

Thus non-empty sets $\mathscr{R}$ and $\mathscr{S}$ are separated if they are not only disjoint but also neither contains a boundary point (or equivalently a limit point) of the other.

We now define connectedness as a negation of separatedness.

DEFINITION (Connected set). A set is *connected* if it is not the union of two separated sets.

The situation presented by this definition is one that is not uncommon in mathematics. That is, although the definition is satisfactory from a logical point of view, it is difficult to apply in testing whether a given set is connected. However, the question of what sets are connected need not be settled immediately. Using the definition of connectedness, we can lay the groundwork for a future solution by showing that if each two points of a set belong to a connected subset, then the set itself must be connected. Later we show that segments are connected sets, and this, with the property just mentioned, will establish that all convex sets are connected, a result one would surely anticipate.

In the theorems to come, the following simple fact is useful.

THEOREM 3. If $\mathscr{R} \subset \mathscr{S}$, then $\mathrm{Cl}(\mathscr{R}) \subset \mathrm{Cl}(\mathscr{S})$. (E)

THEOREM 4. If a connected set is contained in the union of two separated sets, then it is contained in one of those sets.

PROOF. Let $\mathscr{S}$ denote a connected set that is contained in the union of two separated sets $\mathscr{R}_1$ and $\mathscr{R}_2$. Let sets $\mathscr{S}_1$ and $\mathscr{S}_2$ be defined by $\mathscr{S}_1 = \mathscr{S} \cap \mathscr{R}_1$ and $\mathscr{S}_2 = \mathscr{S} \cap \mathscr{R}_2$. Clearly, $\mathscr{S} = \mathscr{S}_1 \cup \mathscr{S}_2$, and, because $\mathscr{S}$ is connected, the sets $\mathscr{S}_1$ and $\mathscr{S}_2$ are not separated. Since $\mathscr{S}_1 \subset \mathscr{R}_1$, Theorem 3 implies that $\mathrm{Cl}(\mathscr{S}_1) \subset \mathrm{Cl}(\mathscr{R}_1)$. Similarly, $\mathscr{S}_2 \subset \mathscr{R}_2$ implies that $\mathrm{Cl}(\mathscr{S}_2) \subset \mathrm{Cl}(\mathscr{R}_2)$. Because $\mathscr{R}_1$ and $\mathscr{R}_2$ are separated, by hypothesis, then $\mathscr{R}_1 \cap \mathrm{Cl}(\mathscr{R}_2) = \varnothing$ and also $\mathscr{R}_2 \cap \mathrm{Cl}(\mathscr{R}_1) = \varnothing$. Therefore $\mathscr{S}_1 \cap \mathrm{Cl}(\mathscr{S}_2) = \varnothing$ and also $\mathscr{S}_2 \cap \mathrm{Cl}(\mathscr{S}_1) = \varnothing$. Thus if neither $\mathscr{S}_1$ nor $\mathscr{S}_2$ were empty, they would be separated sets. Since they are not separated sets, either $\mathscr{S}_1$ is empty, and $\mathscr{S}_2 \subset \mathscr{R}_2$, or else $\mathscr{S}_2$ is empty and $\mathscr{S} \subset \mathscr{R}_1$. □

THEOREM 5. If each two points of a set $\mathscr{S}$ belong to some connected subset of $\mathscr{S}$, then $\mathscr{S}$ itself is a connected set.

PROOF. For an indirect argument, assume that $\mathscr{S}$ is not connected. Then $\mathscr{S}$ is the union of two separated sets $\mathscr{S}_1$ and $\mathscr{S}_2$. By definition, these sets are not empty, so there exists a point $P_1 \in \mathscr{S}_1$ and a point $P_2 \in \mathscr{S}_2$. By the hypothesis of the theorem, P_1 and P_2 belong to some connected subset of $\mathscr{S}$, say the set $\mathscr{R}$. Since $\mathscr{R} \subset \mathscr{S} = \mathscr{S}_1 \cup \mathscr{S}_2$, then, by Theorem 4, either $\mathscr{R} \subset \mathscr{S}_1$ or else $\mathscr{R} \subset \mathscr{S}_2$. But $P_2 \notin \mathscr{S}_1$, so $\mathscr{R} \not\subset \mathscr{S}_1$, and $P_1 \notin \mathscr{S}_2$, so $\mathscr{R} \not\subset \mathscr{S}_2$. The contradiction shows that the assumption is false, hence $\mathscr{S}$ is a connected set. □

THEOREM 6. If two connected sets have a point in common, then their union is a connected set. (E)

THEOREM 7. The closure of a connected set is a connected set. (E)

In defining boundedness and connectedness, we were guided by natural intuitions about these concepts. Our third basic concept—compactness—is more subtle and not readily suggested by intuition. Although the concept has origins in the distant past, it was not clearly formulated nor its importance clearly recognized until the first part of the twentieth century.

Since the notion of compactness, which we are about to define, is perhaps the most difficult and least intuitive concept we will encounter, it may be helpful to give some indication of its importance. First, there is the fact, which we prove a little later, that in $\mathscr{E}^n$ the compact sets are precisely those which are closed and bounded. Since closed and bounded sets form a very important class, especially from the point of view of geometry, an alternate characterization of them is a valuable new insight. But that is only part of the story.

Although in $\mathscr{E}^n$ a closed and bounded set is compact and conversely, it is the properties formulated in compactness that lead to easy proofs for important properties of closed and bounded sets. For example, it is relatively easy to show that the continuous image of a compact set is again a compact set. From this we obtain the same property for closed and bounded sets. In particular, if the continuous function is real valued, then the compact range is a closed and bounded set of real numbers, and the function attains a maximum and a minimum. Since a wide class of geometric figures are compact and many geometric measures on them are real valued, continuous functions, compactness is a powerful tool in establishing extremal properties of such figures. Throughout the text, many apparently difficult problems are solved easily by the use of compactness.

Because the property of a set being closed and bounded *is* an intuitive notion, whereas compactness is not, the former notion initially had the dominant role. However, proofs about closed and bounded sets in $\mathscr{E}^n$ invariably led to some point where the properties (or equivalent ones) of compactness were involved. Thus it gradually became clear, as a practical matter, that singling these properties out and identifying them with a name—compactness—would simplify and unify many proofs. It then turned out that in spaces more abstract than $\mathscr{E}^n$ the concept of "closed and bounded" was less fruitful than that of compactness. Thus compactness became the featured concept.

We now introduce a simple notion required in the definition of compactness, namely, the concept of a cover.

DEFINITION (Cover, open cover, neighborhood cover). A collection of sets $\mathscr{F} = \{\mathscr{S}_\alpha : \alpha \in \Lambda\}$ is a *cover* of set $\mathscr{R}$ if $\mathscr{R}$ is contained in the union of the sets in $\mathscr{F}$, hence if each point of $\mathscr{R}$ belongs to at least one set in $\mathscr{F}$. The cover $\mathscr{F}$ is a *finite* or *infinite cover* according as the number of sets in $\mathscr{F}$ is finite or infinite. The cover is an *open cover* if all the sets in $\mathscr{F}$ are open sets. The cover is a *neighborhood cover* if all the sets in $\mathscr{F}$ are neighborhoods.

DEFINITION (Compactness). A set $\mathscr{S}$ is a *compact set*, or simply compact, if every collection of open sets that covers $\mathscr{S}$ has a finite subcollection that also covers $\mathscr{S}$.

Our first result on compactness can be useful in proving a set to be compact.

THEOREM 8. A set $\mathscr{S}$ is compact if and only if every neighborhood cover of $\mathscr{S}$ has a finite subcollection that covers $\mathscr{S}$.

PROOF. Since neighborhoods are open sets, a neighborhood cover is a special kind of open cover. Thus if we already know that $\mathscr{S}$ is compact, it follows that every neighborhood cover of $\mathscr{S}$ has a finite subcollection that also covers $\mathscr{S}$.

To prove the converse, we must show that every open cover of $\mathscr{S}$ has a finite subcollection that also covers $\mathscr{S}$ given that this is true for the special case of neighborhood covers. Let $\{\mathscr{S}_\alpha : \alpha \in \Lambda\}$ be any open cover of $\mathscr{S}$. For each point $X \in \mathscr{S}$ there is an $\alpha(X) \in \Lambda$ such that $X \in \mathscr{S}_{\alpha(X)}$. Since $\mathscr{S}_{\alpha(X)}$ is an open set, there is a neighborhood $N_X = \mathrm{N}(X, \delta(X))$ such that $X \in N_X \subset \mathscr{S}_{\alpha(X)}$. The collection $\{N_X : X \in \mathscr{S}\}$ is clearly a neighborhood cover of $\mathscr{S}$. By hypothesis, there is a finite subcollection of these neighborhoods, say N_{X_1}, $N_{X_2}, \ldots, N_{X_k}$ whose union still covers $\mathscr{S}$. Thus

$$\mathscr{S} \subset \bigcup_{i=1}^{k} N_{X_i} \subset \bigcup_{i=1}^{k} \mathscr{S}_{\alpha(X_i)},$$

hence $\mathscr{S}_{\alpha(X_1)}, \mathscr{S}_{\alpha(X_2)}, \ldots, \mathscr{S}_{\alpha(X_k)}$ is a finite subcollection of the original open cover, and this finite subcollection also covers $\mathscr{S}$. □

Our next two theorems show that compact sets are necessarily closed and bounded. Happily, the converse is also true in euclidean spaces, but Exercise 16 (Sec. 1.4) shows that the two concepts are not logically equivalent in general metric spaces. Thus compactness is a more restrictive set property than closedness and boundedness.

THEOREM 9. A compact set is a bounded set.

PROOF. Let $\mathscr{S}$ be a compact set. Corresponding to any point A, the sequence of neighborhoods $\mathrm{N}(A, 1), \mathrm{N}(A, 2), \ldots, \mathrm{N}(A, k), \ldots$ is a cover of the whole space $\mathscr{E}$ and hence is a cover of $\mathscr{S}$. Because $\mathscr{S}$ is compact, there is a finite subcollection of these neighborhoods that also covers $\mathscr{S}$. Since they are concentric, their union is just the largest of them. Thus $\mathscr{S}$ is contained in a neighborhood and, by Theorem 1, is a bounded set. □

THEOREM 10. A compact set is a closed set.

PROOF. Let $\mathscr{S}$ be a compact set. We show that $\mathrm{Cp}(\mathscr{S})$ is open and hence that $\mathscr{S}$ is closed. To this end, consider a point $A \in \mathrm{Cp}(\mathscr{S})$. For each X in $\mathscr{S}$, $X \neq A$ implies that $\delta_{AX} = \frac{1}{3}d(A, X)$ is greater than zero. Thus the neighborhoods $\mathrm{N}(X, \delta_{AX})$ and $\mathrm{N}(A, \delta_{AX})$ exist, and clearly they are disjoint. The collection of neighborhoods $\{\mathrm{N}(X, \delta_{AX}): X \in \mathscr{S}\}$ is a cover of $\mathscr{S}$. Since $\mathscr{S}$ is compact, there is a finite subcollection

$$\mathrm{N}(X_1, \delta_{AX_1}), \mathrm{N}(X_2, \delta_{AX_2}), \ldots, \mathrm{N}(X_k, \delta_{AX_k}) \tag{1}$$

that is also a cover of $\mathscr{S}$. Since the corresponding pairs of neighborhoods $\mathrm{N}(X_i, \delta_{AX_i})$ and $\mathrm{N}(A, \delta_{AX_i})$ are disjoint, then the smallest of the concentric neighborhoods $\mathrm{N}(A, \delta_{AX_i})$ does not intersect any of the neighborhoods in (1). Therefore it does not intersect $\mathscr{S}$. It is therefore a neighborhood of A that is contained in $\mathrm{Cp}(\mathscr{S})$. Hence $A \in \mathrm{In}[\mathrm{Cp}(\mathscr{S})]$. Since the points of $\mathrm{Cp}(\mathscr{S})$ are interior points, $\mathrm{Cp}(\mathscr{S})$ is open, hence $\mathscr{S}$ is closed. □

Our next theorem plays a key role in a later proof that closed and bounded sets in $\mathscr{E}^n$ are compact.

THEOREM 11. A closed subset of a compact set is also a compact set.

PROOF. Let $\mathscr{S}$ be a compact set and let $\mathscr{R}$ be a closed subset of $\mathscr{S}$. Consider any collection $\mathscr{F}$ of open sets that is a cover of $\mathscr{R}$. Adjoin to $\mathscr{F}$ the open set $\mathrm{Cp}(\mathscr{R})$ to form a collection $\mathscr{F}_1$.

Now consider any point X in $\mathscr{S}$. If X is in $\mathscr{R}$, then it belongs to some set in $\mathscr{F}$ which is also a set in $\mathscr{F}_1$. If X is not in $\mathscr{R}$, then it belongs to $\mathrm{Cp}(\mathscr{R})$ which is a set in $\mathscr{F}_1$. Hence $\mathscr{F}_1$ is an open cover of $\mathscr{S}$. Because $\mathscr{S}$ is compact, there is a finite subcollection of $\mathscr{F}_1$, say a collection $\mathscr{F}_2$, that is also a cover of $\mathscr{S}$. Since $\mathscr{F}_2$ covers $\mathscr{S}$ it also covers $\mathscr{R} \subset \mathscr{S}$. Now let $\mathscr{F}_3$ be defined as the collection $\mathscr{F}_2$, if $\mathrm{Cp}(\mathscr{R})$ is not in $\mathscr{F}_2$, and as the collection $\mathscr{F}_2$ with $\mathrm{Cp}(\mathscr{R})$ deleted, if

$\text{Cp}(\mathscr{R})$ is in $\mathscr{F}_2$. Then $\mathscr{F}_3$ is a finite subcollection of the original cover $\mathscr{F}$. Moreover, since $\text{Cp}(\mathscr{R})$ contains no points of $\mathscr{R}$, deleting it from $\mathscr{F}_2$, if necessary, still leaves a cover of $\mathscr{R}$. Thus $\mathscr{F}_3$ is a finite subcollection of $\mathscr{F}$ that covers $\mathscr{R}$. Hence $\mathscr{R}$ is compact. □

The proofs of the last two results in this section involve techniques that are of great use throughout the book. We need the following notions from set theory.

DEFINITION (Nested sequences). A sequence of sets $\{\mathscr{S}_i : i = 1, 2, \ldots\}$ is an *increasingly nested* sequence if $\mathscr{S}_i \subset \mathscr{S}_{i+1}$, $i = 1, 2, \ldots$ and is a *decreasingly nested* sequence if $\mathscr{S}_i \supset \mathscr{S}_{i+1}$, $i = 1, 2, \ldots$.

The next theorem is closely related to compactness and is often called the Cantor intersection theorem.

THEOREM 12. If the sets in a decreasingly nested sequence are non-empty and compact, then the sequence has a non-empty intersection set, that is, there is at least one point that belongs to all the sets in the sequence.

PROOF. Let $\mathscr{F} = \{\mathscr{S}_i : i = 1, 2, \ldots\}$ be a decreasingly nested sequence of non-empty compact sets. Assume that the intersection of all the sets in $\mathscr{F}$ is the empty set. Now consider the sequence $\mathscr{G} = \{\text{Cp}(\mathscr{S}_i) : i = 1, 2, \ldots\}$. This is clearly an increasingly nested sequence, and since the sets $\mathscr{S}_i$ are compact, they are closed (Theorem 10), so the sets $\text{Cp}(\mathscr{S}_i)$ are open. If the sequence $\mathscr{G}$ is not a cover of $\mathscr{S}_1$, then at least one point X_o in $\mathscr{S}_1$ is not in any set $\text{Cp}(\mathscr{S}_i)$. It is therefore in all the sets $\mathscr{S}_i$, which contradicts our assumption. On the other hand, if $\mathscr{G}$ is a cover of $\mathscr{S}_1$, it is an open cover of a compact set. Hence some finite subcollection of $\mathscr{G}$ also covers $\mathscr{S}_1$. Since this finite subcollection is finite and increasingly nested, it has a largest set, say $\text{Cp}(\mathscr{S}_n)$. Then $\mathscr{S}_1 \subset \text{Cp}(\mathscr{S}_n)$. But this is impossible because $\mathscr{S}_1 \supset \mathscr{S}_n$. The contradictions show that our assumption must be false, hence there is at least one point common to all the sets in $\mathscr{F}$. □

THEOREM 13. If the intersection set of a decreasingly nested sequence of compact sets is a singleton set $\{P\}$, then every neighborhood of P contains all but a finite number of the sets in the sequence.

PROOF. Let $\{\mathscr{S}_i : i = 1, 2, \ldots\}$ be a decreasingly nested sequence of compact sets whose intersection is the set $\{P\}$. Then none of the sets $\mathscr{S}_i$ can be empty. Let $N = \text{N}(P, \delta)$ be an arbitrary neighborhood of P. If there is any positive integer i_o such that $\mathscr{S}_{i_o} \subset N$, then for all $i > i_o$ we have $N \supset \mathscr{S}_{i_o} \supset \mathscr{S}_i$, hence only the sets $\mathscr{S}_1, \mathscr{S}_2, \ldots, \mathscr{S}_{i_o - 1}$ could fail to be contained in N. For an indirect proof, assume that there is no positive integer i_o such that $\mathscr{S}_{i_o} \subset N$.

Then $\mathscr{R}_i = \mathscr{S}_i \cap \mathrm{Cp}(N) \neq \varnothing$, for $i = 1, 2, \ldots$. Since N is a neighborhood, it is open, hence $\mathrm{Cp}(N)$ is closed. The sets $\mathscr{S}_i$ are compact, so they are closed. Each $\mathscr{R}_i$ is therefore the intersection of closed sets, hence it is closed. Because $\mathscr{R}_i \subset \mathscr{S}_i$, $\mathscr{R}_i$ is a closed subset of a compact set. Thus, by Theorem 11, $\mathscr{R}_i$ is compact. Because $\mathscr{S}_i \supset \mathscr{S}_{i+1}$, we also have $\mathscr{R}_i \supset \mathscr{R}_{i+1}$. Therefore $\{\mathscr{R}_i : i = 1, 2, \ldots\}$ is a decreasingly nested sequence of non-empty compact sets. By the Cantor intersection theorem, there is at least one point Q belonging to all the sets $\mathscr{R}_i$. Since $Q \in \mathrm{Cp}(N) = \mathrm{Cp}[\mathrm{N}(P, \delta)]$, $Q \neq P$. But since Q belongs to all the sets $\mathscr{R}_i$, it also belongs to all the sets $\mathscr{S}_i$, and therefore $Q = P$. The contradiction shows that our assumption is false. Hence there is an integer i_o such that $\mathscr{S}_{i_o}$, and all succeeding $\mathscr{S}_i$ are contained in $N(P, \delta)$ □

To conclude this section, it is an important fact that the three concepts featured—boundedness, connectedness, and compactness—are all intrinsic with respect to relativization of the metric in the sense made explicit in the following theorems.

THEOREM 14. If $\mathscr{E}$ is a metric space and $\mathscr{E}'$ is a relative metric subspace, then a subset of $\mathscr{E}'$ is bounded in $\mathscr{E}'$ if and only if it is bounded in $\mathscr{E}$. (E)

THEOREM 15. Let $\mathscr{E}$ be a metric space and $\mathscr{E}'$ be a relative subspace.

(i) If sets $\mathscr{R}$ and $\mathscr{T}$ are separated in $\mathscr{E}'$, then they are separated in $\mathscr{E}$.
(ii) If $\mathscr{R}$ and $\mathscr{T}$ are separated sets in $\mathscr{E}$, then $\mathscr{R} \cap \mathscr{E}'$ and $\mathscr{T} \cap \mathscr{E}'$ are separated sets in $\mathscr{E}'$, provided that neither is empty.
(iii) A set $\mathscr{S}$ in $\mathscr{E}'$ is connected in $\mathscr{E}'$ if and only if it is connected in $\mathscr{E}$.

PROOF. We use the prime notation of Sec. 1.3 for the subspace $\mathscr{E}'$.

(i) From (iv) of Theorem 7, Sec. 1.3, the sets $\mathscr{R}$ and $\mathscr{T}$, separated in $\mathscr{E}'$, have closure in $\mathscr{E}'$ expressed by $\mathrm{Cl}(\mathscr{R})' = \mathrm{Cl}(\mathscr{R}) \cap \mathscr{E}'$ and $\mathrm{Cl}(\mathscr{T})' = \mathrm{Cl}(\mathscr{T}) \cap \mathscr{E}'$. Because $\mathscr{R}$ and $\mathscr{T}$ are separated in $\mathscr{E}'$, $\mathscr{R} \cap \mathrm{Cl}(\mathscr{T})' = \varnothing$ and $\mathscr{T} \cap \mathrm{Cl}(\mathscr{R})' = \varnothing$, hence $\mathscr{R} \cap [\mathrm{Cl}(\mathscr{T}) \cap \mathscr{E}'] = \varnothing$ and $\mathscr{T} \cap [\mathrm{Cl}(\mathscr{R}) \cap \mathscr{E}'] = \varnothing$. From the commutativity of intersections, it follows that $[\mathscr{R} \cap \mathscr{E}'] \cap \mathrm{Cl}(\mathscr{T}) = \varnothing$ and $[\mathscr{T} \cap \mathscr{E}'] \cap \mathrm{Cl}(\mathscr{R}) = \varnothing$. But $\mathscr{R}$ and $\mathscr{T}$ are subsets of $\mathscr{E}'$, so $\mathscr{R} \cap \mathscr{E}' = \mathscr{R}$ and $\mathscr{T} \cap \mathscr{E}' = \mathscr{T}$. Therefore $\mathscr{R} \cap \mathrm{Cl}(\mathscr{T}) = \varnothing$ and $\mathscr{T} \cap \mathrm{Cl}(\mathscr{R}) = \varnothing$, so $\mathscr{R}$ and $\mathscr{T}$ are separated in $\mathscr{E}$.

(ii) Let $\mathscr{R}$ and $\mathscr{T}$ denote separated sets in $\mathscr{E}$ such that $\mathscr{R} \cap \mathscr{E}' \neq \varnothing$ and $\mathscr{T} \cap \mathscr{E}' \neq \varnothing$. We want to show that

$$(\mathscr{R} \cap \mathscr{E}') \cap \mathrm{Cl}(\mathscr{T} \cap \mathscr{E}')' = \varnothing,$$

and (1)

$$(\mathscr{T} \cap \mathscr{E}') \cap \mathrm{Cl}(\mathscr{R} \cap \mathscr{E}')' = \varnothing.$$

From (iv) of Theorem 7, Sec. 1.3, $\mathrm{Cl}(\mathcal{T} \cap \mathcal{E}')' = \mathrm{Cl}(\mathcal{T} \cap \mathcal{E}') \cap \mathcal{E}'$ and $\mathrm{Cl}(\mathcal{R} \cap \mathcal{E}')' = \mathrm{Cl}(\mathcal{R} \cap \mathcal{E}') \cap \mathcal{E}'$. Thus the relations in (1) are equivalent to

$$\mathcal{R} \cap \mathcal{E}' \cap \mathrm{Cl}(\mathcal{T} \cap \mathcal{E}') = \varnothing, \text{ and } \mathcal{T} \cap \mathcal{E}' \cap \mathrm{Cl}(\mathcal{R} \cap \mathcal{E}') = \varnothing. \tag{2}$$

From the separatedness of $\mathcal{R}$ and $\mathcal{T}$ in $\mathcal{E}$, we know that

$$\mathcal{R} \cap \mathrm{Cl}(\mathcal{T}) = \varnothing \text{ and } \mathcal{T} \cap \mathrm{Cl}(\mathcal{R}) = \varnothing. \tag{3}$$

Since $\mathcal{R} \cap \mathcal{E}' \subset \mathcal{R}$, and since $\mathcal{T} \cap \mathcal{E}' \subset \mathcal{T}$ implies $\mathrm{Cl}(\mathcal{T} \cap \mathcal{E}') \subset \mathrm{Cl}(\mathcal{T})$, by Theorem 3, the first relation in (3) implies the first relation in (2). Similarly, $\mathcal{T} \cap \mathcal{E}' \subset \mathcal{T}$, and $\mathrm{Cl}(\mathcal{R} \cap \mathcal{E}') \subset \mathrm{Cl}(\mathcal{R})$, so the second relation in (3) implies the second relation in (2). Thus (ii) is proved.

(iii) We want to show that a subset $\mathcal{S}$ of $\mathcal{E}'$ is connected in $\mathcal{E}'$ if and only if it is connected in $\mathcal{E}$. We do so by showing that if a subset of $\mathcal{E}'$ is disconnected in either $\mathcal{E}'$ or $\mathcal{E}$, then it is disconnected in both.

Let $\mathcal{U}$ be a subset of $\mathcal{E}'$ that is disconnected in $\mathcal{E}'$. Then $\mathcal{U} = \mathcal{R} \cup \mathcal{T}$, where $\mathcal{R}$ and $\mathcal{T}$ are separated in $\mathcal{E}'$. By (i), $\mathcal{R}$ and $\mathcal{T}$ are also separated in $\mathcal{E}$, hence $\mathcal{U}$ is disconnected in $\mathcal{E}$.

Next, let $\mathcal{U}$ be a subset of $\mathcal{E}'$ that is disconnected in $\mathcal{E}$. Then sets $\mathcal{R}$ and $\mathcal{T}$ exist that are separated in $\mathcal{E}$ and are such that $\mathcal{U} = \mathcal{R} \cup \mathcal{T}$. In particular, $\mathcal{R}$ and $\mathcal{T}$ are non-empty subsets of $\mathcal{U}$ and hence of $\mathcal{E}'$. Thus $\mathcal{R} = \mathcal{R} \cap \mathcal{E}'$ and $\mathcal{T} = \mathcal{T} \cap \mathcal{E}'$ are non-empty sets, and, from (ii), they are separated in $\mathcal{E}'$. Thus $\mathcal{U} = (\mathcal{R} \cap \mathcal{E}') \cup (\mathcal{T} \cap \mathcal{E}')$ shows that $\mathcal{U}$ is disconnected in $\mathcal{E}'$. □

THEOREM 16. If $\mathcal{E}$ is a metric space and $\mathcal{E}'$ is a relative subspace, then a subset of $\mathcal{E}'$ is compact in $\mathcal{E}'$ if and only if it is compact in $\mathcal{E}$.

PROOF. Let $\mathcal{K}$ denote a subset of $\mathcal{E}'$. We suppose first that $\mathcal{K}$ is compact in $\mathcal{E}'$. Let $\{\mathcal{G}_\alpha : \alpha \in \Lambda\}$ be a cover of $\mathcal{K}$ by open sets of $\mathcal{E}$. Then, because $\mathcal{K} \subset \mathcal{E}'$, the collection $\{\mathcal{G}_\alpha \cap \mathcal{E}' : \alpha \in \Lambda\}$ is a cover of $\mathcal{K}$. By (ii) of Theorem 7, Sec. 1.3, the sets $\mathcal{G}_\alpha \cap \mathcal{E}'$ are open in $\mathcal{E}'$. Since $\mathcal{K}$ is compact in $\mathcal{E}'$, the open cover $\{\mathcal{G}_\alpha \cap \mathcal{E}' : \alpha \in \Lambda\}$ has a finite subcollection, say $\{\mathcal{G}_{\alpha_1} \cap \mathcal{E}', \ldots, \mathcal{G}_{\alpha_m} \cap \mathcal{E}'\}$ that is also a cover of $\mathcal{K}$. Because $\mathcal{G}_{\alpha_i} \cap \mathcal{E}' \subset \mathcal{G}_{\alpha_i}$, $i = 1, 2, \ldots, m$, the collection $\{\mathcal{G}_{\alpha_1}, \mathcal{G}_{\alpha_2}, \ldots, \mathcal{G}_{\alpha_m}\}$ is again a cover of $\mathcal{K}$. Thus every cover of $\mathcal{K}$ by open sets in $\mathcal{E}$ has a finite subcollection that covers $\mathcal{K}$. Hence $\mathcal{K}$ is compact in $\mathcal{E}$.

Next, suppose that $\mathcal{K}$ is compact in $\mathcal{E}$. Now let $\{\mathcal{G}_\alpha : \alpha \in \Lambda\}$ be an arbitrary cover of $\mathcal{K}$ by sets that are open in $\mathcal{E}'$. By (ii) of Theorem 7, Sec. 1.3, because $\mathcal{G}_\alpha$ is open in $\mathcal{E}'$, $\alpha \in \Lambda$, there exists a set $\mathcal{H}_\alpha$ that is open in $\mathcal{E}$ and such that $\mathcal{G}_\alpha = \mathcal{H}_\alpha \cap \mathcal{E}'$, $\alpha \in \Lambda$. Since $\mathcal{K} \subset \mathcal{E}'$, the collection $\{\mathcal{H}_\alpha : \alpha \in \Lambda\}$ is also a cover of $\mathcal{K}$ and is a cover of $\mathcal{K}$ by sets open in $\mathcal{E}$. Because $\mathcal{K}$ is compact in $\mathcal{E}$, this collection has a finite subcollection $\{\mathcal{H}_{\alpha_1}, \mathcal{H}_{\alpha_2}, \ldots, \mathcal{H}_{\alpha_j}\}$ that covers $\mathcal{K}$. Then,

$$\mathcal{K} = (\mathcal{K} \cap \mathcal{E}') \subset \bigcup_{i=1}^{j} (\mathcal{H}_{\alpha_i} \cap \mathcal{E}') = \bigcup_{i=1}^{j} \mathcal{G}_{\alpha_i}.$$

shows that $\{\mathscr{G}_{\alpha_1}, \mathscr{G}_{\alpha_2}, \ldots, \mathscr{G}_{\alpha_j}\}$ is a cover of $\mathscr{K}$. Thus every cover of $\mathscr{K}$ by open sets in $\mathscr{E}'$ has a finite subcollection that covers $\mathscr{K}$. Thus $\mathscr{K}$ is compact in $\mathscr{E}'$. □

Exercises – Section 1.4

1. Prove that a set of points in $\mathscr{E}^1$ is bounded if and only if it is contained in some neighborhood of the origin.
2. Prove Theorem 1 and Theorem 2.
3. Prove that if $\mathscr{S}$ is any set such that $\mathrm{In}(\mathscr{S})$ and $\mathrm{Ex}(\mathscr{S})$ are non-empty, then these are separated sets.
4. In $\mathscr{E}^1$, if $P \neq Q$, what are necessary conditions on $d(P, Q)$ in order for the following pairs to be separated sets?

   a. $\mathrm{N}(P, 2)$, $\mathrm{N}(Q, 3)$
   b. $\mathrm{N}(P, 2)$, $\mathrm{B}(Q, 3)$
   c. $\mathrm{S}(P, 2)$, $\mathrm{N}(Q, 3)$
   d. $\mathrm{S}(P, 2)$, $\mathrm{S}(Q, 3)$

5. Prove Theorem 3.
6. Prove Theorem 6.
7. Prove that if a collection of any number of connected sets is such that each two intersect, then the union of the sets is connected.
8. What are the finite connected sets?
9. Prove Theorem 7.
10. Prove that a set $\mathscr{S}$ must be disconnected if there exists some set $\mathscr{R}$ such that $\mathrm{In}(\mathscr{R}) \cap \mathscr{S} \neq \varnothing$, $\mathrm{Ex}(\mathscr{R}) \cap \mathscr{S} \neq \varnothing$, but $\mathrm{Bd}(\mathscr{R}) \cap \mathscr{S} = \varnothing$.
11. Why is any finite set compact?
12. What is wrong with the following reasoning? Suppose $\mathscr{F}$ is any open cover of the open set $\mathscr{S} = \{x \in \mathscr{E}^1 : 0 < x < 1\}$. The sets $\{x \in \mathscr{E}' : 0 < x < \frac{1}{2}\}$, $\{x \in \mathscr{E}^1 : \frac{1}{4} < x < \frac{3}{4}\}$, and $\{x \in \mathscr{E}^1 : \frac{1}{2} < x < 1\}$ form a finite collection of open sets that cover $\mathscr{S}$. Therefore $\mathscr{S}$ is compact.
13. Can *every* set be covered by a finite number of open sets? By a single open set?
14. Let $\mathscr{F} = \{\mathscr{S}_i : i = 1, 2, \ldots\}$, where $\mathscr{S}_i = \{x \in \mathscr{E}^1 : 1/i < x < 1 - 1/i\}$. Show that $\mathscr{F}$ is an open cover of $\mathscr{S} = \{x \in \mathscr{E}^1 : 0 < x < 1\}$. Do the sets $\mathscr{S}_1, \mathscr{S}_2, \ldots, \mathscr{S}_{25}$ form a cover of $\mathscr{S}$? Does any finite subcollection of $\mathscr{F}$ cover $\mathscr{S}$? If not, does that imply that $\mathscr{S}$ is not compact?
15. Let $\mathscr{S} = \{x \in \mathscr{E}^1 : 0 \leq x \leq 1\}$. Let the collection $\mathscr{F}$ consist of the neighborhoods $\mathrm{N}(P, d(P, Q))$, $\mathrm{N}(Q, d(P, Q))$ for all P, Q in $\mathscr{S}$. Is $\mathscr{F}$ a cover of $\mathscr{S}$? If so, is there a finite subcollection of $\mathscr{F}$ that also covers $\mathscr{S}$? If there is, does this prove that $\mathscr{S}$ is compact?

16. For $x, y \in \mathscr{E}^1$, define $d_1(x, y) = \min\{1, |x - y|\}$.

    a. Verify that d_1 is a metric on $\mathscr{E}^1$.
    b. Prove that a set is open with respect to neighborhoods defined in terms of d_1 if and only if it is open with respect to neighborhoods defined by the euclidean metric.
    c. Prove that precisely the same sets are compact in both metric spaces by making use of (b) and the definition of compactness.
    d. Explain why the concept of "closed and bounded" need not be equivalent to "compact" in every metric space.

17. What is wrong with the argument used for Theorem 9 if we apply it with $\mathscr{E} = \mathscr{E}^1$, $\mathscr{S} = \mathscr{E}^1$, and $A = 0$?
18. Using Theorem 10, prove that the set $\{x \in \mathscr{E}^1 : 0 < x < 1\}$ is not compact.
19. Under what conditions will the collection of closed sets in a metric space coincide with the collection of compact sets?
20. Give an example of a decreasingly nested sequence of non-empty, bounded, open sets whose intersection set is empty. What does your example prove about open sets and compactness?
21. Give an example of a decreasingly nested sequence of non-empty closed sets whose intersection is empty. What does your example prove about closed sets and compactness?
22. Prove that the union of a finite number of compact sets is compact.
23. Give an example to show that the union of infinitely many compact sets need not be compact.
24. Using Theorem 10 and Theorem 11, prove that the intersection of any number of compact sets is compact.
25. If $\mathscr{S}$ is a compact set and point A is not in $\mathscr{S}$, prove that there are disjoint open sets $\mathscr{U}$ and $\mathscr{V}$ such that $A \in \mathscr{U}$ and $\mathscr{S} \subset \mathscr{V}$. (Hint: examine the proof of Theorem 10.)
26. Let $\mathscr{S}_1, \mathscr{S}_2, \ldots, \mathscr{S}_n$ be a finite number of connected sets such that each intersects its successor, that is, $\mathscr{S}_i \cap \mathscr{S}_{i+1} \neq \varnothing$, $i = 1, 2, \ldots, n - 1$. Prove that the union of the sets is connected.
27. Prove Theorem 14.

SECTION 1.5. CONTINUOUS MAPPINGS OF METRIC SETS

An intuitive feeling for continuity derives, no doubt, from our experiences with a kinematic world. This intuition is extremely valuable in mathematics and is especially so in geometry, where it often allows us to "see" the answer to a problem almost instantly. For example, consider the question of the

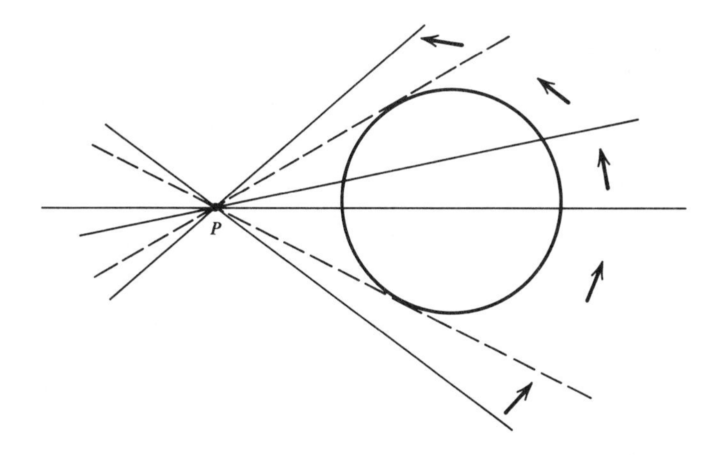

number of tangents to a circle from an outside point P. If we visualize a line turning "continuously" about P, we can "see" that it will pass through the circle. As it changes from a nonintersector of the circle to an intersector and then again to a nonintersector, there must be positions of first and last contact with the circle, and these correspond to two tangents.

To turn the argument just given into a proof, we need a mathematical formulation for the rotation and also for the continuity of the rotation, since it is this which implies that the motion does not "jump" over the first and last contact positions. It is because of these difficulties that elementary texts give a static proof that is less natural and far more complicated.

Although the example just discussed is a simple one, it carries a suggestion of great importance. If we could formulate the motion proof in mathematical terms, then perhaps we could turn our continuity intuitions into actual proofs, not just in this case, but over a wide range of important problems. Such a program is possible, and it is what we embark on in this section. Later, we see its exploitation in geometric problems.

Before we define continuity, it is helpful to have some agreements about language and notations related to functions.

CONVENTIONS. We are interested in functions whose domain of definition is contained in a metric space, referred to as the *domain space*, and whose range is also contained in a metric space, called the *range space*. These may be the same space or different spaces. In keeping with our previous convention, we use the symbol d for the metric of both spaces with the understanding that these need not be the same metric.

We ordinarily use $\mathscr{D}$ and $\mathscr{R}$ for the domain and range sets of a function f and denote the function by $f : \mathscr{D} \to \mathscr{R}$. This notation not only displays the three static elements, f, $\mathscr{D}$, $\mathscr{R}$, of the function, but also indicates the dynamic view of a function in which it is seen as the mapping that carries the domain to the range. In fact, the function $f : \mathscr{D} \to \mathscr{R}$ is said to be a mapping of $\mathscr{D}$ *onto* $\mathscr{R}$ and *into* any set that contains $\mathscr{R}$. The point $f(X)$ is *the image* of X under the mapping, and X is *a pre-image* of $f(X)$. The function is one to one, or 1-1, if each point in the range has a unique pre-image, that is, if $X \neq Y$ implies $f(X) \neq f(Y)$. If f is 1-1, then the *inverse function* $f^{-1} : \mathscr{R} \to \mathscr{D}$ is well defined by "$f^{-1}(Y) = X$ if and only if $Y = f(X)$."

If $\mathscr{S}$ is any set *in the domain* of f, we use the abbreviation $f(\mathscr{S})$ for the *image of set* $\mathscr{S}$, that is,

$$f(\mathscr{S}) = \{f(X): X \in \mathscr{S}\}.$$

If $\mathscr{T}$ is any set *in the range space* of f, we denote the *inverse image of set* $\mathscr{T}$ by $f^{-1}(\mathscr{T})$, that is,

$$f^{-1}(\mathscr{T}) = \{X : f(X) \in \mathscr{T}\}.$$

Comments. The definition for $f^{-1}(\mathscr{T})$ does not imply that f is 1-1. Here f^{-1} has the meaning of the inverse *relation* to f. Note especially that the definitions for $f(\mathscr{S})$ and $f^{-1}(\mathscr{T})$ are not symmetric with respect to the domain and range, respectively. Thus $f(\mathscr{S})$, by definition, requires $\mathscr{S} \subset \mathscr{D}$, so $f(\mathscr{S})$ is a subset of the range. But the definition of $f^{-1}(\mathscr{T})$ does not require $\mathscr{T}$ to be a subset of the range, but only a subset of the range space. Thus, in the event that $\mathscr{T}$ and $\mathscr{R}$ are disjoint, there is no point $f(X) \in \mathscr{T}$, so $f^{-1}(\mathscr{T}) = \varnothing$.

Analogous to a space having relative subspaces, a function has relative subfunctions, called "restrictions," and defined as follows.

DEFINITION (Restriction of a function). If $\mathscr{D}$ is the domain of a function $f : \mathscr{D} \to \mathscr{R}$, and if $\mathscr{D}_1 \subset \mathscr{D}$, the *restriction of f to $\mathscr{D}_1$* is the function with domain $\mathscr{D}_1$ that has the same range element as f at each point of $\mathscr{D}_1$. Thus if f_1 denotes the restriction, then $f_1(X) = f(X)$ for each $X \in \mathscr{D}_1$.

We now introduce the central concept of this section.

DEFINITION (Continuous function). Let $f : \mathscr{D} \to \mathscr{R}$ be a function whose domain and range are subsets of metric spaces. The function f is *continuous at A in the domain* if and only if to each real number $\varepsilon > 0$ there corresponds at least one real number $\delta > 0$ such that $X \in \mathscr{D}$ and $d(A, X) < \delta$ imply that $d(f(A), f(X)) < \varepsilon$. Equivalently, in terms of the neighborhoods of the spaces, $X \in \mathrm{N}(A, \delta) \cap \mathscr{D}$ implies that $f(X) \in \mathrm{N}(f(A), \varepsilon)$. The function f is a *continuous function* if it is continuous at each point in its domain.

By the definition of $f : \mathscr{D} \to \mathscr{R}$, the sets $\mathscr{D}$ and $\mathscr{R}$ are subsets of metric spaces, hence they are relative metric spaces. Thus neighborhoods in the domain and range spaces have associated relative neighborhoods in $\mathscr{D}$ and

$\mathscr{R}$, respectively. In the definition of continuity of f at A, the point A and the number ε determine the neighborhood $N(f(A), \varepsilon)$. The corresponding neighborhood $N(A, \delta)$ in the definition is in the domain space and so determines the relative neighborhood $N(A, \delta) \cap \mathscr{D}$. Since this neighborhood is contained in $\mathscr{D}$, it has an image $f[N(A, \delta) \cap \mathscr{D}]$. The continuity condition "$X \in N(A, \delta) \cap \mathscr{D}$ implies that $f(X) \in N(f(A), \varepsilon)$" is just the condition "$f[N(A, \delta) \cap \mathscr{D}] \subset N(f(A), \varepsilon)$." Thus the next theorem is simply a translation of the continuity definition into the language of relative spaces.

THEOREM 1. The function $f : \mathscr{D} \to \mathscr{R}$ is continuous at $A \in \mathscr{D}$ if and only if to each neighborhood of $f(A)$ there corresponds at least one relative neighborhood of A whose image is contained in that neighborhood of $f(A)$.

Now, let $\mathscr{M}$ denote a neighborhood of $f(A)$, let $\mathscr{N}'$ denote a relative neighborhood of A, and let Z be an arbitrary point of $\mathscr{N}'$. Suppose that $f(\mathscr{N}') \subset \mathscr{M}$. Since $Z \in \mathscr{N}'$, then $f(Z) \in f(\mathscr{N}')$, so $f(Z) \in \mathscr{M}$. By definition, $f^{-1}(\mathscr{M}) = \{X : f(X) \in \mathscr{M}\}$. Therefore $Z \in f^{-1}(\mathscr{M})$. Thus $f(\mathscr{N}') \subset \mathscr{M} \Rightarrow \mathscr{N}' \subset f^{-1}(\mathscr{M})$. Conversely, suppose that $\mathscr{N}' \subset f^{-1}(\mathscr{M})$. Then $Z \in f^{-1}(\mathscr{M})$, which is to say that $f(Z) \in \mathscr{M}$. Thus $\mathscr{N}' \subset f^{-1}(\mathscr{M}) \Rightarrow f(\mathscr{N}') \subset \mathscr{M}$. The opposite implications show that the image of $\mathscr{N}'$ is contained in $\mathscr{M}$ if and only if the pre-image of $\mathscr{M}$ contains $\mathscr{N}'$. Thus we have the following form of Theorem 1.

COROLLARY 1. The function $f : \mathscr{D} \to \mathscr{R}$ is continuous at $A \in \mathscr{D}$ if and only if the pre-image of each neighborhood of $f(A)$ contains at least one relative neighborhood of A.

The proof of the next theorem illustrates how Theorem 1 can be applied.

THEOREM 2. Every function is continuous at the isolated points of its domain.

PROOF. Let A denote an isolated point in the domain $\mathscr{D}$ of the function $f : \mathscr{D} \to \mathscr{R}$. Since A is isolated, there exists a deleted neighborhood $N_o(A, \delta)$ such that $N_o(A, \delta) \cap \mathscr{D} = \varnothing$. Thus $N(A, \delta) \cap \mathscr{D} = \{A\}$, that is, the singleton set $\{A\}$ is itself a relative neighborhood of A. Since the image of this neighborhood, namely, $\{f(A)\}$, is contained in every neighborhood of $f(A)$, it follows from Theorem 1 that f is continuous at A. □

Because of the importance of continuity, it is desirable to have different ways of describing continuous functions. We now give two such descriptions, one in terms of open sets and one in terms of closed sets.

THEOREM 3. A function $f : \mathscr{D} \to \mathscr{R}$ is continuous if and only if every open set in the range space has a pre-image that is a relative open subset of the domain $\mathscr{D}$.

PROOF. First, suppose that f is continuous and that $\mathscr{S}$ is an open set in the range space. Since the empty set is an open subset of every set, it follows trivially that if $f^{-1}(\mathscr{S}) = \varnothing$ then $f^{-1}(\mathscr{S})$ is open in $\mathscr{D}$, and we suppose $f^{-1}(\mathscr{S}) \neq \varnothing$. Now consider $A \in f^{-1}(\mathscr{S})$. Then $f(A) \in \mathscr{S}$. Because $\mathscr{S}$ is open, there exists a neighborhood of $f(A)$, say $\mathscr{M} = \mathrm{N}(f(A), \varepsilon)$, such that $\mathscr{M} \subset \mathscr{S}$. Since f is continuous at A, then, by Corollary 1, $f^{-1}(\mathscr{M})$ must contain some relative neighborhood of A, say $\mathscr{N}' = \mathrm{N}(A, \delta) \cap \mathscr{D}$. Because $\mathscr{M} \subset \mathscr{S}$, then $f^{-1}(\mathscr{M}) \subset f^{-1}(\mathscr{S})$. Therefore we have $\mathscr{N}' \subset f^{-1}(\mathscr{M}) \subset f^{-1}(\mathscr{S})$. Since each point A in $f^{-1}(\mathscr{S})$ has a relative neighborhood that is contained in $f^{-1}(\mathscr{S})$, then $f^{-1}(\mathscr{S})$ is open in $\mathscr{D}$.

Conversely, suppose that $\mathscr{S}$ open in the range space implies that $f^{-1}(\mathscr{S})$ is a relative open subset of $\mathscr{D}$. Consider any point $A \in \mathscr{D}$, and let $\mathscr{M} = \mathrm{N}(f(A), \varepsilon)$ be an arbitrary neighborhood of $f(A)$. Then $\mathscr{M}$ is an open set in the range space and so, by hypothesis, $f^{-1}(\mathscr{M})$ is a relative open subset of $\mathscr{D}$. Therefore A in $f^{-1}(\mathscr{M})$ is a relative interior point of $f^{-1}(\mathscr{M})$, so there exists a relative neighborhood of A that is contained in $f^{-1}(\mathscr{M})$. Thus each neighborhood of $f(A)$ has a pre-image that contains a relative neighborhood of A and so, by Corollary 1, f is continuous at A. □

We can state Theorem 3 in the following more explicit form.

COROLLARY 3.1. A function $f : \mathscr{D} \to \mathscr{R}$ is continuous if and only if for each open subset $\mathscr{S}$ of the range space there exists an open subset $\mathscr{T}$ of the domain space such that

$$\{X : f(X) \in \mathscr{S}\} = \mathscr{D} \cap \mathscr{T}. \qquad \text{(E)}$$

COROLLARY 3.2. If $f : \mathscr{D} \to \mathscr{R}$ is continuous, and if $\mathscr{D}$ is open in the domain space (in particular, if $\mathscr{D}$ is the domain space), then for every open set $\mathscr{S}$ in the range space, the set $\{X : f(X) \in \mathscr{S}\}$ is also open. (E)

THEOREM 4. A function $f : \mathscr{D} \to \mathscr{R}$ is continuous if and only if every closed set in the range space has a pre-image that is a relative closed subset of the domain $\mathscr{D}$.

PROOF. We note first that if $\mathscr{S}$ and $\mathscr{T}$ are complements in the range space, then $f^{-1}(\mathscr{S})$ and $f^{-1}(\mathscr{T})$ are relative complements in $\mathscr{D}$. First, by definition, $f^{-1}(\mathscr{S})$ and $f^{-1}(\mathscr{T})$ are subsets of $\mathscr{D}$. They are disjoint subsets because $X \in f^{-1}(\mathscr{S}) \cap f^{-1}(\mathscr{T})$ would imply that $f(X) \in \mathscr{S} \cap \mathscr{T}$, contradicting $\mathscr{S} \cap \mathscr{T} \neq \varnothing$. Finally, if $X \in \mathscr{D}$, then $f(X) \in \mathscr{R} \subset \mathscr{S} \cup \mathscr{T}$. Since $f(X) \in \mathscr{T}$ implies that $X \in f^{-1}(\mathscr{T})$ and $f(X) \in \mathscr{S}$ implies that $X \in f^{-1}(\mathscr{S})$, it follows that $X \in f^{-1}(\mathscr{T}) \cup f^{-1}(\mathscr{S})$. Thus $\mathscr{D} = f^{-1}(\mathscr{T}) \cup f^{-1}(\mathscr{S})$.

Now suppose that f is continuous and that $\mathscr{S}$ is a closed set in the range space. Then $\text{Cp}(\mathscr{S})$ is an open set in the range space. By Theorem 3, $f^{-1}[\text{Cp}(\mathscr{S})]$ is a relative open subset of $\mathscr{D}$, and so its relative complement is a relative closed subset of $\mathscr{D}$. But, as noted above, the relative complement of $f^{-1}[\text{Cp}(\mathscr{S})]$ is $f^{-1}(\mathscr{S})$, so $f^{-1}(\mathscr{S})$ is closed in $\mathscr{D}$.

Conversely, suppose that the pre-image of any set closed in the range space is a relative closed subset of $\mathscr{D}$. Let $\mathscr{S}$ be any open set in the range space. Then $\text{Cp}(\mathscr{S})$ is a closed set in the range space and so, by hypothesis, $f^{-1}[\text{Cp}(\mathscr{S})]$ is a relative closed subset of $\mathscr{D}$. Therefore its relative complement is a relative open subset of $\mathscr{D}$. But its relative complement is $f^{-1}(\mathscr{S})$. Thus open subsets of the range space have pre-images that are relatively open in $\mathscr{D}$, and so, by Theorem 3, f is a continuous function. □

COROLLARY 4.1. A function $f: \mathscr{D} \to \mathscr{R}$ is continuous if and only if for each closed subset $\mathscr{S}$ of the range space there exists a closed subset $\mathscr{T}$ of the domain space such that

$$\{X : f(X) \in \mathscr{S}\} = \mathscr{D} \cap \mathscr{T}. \tag{E}$$

COROLLARY 4.2. If $f: \mathscr{D} \to \mathscr{R}$ is continuous, and if $\mathscr{D}$ is closed in the domain space (in particular, if $\mathscr{D}$ is the domain space), then for every closed set $\mathscr{S}$ in the range space the set $\{X : f(X) \in \mathscr{S}\}$ is also closed. (E)

Corollaries 3.2 and 4.2 are of great practical importance in showing sets to be open or closed. For example, if a real valued function f is continuous on a closed set, and c is any real number, then the points X in the domain at which $f(X) = c$ form a closed set. This is so because $\mathscr{S} = \{c\}$ is a closed set in the range space. Then, by Corollary 4.2, $\{X : f(X) = c\} = \{X : f(X) \in \mathscr{S}\}$ is closed. Similarly, if f is continuous on an open set, then Corollary 3.2 implies the openess of such sets as $\{X : a < f(X) < b\}$ or $\{X : f(X) > c\}$.

Previously we emphasized the fact that connectedness and compactness are natural properties of what we called object sets. We now prove that these properties are preserved by continuous processes. This shows that we can often expect the continuous image of an object set to be again an object set, although it might be an object set of different dimension (for example, the first might be an object set in three space and its image an object set in the plane).

We want to show that if the domain of a continuous function is connected, then so is the range. A proof can be facilitated by the following lemma which is interesting in its own right.

LEMMA. If the function $f: \mathscr{D} \to \mathscr{R}$ is continuous, and if $\mathscr{T}$ is any subset of the range space, then

$$\mathscr{D} \cap \text{Cl}[f^{-1}(\mathscr{T})] \subset f^{-1}[\text{Cl}(\mathscr{T})].$$

PROOF. We must show that if $A \in \mathscr{D} \cap \mathrm{Cl}[f^{-1}(\mathscr{T})]$, then $f(A) \in \mathrm{Cl}(\mathscr{T})$. Let $\mathscr{M} = \mathrm{N}(f(A), \varepsilon)$ be an arbitrary neighborhood of $f(A)$. Since $A \in \mathscr{D}$ and f if continuous, then f is continuous at A, and so, by Corollary 1, $f^{-1}(\mathscr{M})$ contains some relative neighborhood of A, say $\mathrm{N}(A, \delta)'$. By (iv) of Theorem 7, Sec. 1.3, $\mathscr{D} \cap \mathrm{Cl}[f^{-1}(\mathscr{T})]$ is the relative closure of $f^{-1}(\mathscr{T})$. Thus, by Theorem 6, Sec. 1.3, there exists $Y \in \mathrm{N}(A, \delta)' \cap f^{-1}(\mathscr{T})$. This implies that $f(Y) \in \mathscr{T}$. Also, $Y \in \mathrm{N}(A, \delta)' \subset f^{-1}(\mathscr{M})$ implies that $f(Y) \in \mathscr{M}$. Therefore $f(Y) \in \mathscr{M} \cap \mathscr{T}$. Thus every neighborhood of $f(A)$ intersects $\mathscr{T}$ and so, by Theorem 6, Sec. 1.3, $f(A) \in \mathrm{Cl}(\mathscr{T})$. □

THEOREM 5. If the function $f : \mathscr{D} \to \mathscr{R}$ is continuous, and if $\mathscr{D}$ is connected, then $\mathscr{R}$ is connected. That is, the continuous image of a connected set is connected.

PROOF. The theorem follows if we can prove that the continuity of f and the disconnectedness of the range $\mathscr{R}$ imply the disconnectedness of $\mathscr{D}$. Let $\mathscr{R}$ be a disconnected range. Then $\mathscr{R} = \mathscr{R}_1 \cup \mathscr{R}_2$, where $\mathscr{R}_1$ and $\mathscr{R}_2$ are separated sets. Now define $\mathscr{D}_1 = f^{-1}(\mathscr{R}_1)$ and $\mathscr{D}_2 = f^{-1}(\mathscr{R}_2)$. From the relations $\mathscr{R} = \mathscr{R}_1 \cup \mathscr{R}_2, \mathscr{R}_1 \neq \varnothing$, and $\mathscr{R}_2 \neq \varnothing$, it follows that

$$\mathscr{D} = \mathscr{D}_1 \cup \mathscr{D}_2, \mathscr{D}_1 \neq \varnothing, \mathscr{D}_2 \neq \varnothing. \tag{1}$$

From the lemma, we know that

$$\mathscr{D} \cap \mathrm{Cl}(\mathscr{D}_1) \subset f^{-1}[\mathrm{Cl}(\mathscr{R}_1)]. \tag{2}$$

Assume that there exists a point $X \in \mathscr{D}_2 \cap \mathrm{Cl}(\mathscr{D}_1)$. Then, from $X \in \mathscr{D}_2$, we have $f(X) \in \mathscr{R}_2$. Since $\mathscr{D}_2 \subset \mathscr{D}$, then $X \in \mathscr{D}_2 \cap \mathrm{Cl}(\mathscr{D}_1)$ also implies that $X \in \mathscr{D} \cap \mathrm{Cl}(\mathscr{D}_1)$. This, with (2), shows that $f(X) \in \mathrm{Cl}(\mathscr{R}_1)$. Thus $f(X) \in \mathscr{R}_2 \cap \mathrm{Cl}(\mathscr{R}_1)$, which contradicts the separatedness of $\mathscr{R}_1$ and $\mathscr{R}_2$. Hence there is no point X in $\mathscr{D}_2 \cap \mathrm{Cl}(\mathscr{D}_1)$, so

$$\mathscr{D}_2 \cap \mathrm{Cl}(\mathscr{D}_1) = \varnothing. \tag{3}$$

By an entirely similar argument,

$$\mathscr{D}_1 \cap \mathrm{Cl}(\mathscr{D}_2) = \varnothing. \tag{4}$$

From (1), (3), and (4) it follows that $\mathscr{D}_1$ and $\mathscr{D}_2$ are separated sets and hence that $\mathscr{D}$ is disconnected. □

THEOREM 6. If the function $f : \mathscr{D} \to \mathscr{R}$ is continuous and if $\mathscr{D}$ is compact, then $\mathscr{R}$ is compact. That is, the continuous image of a compact set is compact.

PROOF. Let $\mathscr{F} = \{\mathscr{S}_\alpha : \alpha \in \Lambda\}$ be a collection of sets $\mathscr{S}_\alpha$ that are open in the range space and such that $\mathscr{F}$ is a cover of $\mathscr{R}$. The theorem follows if we can show that there is a finite subcollection of $\mathscr{F}$ that is also a cover of $\mathscr{R}$. We start with Theorem 3. Since f is continuous and since each of the sets $\mathscr{S}_\alpha$ is open in the range space, then, by Theorem 3, each of the sets $f^{-1}(\mathscr{S}_\alpha)$ is a relative open set in $\mathscr{D}$. Moreover, the collection

$$\mathscr{G} = \{f^{-1}(\mathscr{S}_\alpha) : \alpha \in \Lambda\} \tag{1}$$

is a cover of $\mathscr{D}$. For if $A \in \mathscr{D}$, then $f(A) \in \mathscr{R}$ and, since $\mathscr{F}$ covers $\mathscr{R}, f(A) \in \mathscr{S}_\beta$ for at least one value of $\beta \in \Lambda$. Then $A \in f^{-1}(\mathscr{S}_\beta)$ and $f^{-1}(\mathscr{S}_\beta) \in \mathscr{G}$. Since $\mathscr{D}$ is compact in the domain space, then, by Theorem 16, Sec. 1.4, $\mathscr{D}$ is relatively compact. Therefore the collection $\mathscr{G}$ of relatively open sets covering $\mathscr{D}$ has a finite subcollection, say

$$\mathscr{G}_1 = \{f^{-1}(\mathscr{S}_{\alpha_1}), f^{-1}(\mathscr{S}_{\alpha_2}), \ldots, f^{-1}(\mathscr{S}_{\alpha_k})\} \tag{2}$$

that is also a cover of $\mathscr{D}$. Corresponding to $\mathscr{G}_1$, let

$$\mathscr{F}_1 = \{\mathscr{S}_{\alpha_1}, \mathscr{S}_{\alpha_2}, \ldots, \mathscr{S}_{\alpha_k}\}, \tag{3}$$

If $Z \in \mathscr{R}$, there is at least one point $X \in \mathscr{D}$ and such that $f(X) = Z$. Because $X \in \mathscr{D}$, and $\mathscr{G}_1$ is a cover of $\mathscr{D}$, there is at least one integer i, $1 \le i \le k$, such that $X \in f^{-1}(\mathscr{S}_{\alpha_i})$. Then $f(X) = Z \in \mathscr{S}_{\alpha_i}$. Thus $\mathscr{F}_1$ is a cover of $\mathscr{R}$, and $\mathscr{F}_1$ is a finite subcollection of $\mathscr{F}$. □

As we noted before, when f is a one-to-one function, then the inverse relation f^{-1} is also a function. However, even when both f and f^{-1} are functions, the continuity of one need not imply the continuity of the other. It is an important property of compactness that if f and f^{-1} are both functions and either is continuous on a compact domain, then both are continuous. In the proof we give for this, we need the following basic and useful fact.

THEOREM 7. Every restriction of a continuous function is continuous. (E)

THEOREM 8. If the function $f : \mathscr{D} \to \mathscr{R}$ is continuous and one to one and $\mathscr{D}$ is compact, then the inverse function is continuous.

PROOF. The argument is easier to follow if we let $\mathscr{D}^*$ and $\mathscr{R}^*$ denote the domain and range spaces of the function $f : \mathscr{D} \to \mathscr{R}$ and let g be another designation of f^{-1}. Then $\mathscr{R}^*$ and $\mathscr{D}^*$ are, respectively, the domain and range spaces of the function $g = f^{-1} : \mathscr{R} \to \mathscr{D}$. If $Y = f(X)$, then $X = f^{-1}(Y) = g(Y)$.

Now let $\mathscr{S}$ be an arbitrary closed set in the range space of g. By Theorem 4, the continuity of g is established if we can show that $g^{-1}(\mathscr{S})$ is a relative closed set in the domain of g. That is, $\mathscr{S}$ is a closed set in $\mathscr{D}^*$, and we must show that $g^{-1}(\mathscr{S})$ is a relative closed subset of $\mathscr{R}$.

By definition, $g^{-1}(\mathscr{S})$ is the set of those points Y in $\mathscr{R}^*$ that have g images in $\mathscr{S}$. For Y in $\mathscr{R}^*$ to have a g image, it must be in $\mathscr{R}$, in which case $g(Y) = X$ is in $\mathscr{D}$. Thus $g(Y) \in \mathscr{S}$ for those Y such that $g(Y) = X$ is in $\mathscr{S} \cap \mathscr{D}$. Hence

$$g^{-1}(\mathscr{S}) = \{f(X): X \in \mathscr{S} \cap \mathscr{D}\} = f(\mathscr{S} \cap \mathscr{D}).$$

In $\mathscr{D}^*$, by hypothesis, $\mathscr{D}$ is compact and $\mathscr{S}$ is closed. Therefore $\mathscr{S} \cap \mathscr{D}$ is a closed subset of a compact set, and so, by Theorem 11, Sec. 1.4, $\mathscr{S} \cap \mathscr{D}$ is compact in $\mathscr{D}^*$. Since f is continuous on $\mathscr{D}$, its restriction to $\mathscr{S} \cap \mathscr{D}$ is continuous, by Theorem 7, and so $f(\mathscr{S} \cap \mathscr{D})$ is compact in $\mathscr{R}^*$ (Theorem 11, Sec. 1.4). Thus, by Theorem 10, Sec. 1.4, $f(\mathscr{S} \cap \mathscr{D})$ is closed in $\mathscr{R}^*$. But, by (iii) of Theorem 7, Sec. 1.4, any set that is a subset of $\mathscr{R}$ and is closed in $\mathscr{R}^*$ is also closed in $\mathscr{R}$. Thus $f(\mathscr{S} \cap \mathscr{D})$, which is a subset of $\mathscr{R}$ and is closed in $\mathscr{R}^*$, is also a relative closed set in $\mathscr{R}$. Hence $g = f^{-1}$ is continuous. □

We conclude this section with one important method of generating continuous functions, namely, by composing continuous functions.

DEFINITION (Function of a function). If the functions $f: \mathscr{D}_1 \to \mathscr{R}_1$ and $g: \mathscr{D}_2 \to \mathscr{R}_2$ are such that $\mathscr{R}_1 \subset \mathscr{D}_2$, then the *composite function* "g of f," denoted by $g(f)$, is defined by

$$g(f)(X) = g(f(X)),\ X \in \mathscr{D}_1$$

and is a mapping of $\mathscr{D}_1$ into $\mathscr{R}_2$.

It is a familiar fact of elementary mathematics that for real valued functions of a real variable, a continuous function of a continuous function is again a continuous function. The proof is essentially the same for general functions.

THEOREM 9. If the functions f and g are continuous, and if the function $g(f)$ exists, then $g(f)$ is continuous. (E)

Exercises – Section 1.5

1. Let f be the real valued function defined on $\mathscr{E}^1$ by $f(x) = x^2$. Give explicit representation for the sets

   a. $f(\{x: 0 \le x \le 1\})$
   b. $f(\{x: -1 \le x \le 1\})$
   c. $f^{-1}(\{x: 0 \le x \le 4\})$
   d. $f^{-1}(\{x: -3 < x < -2\})$

2. Let P denote a fixed point of $\mathscr{E}^2$ and let f be defined by $f(X) = d(P, X)$ for $X \in \mathscr{E}^2$.

   a. Is f a mapping of $\mathscr{E}^2$ onto or into $\mathscr{E}^1$?
   b. Given two points A, X in $\mathscr{E}^2$, apply the triangle inequality in two ways to the points P, A, X to obtain $|d(P, X) - d(P, A)| \leq d(A, X)$.
   c. Prove that f is continuous.

3. If f is defined on $\mathscr{E}^2$ by $f(X) = |x| = \sqrt{x_1^2 + x_2^2}$, how does the continuity of f follow from Exercise 2?
4. The space $\mathscr{E}^2$ can be denoted by $(\mathscr{V}^2, d)$, where d is the euclidean metric on the pairs of points in the vector space $\mathscr{V}^2$. Now let d' be defined on the pairs of $\mathscr{V}^2$ by $d'(X, Y) = \max\{|x_1 - y_1|, |x_2 - y_2|\}$. It can be shown that d' is a metric, hence that $(\mathscr{V}^2, d')$ is a metric space. Using N′ for neighborhoods in this space,

$$\mathrm{N}'(A, \varepsilon) = \{X : a_1 - \varepsilon < x_1 < a_1 + \varepsilon, a_2 - \varepsilon < x_2 < a_2 + \varepsilon\}.$$

   A cartesian picture of $\mathrm{N}'(A, \varepsilon)$ is a square centered at (a_1, a_2) with sides of length 2ε and parallel to the axes.

   a. Prove that for real numbers a, b,

$$\max\{|a|, |b|\} \leq \sqrt{a^2 + b^2} \leq \sqrt{2} \max\{|a|, |b|\}.$$

   b. Use (a) to prove that $\mathrm{N}'(A, \varepsilon) \subset \mathrm{N}(A, \sqrt{2}\, \varepsilon)$ and that $\mathrm{N}(A, \varepsilon) \subset \mathrm{N}'(A, \varepsilon)$.
   c. Show that if a set is open in either of the spaces $(\mathscr{V}^2, d)$ and $(\mathscr{V}^2, d')$ then it is open in both.
   d. Show that if $\mathscr{D} \subset \mathscr{V}^2$ and if f is continuous on $\mathscr{D}$ with respect to either d or d', then it is continuous with respect to the other. Thus the spaces have the same continuous functions.

5. Addition of real numbers can be thought of as a function $+ : \mathscr{V}^2 \to \mathscr{V}^1$ defined by $+(X) = x_1 + x_2$. Using d', defined in Exercise 4, show that $+$ is a continuous mapping of $(\mathscr{V}^2, d')$ onto $(\mathscr{V}^1, d) = \mathscr{E}^1$. Apply Exercise 4 to show that $+ : \mathscr{E}^2 \to \mathscr{E}^1$ is a continuous function.
6. Let f and g be continuous, real valued functions on a common domain $\mathscr{D}$ in $\mathscr{E}^1$. Let a function h, mapping $\mathscr{D}$ into $\mathscr{E}^2$, be defined on $\mathscr{D}$ by $h(X) = (f(x), g(x))$. Prove that h is a continuous function.
7. Prove Corollaries 3.1 and 3.2.
8. Prove Corollaries 4.1 and 4.2.
9. Prove Theorem 7.
10. Use Exercise 2 to prove that balls and spheres in $\mathscr{E}^2$ are closed sets.

11. Give an example of a continuous function and an open set in its domain whose image is not open.
12. Prove Theorem 9.
13. Referring to Exercises 5 and 6, compute the composite function $+(h)$. Is it continuous?
14. Two sets $\mathscr{R}$ and $\mathscr{S}$ are "homeomorphic" if there exists a function f such that f and f^{-1} are continuous functions and $\mathscr{S} = f(\mathscr{R})$ and $\mathscr{R} = f^{-1}(\mathscr{S})$. In $\mathscr{E}^2$, if the points of a set $\mathscr{R}$ form the letter "A" and the points of a set $\mathscr{S}$ form the letter "T", explain why $\mathscr{R}$ and $\mathscr{S}$ are not homeomorphic.
15. Give an example of a continuous function $f: \mathscr{D} \to \mathscr{R}$, in which $\mathscr{D}$ is not closed, not bounded, and not connected, but $\mathscr{R}$ is closed, bounded, and connected.

SECTION 1.6. CONNECTEDNESS IN $\mathscr{E}^1$

As we mentioned in Sec. 1.4, it is difficult to apply the definition of connectedness directly to the determination of whether a given set in $\mathscr{E}^n$ is connected. This is true even in $\mathscr{E}^2$. However, in $\mathscr{E}^1$ it is possible to characterize the connected sets and to verify our intuitive expectations.

We first review some basic concepts involved with real numbers. In $\mathscr{E}^1$, a point P is just the number p, and we use the lower case notation. The euclidean metric also simplifies to $d(x, y) = |x - y|$.

The basic sets of $\mathscr{E}^1$ are the intervals.

DEFINITION (Intervals). A set of real numbers is an *interval* if it is one of the following sets:

1. *The bounded intervals.* For $a < b$
   $(a, b) = \{x: a < x < b\}$ is an *open interval*
   $[a, b] = \{x: a \leq x \leq b\}$ is a *closed interval*
   $[a, b) = \{x: a \leq x < b\}$ and $(a, b] = \{x: a < x \leq b\}$ are *half open intervals*
2. The *unbounded intervals.*
   $(a, \infty) = \{x: a < x\}$; $[a, \infty) = \{x: a \leq x)$; $(-\infty, a) = \{x: x < a\}$; $(-\infty, a] = \{x: x \leq a\}$; $(-\infty, \infty) = \mathscr{V}^1$
3. The *degenerate intervals.*
   The set $\varnothing$, and for each real number a the singleton set $\{a\}$.

The neighborhood $\mathrm{N}(c, \varepsilon)$ is the open interval $(c - \varepsilon, c + \varepsilon)$. The open interval (a, b) is the neighborhood $\mathrm{N}(c, \varepsilon)$, where $c = (a + b)/2$ and $\varepsilon = (b - a)/2$. The closed interval $[a, b]$ is the ball $\mathrm{B}(c, \varepsilon)$, and the set of interval endpoints $\{a, b\}$ is the one-dimensional sphere $\mathrm{S}(c, \varepsilon)$.

In $\mathscr{E}^1$, a set $\mathscr{S}$ is ordinarily defined to be bounded if it is bounded below and bounded above, that is, if numbers a, b exist such that $a \leq x \leq b$ for all $x \in \mathscr{S}$. This is equivalent to the condition that $\mathscr{S}$ be contained in some ball and thus coincides with the concept of a bounded set in a metric space.

We wish to place special emphasis on a fundamental property of real numbers which will find application throughout the book. *A non-empty set of real numbers which is bounded above has a least upper bound, and a non-empty set of real numbers which is bounded below has a greatest lower bound.*

Comment. An endpoint to an interval is either the least upper bound or greatest lower bound to the interval.

The following are some rather obvious but important facts.

THEOREM 1. If a set $\mathscr{S}$ is bounded above (below), then the least upper bound (greatest lower bound) belongs to the closure of the set.

PROOF. Let $b = \text{lub}(\mathscr{S})$. If $b \notin \text{Cl}(\mathscr{S})$, then b is interior to $\text{Cp}(\mathscr{S})$, hence there exists $\text{N}(b, \delta) = (b - \delta, b + \delta)$ such that $(b - \delta, b + \delta) \cap \mathscr{S} = \varnothing$. because b is an upper bound, $(b, \infty) \cap \mathscr{S} = \varnothing$. Therefore $(b - \delta, \infty) \cap \mathscr{S} = \varnothing$, so $b - \delta$ is an upper bound to $\mathscr{S}$. Since $b - \delta < b$, then $b \neq \text{lub}(\mathscr{S})$. The contradiction shows that $b \notin \text{Cl}(\mathscr{S})$ is false, hence $\text{lub}(\mathscr{S}) \in \text{Cl}(\mathscr{S})$. By an entirely similar argument, $\text{glb}(\mathscr{S}) \in \text{Cl}(\mathscr{S})$. □

THEOREM 2. An endpoint to a (nondegenerate) interval is a limit point to the interval. (E)

THEOREM 3. An open interval is an open set, and a closed interval is a closed set. (E)

For the purposes of this section, we must define the notion of convexity in the special case of the real line.

DEFINITION (Convex set of real numbers). A set $\mathscr{S}$ in $\mathscr{E}^1$ is *convex* if $a, b \in \mathscr{S}$ and $a < x < b$ imply $x \in \mathscr{S}$.

The next two theorems identify intervals and convex sets of numbers as the same class of sets.

THEOREM 4. Every interval is a convex set. (E)

THEOREM 5. Every convex set of real numbers is an interval.

PROOF. Let $\mathscr{S}$ be a convex set of real numbers. If $\mathscr{S} = \varnothing$, then $\mathscr{S}$ is an interval, so we suppose there exists $c \in \mathscr{S}$. Let x be any point of (c, ∞). If $\mathscr{S}$ is unbounded above, then there exists $x_1 \in \mathscr{S}$ and $x_1 > x$. Then from $c, x_1 \in \mathscr{S}$ and $c < x < x_1$, the convexity of $\mathscr{S}$ implies $x \in \mathscr{S}$. Thus if $\mathscr{S}$ is unbounded above, $[c, \infty) \subset \mathscr{S}$. By a similar argument, if $\mathscr{S}$ is unbounded below, then $(-\infty, c] \subset \mathscr{S}$. Thus if $\mathscr{S}$ is unbounded above and below, every number belongs to $\mathscr{S}$, so $\mathscr{S}$ is the interval $\mathscr{V}^1$. We turn then to the case in which $\mathscr{S}$ is bounded either above or below, and, from symmetry, it suffices to consider the case in which $a = \text{glb}(\mathscr{S})$ exists.

Now consider $x \in (a, \infty)$, and let $\delta = x - a$. By Theorem 1, $a \in \text{Cl}(\mathscr{S})$ so, by the lemma, Sec. 1.5, $\text{N}(A, \delta)$ intersects $\mathscr{S}$ at some point x_1. Clearly, $x_1 \in [a, x)$, so $x_1 < x$. If $\mathscr{S}$ is unbounded above, there exists $x_2 \in \mathscr{S}$ and $x_2 > x$. Then $x_1, x_2 \in \mathscr{S}$ and $x_1 < x < x_2$ imply $x \in \mathscr{S}$, by the convexity of $\mathscr{S}$, hence $(a, \infty) \subset \mathscr{S}$. Since $(-\infty, a) \cap \mathscr{S} = \varnothing$, it follows that if $\mathscr{S}$ is unbounded above, it must be either $[a, \infty)$ or (a, ∞).

Finally, suppose that $\mathscr{S}$ is also bounded above. Then $b = \text{lub}(\mathscr{S})$ exists. If $a = b$, then $\mathscr{S} \neq \varnothing$ implies that $\mathscr{S} = \{a\}$, which is a degenerate interval, so we suppose that $a < b$. Now consider $x \in (a, b)$, and let $\delta_1 = x - a$ and $\delta_2 = b - x$. As before, $a \in \text{Cl}(\mathscr{S})$ implies that $\text{N}(A, \delta_1) \cap \mathscr{S} \neq \varnothing$, so there exists $x_1 \in [a, x) \cap \mathscr{S}$. Similarly, $b \in \text{Cl}(\mathscr{S})$ implies that $\text{N}(b, \delta_2) \cap \mathscr{S} \neq \varnothing$, so there exists $x_2 \in (x, b] \cap \mathscr{S}$. Then $x_1, x_2 \in \mathscr{S}$ and $x_1 < x < x_2$ imply $x \in \mathscr{S}$. Hence $(a, b) \subset \mathscr{S}$. Since $(-\infty, a) \cap \mathscr{S} = \varnothing$ and $(b, \infty) \cap \mathscr{S} = \varnothing$, it follows that $\mathscr{S}$ is one of the four possible intervals with endpoints a, b. □

THEOREM 6. Every interval is a connected set.

PROOF. Let I be an interval. For an indirect proof, assume that I is not connected. Then I is the union of two separated sets $\mathscr{L}$ and $\mathscr{R}$. Since $\mathscr{L}$ and $\mathscr{R}$ are non-empty, there exists $a \in \mathscr{L}$ and $b \in \mathscr{R}$, and we may suppose that $a < b$. By Theorem 4, I is convex, so $[a, b] \subset I$.

Now, define $\mathscr{L}_1 = [a, b] \cap \mathscr{L}$ and $\mathscr{R}_1 = [a, b] \cap \mathscr{R}$. Since $a \in \mathscr{L}_1$ and $b \in \mathscr{R}_1$, the sets $\mathscr{L}_1, \mathscr{R}_1$ are non-empty and $[a, b] = \mathscr{L}_1 \cup \mathscr{R}_1$. Since $\mathscr{L}_1$ is a non-empty subset of $\mathscr{L}$ and $\mathscr{R}_1$ is a non-empty subset of $\mathscr{R}$, the sets $\mathscr{L}_1$ and $\mathscr{R}_1$ are separated sets.

Because $\mathscr{L}_1 \subset [a, b]$, it is bounded above by b, so $c = \text{lub}(\mathscr{L}_1)$ exists, and $c \leq b$ implies that $c \in \mathscr{L}_1 \cup \mathscr{R}_1$. By Theorem 1, $c \in \text{Cl}(\mathscr{L}_1)$. Since $\mathscr{R}_1 \cap \text{Cl}(\mathscr{L}_1) = \varnothing$, then $c \in \text{Cl}(\mathscr{L}_1)$ implies that $c \notin \mathscr{R}_1$, hence that $c \in \mathscr{L}_1$. Therefore $c \neq b$, and $c < b$. Because c is an upper bound to $\mathscr{L}_1$ there are no points of $\mathscr{L}_1$ in $(c, b]$, so $(c, b] \subset \mathscr{R}_1$. But, by Theorem 2, c is a limit point to $(c, b]$, hence $c \in \text{Lp}(\mathscr{R}_1) \subset \text{Cl}(\mathscr{R}_1)$. Therefore $c \in \mathscr{L}_1 \cap \text{Cl}(\mathscr{R}_1)$, contradicting the separatedness of $\mathscr{L}_1$ and $\mathscr{R}_1$. This shows that the initial assumption is false, hence I is a connected set. □

THEOREM 7. Every connected set of real numbers is a convex set.

PROOF. We prove the theorem by showing that every nonconvex set is disconnected. Let $\mathscr{S}$ be a set that is not convex. Then there must exist numbers a, b, c with $a, b \in \mathscr{S}$, $a < c < b$, and $c \notin \mathscr{S}$. Define $\mathscr{L} = (-\infty, c) \cap \mathscr{S}$ and define $\mathscr{R} = \mathscr{S} \cap (c, \infty)$. Since $a \in \mathscr{L}$, and $b \in \mathscr{R}$, and $(-\infty, c) \cap (c, \infty) = \varnothing$, then $\mathscr{L}$ and $\mathscr{R}$ are non-empty disjoint sets. Clearly, $\mathscr{S} = \mathscr{L} \cup \mathscr{R}$. Furthermore, $\mathrm{Cl}(\mathscr{L}) = (-\infty, c]$ and $\mathrm{Cl}(\mathscr{R}) = [c, \infty)$, so $\mathscr{R} \cap \mathrm{Cl}(\mathscr{L}) = \varnothing$ and $\mathscr{L} \cap \mathrm{Cl}(\mathscr{R}) = \varnothing$. Therefore $\mathscr{L}$ and $\mathscr{R}$ are separated sets, which shows that $\mathscr{S}$ is disconnected. □

A summary of our results gives the characterization of connected sets of real numbers promised at the beginning of this section.

THEOREM 8. A set of real numbers is connected if and only if it is a convex set and, in turn, if and only if it is an interval. (E)

In Sec. 1.5 we proved that the continuous image of a connected set is connected. The next theorem gives a concrete realization of that fact and also establishes one of the most important results in analysis.

THEOREM 9. The image of a connected subset of a metric space under a continuous, real valued function is an interval. In particular, if $f : \mathscr{D} \to \mathscr{R}$ is real valued and continuous and if a and b are any two function values, then f assumes every value intermediate to a and b. (E)

COROLLARY 9. A real valued, continuous function defined on an interval of real numbers assumes all values intermediate to any two function values. (E)

Exercises – Section 1.6

1. Prove Theorem 2.
2. Prove Theorem 3.
3. From Theorem 1 it follows that if b is the least upper bound to $\mathscr{S}$ and $b \notin \mathscr{S}$, then $b \in \mathrm{Lp}(\mathscr{S})$. Give examples in which

   a. $b \notin \mathscr{S}$
   b. $b \in \mathscr{S}$ and $b \notin \mathrm{Lp}(\mathscr{S})$
   c. $b \in \mathscr{S} \cap \mathrm{Lp}(\mathscr{S})$.

4. Prove Theorem 4.
5. If $a = \mathrm{glb}(\mathscr{S})$, where $\mathscr{S}$ is a set of real numbers, and if x is any number greater than a, explain in detail why $[a, x) \cap \mathscr{S} \neq \varnothing$. (This fact was used in the proof of Theorem 5.)

6. If $\mathscr{L}_1$ and $\mathscr{R}_1$ are non-empty subsets of $\mathscr{L}$ and $\mathscr{R}$, respectively, and if $\mathscr{L}$ and $\mathscr{R}$ are separated, explain why $\mathscr{L}_1$ and $\mathscr{R}_1$ are separated. (This fact was used in the proof of Theorem 6.)
7. If $\mathscr{R}$ is a connected set and if $\mathscr{R} \cup \mathscr{S}$ is a connected set, must $\mathscr{S}$ be connected? Give an example to support your answer.
8. Prove Theorem 8.
9. Prove Theorem 9 and Corollary 9.
10. Let $a_n x^n + a_{n-1}x^{n-1} + \cdots + a_1 x + a_o = 0$, $a_n \neq 0$ be a polynomial equation with real coefficients. Explain how Corollary 9 can be used to show that if n is odd, then every such equation has at least one real root.

SECTION 1.7. COMPACTNESS IN $\mathscr{E}^n$

We prove in this section that a closed, bounded set in euclidean space is compact. Since we already know that a compact set in any metric space is closed and bounded, the proof gives us a characterization of the compact subsets of $\mathscr{E}^n$. We can then apply the fruitful idea of compactness to closed and bounded sets and in particular to many object sets of importance.

To obtain the characterization just mentioned, we need some theorems which are interesting and useful in their own right.

THEOREM 1. In $\mathscr{E}^1$ a decreasingly nested sequence of closed, bounded intervals has a non-empty intersection set.

PROOF. Let $\{[a_n, b_n] : n = 1, 2, \ldots\}$ be a decreasingly nested sequence of closed intervals. Then for positive integers n, m such that $n < m$, $[a_n, b_n] \supset [a_m, b_m]$ implies the inequalities $a_n \leq a_m \leq b_m \leq b_n$. If we let $\mathscr{S}$ denote the set $\{a_n : n = 1, 2, \ldots\}$ of all left interval endpoints, then $\mathscr{S}$ is a non-empty set of real numbers that is bounded above. In fact, because of the nestedness, every right endpoint b_n is an upper bound to $\mathscr{S}$. Let $a = \text{lub}(\mathscr{S})$. Then, as we just observed, $a \leq b_n$, $n = 1, 2, \ldots$. Also, by the definition of a, $a_n \leq a$, $n = 1, 2, \ldots$. Hence $a_n \leq a \leq b_n$, $n = 1, 2, \ldots$, which is just to say that $a \in [a_n, b_n]$, $n = 1, 2, \ldots$. Thus the intersection of the intervals is not empty. □

If we introduce a kind of n-dimensional analog of a closed interval, we can obtain an n-dimensional analog of Theorem 1.

DEFINITION (n cell). In $\mathscr{E}^n$, a set defined by n inequalities,

$$\mathscr{S} = \{x : a_i \leq x_i \leq b_i, i = 1, 2, \ldots n\},$$

determined by real numbers $a_i < b_i$, $i = 1, 2, \ldots, n$ is an *n cell*, and the numbers $b_1 - a_1, b_2 - a_2, \ldots, b_n - a_n$ are the *side lengths* of the *n cell*. When all the side lengths are the same positive number, the cell is a special kind of *solid cube* and is called a *cube n cell*.

THEOREM 2. In $\mathscr{E}^n$, a decreasingly nested sequence of n cells has a non-empty intersection set.

PROOF. Let $\{\mathscr{S}_j : j = 1, 2, \ldots\}$ denote a decreasingly nested sequence of n cells, where the n cell $\mathscr{S}_j$ is represented by

$$\mathscr{S}_j = \{x \in \mathscr{E}^n : a_{j,i} \leq x_i \leq b_{j,i}, i = 1, =, \ldots, n\}. \tag{1}$$

Now, consider a fixed positive integer i, $1 \leq i \leq n$. Then the ith inequalities in the definitions for $\mathscr{S}_j$ and $\mathscr{S}_{j+1}$ are

$$a_{j,i} \leq x_i \leq b_{j,i} \text{ and } a_{j+1,i} \leq x_i \leq b_{j+1,i}. \tag{2}$$

If $x = (x_1, \ldots, x_i, \ldots, x_n)$ is a point in $\mathscr{S}_{j+1}$, then x_i satisfies the second inequality in (2). But, since $\mathscr{S}_j \supset \mathscr{S}_{j+1}$, then x is also in $\mathscr{S}_j$, so x_i satisfies the first inequality in (2). That is, $\mathscr{S}_j \supset \mathscr{S}_{j+1}$ implies $[a_{j,i}, b_{j,i}] \supset [a_{j+1,i}, b_{j+1,i}]$, for each i, $1 \leq i \leq n$. Thus, as j runs through the integers $1, 2, \ldots$, there are n corresponding sequences

$$\{[a_{j,1}, b_{j,1}]\}, \{[a_{j,2}, b_{j,2}]\}, \ldots, \{[a_{j,n}, b_{j,n}]\} \tag{3}$$

$$j = 1, 2, \ldots$$

and each of these is a decreasingly nested sequence of closed intervals. By Theorem 1, each of these sequences has a non-empty intersection set. Thus there is a number c_1 that belongs to all the intervals in the first sequence in (3), a number c_2 that belongs to all the intervals in the second sequence, and so on up to a number c_n that is in all the intervals of the nth sequence in (3). Thus the point $c = (c_1, c_2, \ldots, c_n)$ is in every n cell $\mathscr{S}_j$, so the intersection set of the cells is non-empty. □

The key theorem we need to establish the major results in this section is that n cells in $\mathscr{E}^n$ are compact, and it will suffice, in fact, to prove that cube n cells are compact. However, since the metric properties of $\mathscr{E}^n$ are described in terms of neighborhoods, we require the following theorem relating n cells and neighborhoods (note that in $\mathscr{E}^1$, 1 cells and balls are the same). Similar ideas for $\mathscr{E}^2$ were stated in Exercise 4 of Sec. 1.5.

THEOREM 3. In $\mathscr{E}^n$, let

$$\mathscr{S} = \{x \in \mathscr{E}^n : a_i - \tfrac{1}{2}\delta \leq x_i \leq a_i + \tfrac{1}{2}\delta, i = 1, 2, \ldots, n\}$$

be a cube n cell of side length $\delta > 0$, and let P be any point of $\mathscr{S}$.

(i) If $\varepsilon > 0$ and $\delta < (1/\sqrt{n})\varepsilon$, then $\mathscr{S} \subset \mathrm{N}(P, \varepsilon)$.
(ii) If $a = (a_1, a_2, \ldots, a_n)$, then $\mathrm{B}(A, \tfrac{1}{2}\delta) \subset \mathscr{S}$. (Point A is the "center" of the cell.)

PROOF. To prove (i), let ε and δ be positive numbers such that $\delta < (1/\sqrt{n})\varepsilon$. Let X be an arbitrary point of $\mathscr{S}$. From the definition of $\mathscr{S}$, and $X, P \in \mathscr{S}$, we have

$$a_i - \tfrac{1}{2}\delta \leq x_i \leq a_i + \tfrac{1}{2}\delta, i = 1, 2, \ldots, n, \tag{1}$$

and

$$a_i - \tfrac{1}{2}\delta \leq p_i \leq a_i + \tfrac{1}{2}\delta, i = 1, 2, \ldots, n. \tag{2}$$

From (1) and (2), it follows that

$$x_i, p_i \in [a_i - \tfrac{1}{2}\delta, a_i + \tfrac{1}{2}\delta], i = 1, 2, \ldots, n, \tag{3}$$

hence that

$$|x_i - p_i| \leq \delta, i = 1, 2, \ldots, n. \tag{4}$$

Therefore

$$d(P, X)^2 = \sum_1^n (x_i - p_i)^2 \leq n\delta^2, \tag{5}$$

so

$$d(P, X) \leq \sqrt{n}\delta < \varepsilon. \tag{6}$$

Thus if $X \in \mathscr{S}$, then $X \in \mathrm{N}(P, \varepsilon)$, which establishes (i).

Part (ii) is more immediate. If $X \in \mathrm{B}(A, \tfrac{1}{2}\delta)$, then $d(A, X) \leq \tfrac{1}{2}\delta$, hence $d(A, X)^2 \leq \tfrac{1}{4}\delta^2$, which is just to say that

$$\sum_1^n (a_i - x_i)^2 \leq \tfrac{1}{4}\delta^2. \tag{7}$$

Each term of the summation in (7) is nonnegative, so each term is certainly equal to or less than the sum of the terms. That is, $(a_i - x_i)^2 \le \frac{1}{4}\delta^2$, for each i, so $|a_i - x_i| \le \frac{1}{2}\delta$, for each i, which is just to say that

$$a_i - \tfrac{1}{2}\delta \le x_i \le a_i + \tfrac{1}{2}\delta, i = 1, 2, \ldots, n, \tag{8}$$

and hence that $X \in \mathscr{S}$. Thus $\mathrm{B}(A, \frac{1}{2}\delta) \subset \mathscr{S}$. □

COROLLARY 3. Every bounded set in $\mathscr{E}^n$ is contained in some cube n cell.

We can now make clear how we intend to show that any closed, bounded set $\mathscr{R}$ in $\mathscr{E}^n$ is compact. First, since $\mathscr{R}$ is bounded, there must exist some ball $\mathrm{B}(A, \frac{1}{2}\delta)$ that contains $\mathscr{R}$. In Theorem 3, we have an explicit definition of a cube cell $\mathscr{S}$ that contains $\mathrm{B}(A, \frac{1}{2}\delta)$. Thus $\mathscr{R}$ is a closed subset of $\mathscr{S}$. Since closed subsets of a compact set are compact, then $\mathscr{R}$ will be compact if $\mathscr{S}$ is compact.

In the proof that a cube cell $\mathscr{S}$ is compact, we want to subdivide $\mathscr{S}$ into a finite number of congruent subcube cells. That this is possible in $\mathscr{E}^1$ is obvious, since clearly we can divide a segment of length δ into two subsegments of length $\frac{1}{2}\delta$. Similarly, in $\mathscr{E}^2$ we can divide a full square with side δ into four full squares, each with side $\frac{1}{2}\delta$, and in $\mathscr{E}^3$ we can divide a solid cube with side δ into eight solid cubes, each with side $\frac{1}{2}\delta$. Our next theorem simply establishes that this intuitive process generalizes to cube cells in $\mathscr{E}^n$.

THEOREM 4. If $\mathscr{S}$ is a cube cell in $\mathscr{E}^n$, with side length $\delta > 0$, then $\mathscr{S}$ is the union of a finite number of cube cells with side length $\frac{1}{2}\delta$.

PROOF. Let $\mathscr{S}$ be defined by

$$\mathscr{S} = \{x \in \mathscr{E}^n : a_i \le x_i \le a_i + \delta, i = 1, 2, \ldots, n\}. \tag{1}$$

If we let $\mathscr{I}_i = [a_i, a_i + \delta]$, then (1) can also be expressed as

$$\mathscr{S} = \{x : x_i \in \mathscr{I}_i, i = 1, 2, \ldots, n\}. \tag{2}$$

Corresponding to each of the n closed intervals $\mathscr{I}_i$, of length δ, there are two closed subintervals $\mathscr{I}_{i,1}$ and $\mathscr{I}_{i,2}$, of length $\frac{1}{2}\delta$, defined by

$$\mathscr{I}_{i,1} = [a_i, a_i + \tfrac{1}{2}\delta], \mathscr{I}_{i,2} = [a_i + \tfrac{1}{2}\delta, a_i + \delta], \tag{3}$$

$i = 1, 2, \ldots, n$, and clearly

$$\mathscr{I}_i = \mathscr{I}_{i,1} \cup \mathscr{I}_{i,2}, i = 1, 2, \ldots, n. \tag{4}$$

Now consider an n-tuple $u = (u_1, u_2, \ldots, u_n)$ in which each coordinate u_i is either 1 or 2. Since there are two choices for each of n coordinates, there are clearly 2^n such n-tuples. Let $\mathscr{U}$ denote the set of these 2^n points. Then, corresponding to each $u \in \mathscr{U}$, there is a cube cell $\mathscr{S}_u$ defined by

$$\mathscr{S}_u = \{x : x_i \in \mathscr{I}_{i,u_i}, i = 1, 2, \ldots, n\}, u \in \mathscr{U}. \tag{5}$$

We contend that $\mathscr{S}$ is the union of the 2^n sets $\mathscr{S}_u$.

First, consider $X \in \mathscr{S}_{u'}$, for some $u' \in \mathscr{U}$. From (4), and the fact that each coordinate u_i' is either 1 or 2, we have

$$x_i \in \mathscr{I}_{i,u_i'} \subset \mathscr{I}_{i,1} \cup \mathscr{I}_{i,2} = \mathscr{I}_i, i = 1, 2, \ldots, n, \tag{6}$$

hence $X \in \mathscr{S}$, and so $\mathscr{S}_{u'} \subset \mathscr{S}$.

Conversely, suppose that $X \in \mathscr{S}$. Then,

$$x_i \in \mathscr{I}_i = \mathscr{I}_{i,1} \cup \mathscr{I}_{i,2}, i = 1, 2, \ldots, n. \tag{7}$$

Define u_i' to be 1 if $x_i \in \mathscr{I}_{i,1}$ and u_i' to be 2 if $x_i \notin I_{i,1}$ (hence if $x_i \in \mathscr{I}_{i,2} \cap \mathrm{Cp}(\mathscr{I}_{i,1})$), $i = 1, 2, \ldots, n$. Clearly, $u' \in \mathscr{U}$ and $x \in \mathscr{S}_{u'}$. Therefore $\mathscr{S} \subset \bigcup_{u \in U} \mathscr{S}_u$.

From the two parts of the argument, $\mathscr{S}$ is the union of the sets $\mathscr{S}_u$. □

THEOREM 5. A cube cell in $\mathscr{E}^n$ is a compact set.

PROOF. Let $\mathscr{S}$ denote an arbitrary cube cell with side length $\delta > 0$. For an indirect argument,

$$\text{assume that } \mathscr{S} \text{ is not compact.} \tag{*}$$

By assumption, then, there must exist some infinite open cover $\mathscr{F}$ of $\mathscr{S}$ such that *no* finite subcollection of $\mathscr{F}$ covers $\mathscr{S}$.

By Theorem 4, $\mathscr{S}$ is the union of a finite number of cube cells, each of side length $\frac{1}{2}\delta$. If each of these subcubes could be covered by a finite subcollection from $\mathscr{F}$, then the union of all these subcollections would also be a finite subcollection from $\mathscr{F}$, and would be one that covered $\mathscr{S}$. Thus, since we are assuming that there is no such cover of $\mathscr{S}$, then at least one of the subcubes cannot be covered by *any* finite subcollection from $\mathscr{F}$. Let $\mathscr{S}_1$ denote such a subcube. Its side length, of course, is $\delta_1 = \frac{1}{2}\delta$.

We have shown that if the cube cell $\mathscr{S}$, of side length δ, has no finite cover from $\mathscr{F}$, then $\mathscr{S}$ has a subcube cell $\mathscr{S}_1$, of side length $\delta_1 = \frac{1}{2}\delta$, that also has no finite cover from $\mathscr{F}$. By exactly the same argument, since $\mathscr{S}_1$, of side length δ_1, has no finite cover from $\mathscr{F}$, then $\mathscr{S}_1$ has a subcube cell $\mathscr{S}_2$, of side length

$\delta_2 = \frac{1}{2}\delta_1 = \frac{1}{2}\delta$, that has no finite cover from $\mathscr{F}$. Thus repetition of the argument implies the existence of a sequence of cube cells

$$\mathscr{S} \supset \mathscr{S}_1 \supset \mathscr{S}_2 \supset \cdots \supset \mathscr{S}_m \supset \cdots, \tag{1}$$

such that no one of these cube cells can be covered by *any* finite subcollection from $\mathscr{F}$, and $\mathscr{S}_m$ has side length $\delta_m = (1/2^m)\delta$.

Since the cube cells in the sequence (1) are non-empty and decreasingly nested, then, by Theorem 2, there is a point P that belongs to all of them. Since P is in $\mathscr{S}$, and $\mathscr{S}$ is covered by $\mathscr{F}$, then there is some open set $\mathscr{T}$ in the collection $\mathscr{F}$ such that $P \in \mathscr{T}$. Because $P \in \mathscr{T}$ implies that P is interior to $\mathscr{T}$, there exists $\mathrm{N}(P, \varepsilon) \subset \mathscr{T}$. Because δ, ε, and n are specific, fixed, positive numbers, there exists a positive integer m such that $\delta_m = (1/2^m)\delta < (1/\sqrt{n})\varepsilon$. We now have $P \in \mathscr{S}_m$, $\varepsilon > 0$, and $\delta_m < (1/\sqrt{n})\varepsilon$. Thus, by (i) of Theorem 3, $\mathscr{S}_m \subset \mathrm{N}(P, \varepsilon)$, and so $\mathscr{S}_m \subset \mathscr{T}$. Since $\mathscr{T}$, by itself, is a finite subcollection from $\mathscr{F}$, the cube cell $\mathscr{S}_m$ has a finite cover from $\mathscr{F}$. The contradiction that $\mathscr{S}_m$ does and does not have a finite cover from $\mathscr{F}$ shows that the assumption (*) is false and hence that $\mathscr{S}$ is compact. □

COROLLARY 5. A closed, bounded interval in $\mathscr{E}^1$ is compact.

A number of concrete results of central importance in the text (and in analysis in general) now follow easily. In particular, we observe that Theorems 1, 2, and 5 are special cases of the following more general Theorems 6 and 8.

THEOREM 6. In $\mathscr{E}^n$, a set is compact if and only if it is closed and bounded.

PROOF. Let $\mathscr{S}$ denote a subset of $\mathscr{E}^n$. If $\mathscr{S}$ is compact, then, by Theorems 9 and 10, Sec. 1.4, $\mathscr{S}$ is closed and bounded.

If $\mathscr{S}$ is closed and bounded, then, by Corollary 3, $\mathscr{S}$ is contained in some cube n cell. By Theorem 5, this cube n cell is compact. Thus $\mathscr{S}$ is a closed subset of a compact set, hence, by Theorem 11, Sec. 1.4, $\mathscr{S}$ is compact. □

THEOREM 7. A set of real numbers is compact and connected if and only if it is a closed, bounded interval. (E)

THEOREM 8. A decreasingly nested sequence of non-empty, closed, bounded sets in $\mathscr{E}^n$ has a non-empty intersection set. (E)

THEOREM 9. Let $f : \mathscr{D} \to \mathscr{R}$ be continuous with $\mathscr{D}, \mathscr{R} \subset \mathscr{E}^n$. If $\mathscr{D}$ is a compact set, then $\mathscr{R}$ is a compact set, and both $\mathscr{D}$ and $\mathscr{R}$ are closed, bounded sets. (E)

THEOREM 10. Let $f : \mathscr{D} \to \mathscr{R}$ be a continuous function. If $\mathscr{D}$ is a closed and bounded subset of $\mathscr{E}^n$, and if f maps $\mathscr{D}$ into a metric space, then $\mathscr{R}$ is compact. (E)

THEOREM 11 (Min–Max Theorem). If $f : \mathscr{D} \to \mathscr{R}$ is a continuous, real valued function, and $\mathscr{D}$ is a compact subset of a metric space, then f is a bounded function that attains a minimum value and a maximum value. (E)

THEOREM 12. If $f : \mathscr{D} \to \mathscr{R}$ is a continuous, real valued function, and $\mathscr{D}$ is a compact, connected subset of a metric space, then f achieves a minimum value, a maximum value, and every value intermediate to these. (E)

Exercises – Section 1.7

1. Give an infinite cover of the closed interval $[a, b]$ by closed sets that cannot be reduced to a finite cover.
2. Give an example in $\mathscr{E}^1$ of a decreasingly nested sequence of sets $\mathscr{S}_i$ whose intersection set is empty and

   a. each set $\mathscr{S}_i$ is non-empty and open;
   b. each set $\mathscr{S}_i$ is non-empty and closed.

3. Illustrate Theorem 3 pictorially in $\mathscr{E}^1$ and $\mathscr{E}^2$ with $\delta = 2$ and with $A = P = 0$.
4. In the proof of Theorem 4, following $x_i \in \mathscr{I}_{i,1} \cup \mathscr{I}_{i,2}$, what is wrong with defining u_i' to be 1 if $x_i \in \mathscr{I}_{i,1}$ and to be 2 if $x_i \in \mathscr{I}_{i,2}$?
5. Write a proof modeled on that for Theorem 5 to show that a closed, bounded interval in $\mathscr{E}^1$ is compact.
6. A little later we establish a result that one would expect, namely, that Corollary 5 implies that closed, bounded segments in $\mathscr{E}^n$ are compact sets. Why will this result, in turn, imply that the polygons of plane euclidean geometry are compact sets?
7. Prove Theorem 7.
8. Prove Theorem 8.
9. Prove Theorem 9.
10. Prove Theorem 10.
11. Prove Theorem 11.
12. Prove Theorem 12.
13. A *curve* in $\mathscr{E}^n$ is sometimes defined as a set that can be represented as the range $\mathscr{R}$ of a continuous function $f : \mathscr{D} \to \mathscr{R}$, where $\mathscr{D}$ is the unit interval $[0, 1]$ and $\mathscr{R} \subset \mathscr{E}^n$. Why is such a curve always a compact, connected set?
14. Let $\mathscr{S}$ be a closed, bounded, connected set in $\mathscr{E}^2$. If P is a point of $\mathscr{E}^2$ and if A, B are points of $\mathscr{S}$, prove that there exists a point C in $\mathscr{S}$ such that $d(P, C) = \frac{1}{2}[d(P, A) + d(P, B)]$. (Hint: see Exercise 2, Sec. 1.5.)
15. Let f be a real valued, continuous function defined on an interval $[a, b]$. If f is strictly increasing ($x < y \Rightarrow f(x) < f(y)$), prove that the inverse

relation f^{-1} is a function and is also a continuous function. The same result holds if f is strictly decreasing.

16. If the function f is real valued and continuous on an open interval (a, b) and if for each open interval $(c, d) \subset (a, b)$ the set $\{f(x): x \in (c, d)\}$ is open, prove that f has neither a minimum nor a maximum value on (a, b).

SECTION 1.8. BASIC DISTANCE VARIATIONS, FOOT IN A SET, DIAMETER, NEARNESS OF SETS

Although we will not be studying abstract metric geometry, which is an extensive, special field of mathematics, we can, in an efficient way, convey something of the spirit and method of the subject. There are a number of basic geometric properties that are indispensable in our later work with euclidean geometry. We are now in a position to show that they are also valid in general metric spaces. In this section we consider a few of these, both for future needs and as an illustration of mathematical method.

We begin with the idea of the distance of a point P from a set $\mathscr{S}$. The natural interpretation is that this means the distance from P to the point F in $\mathscr{S}$ that is nearest to P. But we cannot be sure that such a point F exists; in fact, a little reflection shows that it need not, so we introduce a distance that does not depend on a nearest point.

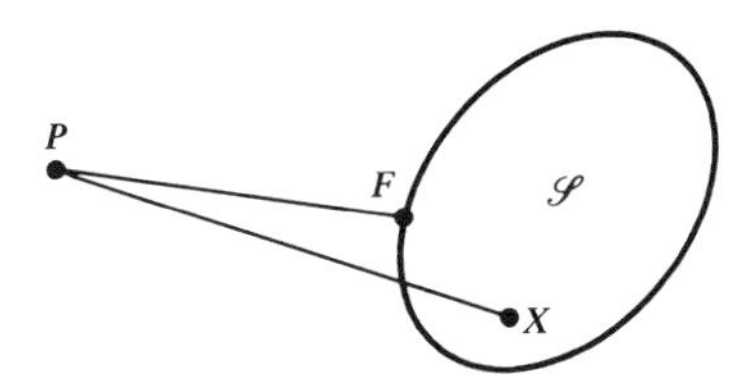

DEFINITION (Distance from a point to a set). The distance from a point P to a non-empty set $\mathscr{S}$ is a number, denoted by $d(P, \mathscr{S})$, defined by

$$d(P, \mathscr{S}) = \text{glb}\{d(P, X): X \in \mathscr{S}\}.$$

Because distances are nonnegative, every set of distances is bounded below by zero and hence has a greatest lower bound. Thus the following fact is immediate.

THEOREM 1. Each point of a metric space $\mathscr{E}$ has a unique distance from each non-empty subset of $\mathscr{E}$.

Given a point P and a set $\mathscr{S}$, the concepts of a point in $\mathscr{S}$ nearest to P and a point in $\mathscr{S}$ farthest from P are really self-explanatory. However, the first of these concepts plays such an important role in geometry that special language is used in connection with it.

DEFINITION (Foot in a set). A point P has a *foot* F in a set $\mathscr{S}$ if $F \in \mathscr{S}$ and $d(P, F) = d(P, \mathscr{S})$. Thus F is a point of $\mathscr{S}$ nearest to P and $d(P, F) \leq d(P, X)$ for all $X \in \mathscr{S}$.

DEFINITION (Farthest point in a set). Corresponding to a point P and a set $\mathscr{S}$, a point F is a *farthest point* of $\mathscr{S}$ from P if $F \in \mathscr{S}$ and $d(P, F) \geq d(P, X)$ for all $X \in \mathscr{S}$.

Obviously, a set $\mathscr{S}$ may contain more than one foot of a point P and more than one point farthest from P. For example, each point of the sphere $\mathrm{S}(P, \delta)$ is a foot of P in the sphere and also a farthest point from P. If P belongs to $\mathscr{S}$, then from $d(P, P) = 0$ it follows that $d(P, \mathscr{S}) = 0$ and that P is its own foot in $\mathscr{S}$. However, it is not true that $d(P, \mathscr{S}) = 0$ implies that P is in $\mathscr{S}$ or even that P has a foot in $\mathscr{S}$. The exact implication of $d(P, \mathscr{S}) = 0$ is the content of the next theorem.

THEOREM 2. If $\mathscr{S}$ is a non-empty set, then $d(P, \mathscr{S}) = 0$ if and only if $P \in \mathrm{Cl}(\mathscr{S})$.

PROOF. We show that "$d(P, \mathscr{S}) > 0$" and "$P \notin \mathrm{Cl}(\mathscr{S})$" are equivalent statements. First, $P \notin \mathrm{Cl}(\mathscr{S})$ is equivalent to $P \in \mathrm{Ex}(\mathscr{S})$. In turn, $P \in \mathrm{Ex}(\mathscr{S})$ is equivalent to the existence of $\mathrm{N}(P, \delta)$ such that $\mathrm{N}(P, \delta) \cap \mathscr{S} = \varnothing$. Thus, in terms of distance, it follows that $P \notin \mathrm{Cl}(\mathscr{S})$ if and only if there exists $\delta > 0$ such that $d(P, X) \geq \delta > 0$ for all $X \in \mathscr{S}$. Therefore, $P \notin \mathrm{Cl}(\mathscr{S})$ if and only if $\mathrm{glb}\{d(P, X) : X \in \mathscr{S}\} \geq \delta$, hence if and only if $d(P, \mathscr{S}) \geq \delta > 0$. □

COROLLARY 2. A non-empty set $\mathscr{S}$ is closed if and only if $d(P, \mathscr{S}) = 0$ implies $P \in \mathscr{S}$. (E)

Corresponding to a set $\mathscr{S}$ and a point P, it is natural to wonder what determines whether or not $\mathscr{S}$ contains a foot of P or a point farthest from P. Because P is fixed, $d(P, X)$ is a function of X, so the question turns on whether or not this function achieves minimum and maximum function values on the domain $\mathscr{S}$. We can get some answers to this by generalizing a problem given in Sec. 1.5.

THEOREM 3. If P is a point of the metric space $\mathscr{E}$ and $\mathscr{S}$ is a non-empty subset of $\mathscr{E}$, then the function f defined by

$$f(X) = d(P, X)$$

is continuous on $\mathscr{E}$.

PROOF. Corresponding to $X_o \in \mathscr{E}$ and an arbitrary positive number ε, let $\delta = \varepsilon$, and consider $X \in \mathrm{N}(X_o, \delta)$. From the definition of X and the triangle inequality for distance, it follows that

$$f(X) = d(P, X) \leq d(P, X_o) + d(X_o, X). \tag{1}$$

hence that

$$f(X) - f(X_o) \leq d(X_o, X). \tag{2}$$

But, similarly,

$$f(X_o) = d(P, X_o) \leq d(P, X) + d(X, X_o), \tag{3}$$

hence

$$f(X_o) - f(X) \leq d(X, X_o). \tag{4}$$

From (2) and (4), we have

$$|f(X) - f(X_o)| \leq d(X, X_o) < \varepsilon.$$

Thus f is continuous at X_o and hence at all points of $\mathscr{E}$. □

COROLLARY 3.1. Each point of space has a foot in each non-empty, compact set.

PROOF. Let P be a point of $\mathscr{E}$ and let $\mathscr{S}$ be a non-empty, compact subset of $\mathscr{E}$. By Theorem 3, the function $f(X) = d(P, X)$ is continuous at all points of $\mathscr{E}$, hence its restriction to the domain $\mathscr{S}$ is also continuous. That is, $f(X) = d(P, X)$, $X \in \mathscr{S}$ is a continuous, real valued function on the compact domain $\mathscr{S}$. Hence by the min–max theorem (Theorem 11, Sec. 1.7), f achieves a minimum value at some point X_o in $\mathscr{S}$. Thus $f(X_o) = d(P, X_o) \leq d(P, X)$ for all X in $\mathscr{S}$, hence P has foot X_o in $\mathscr{S}$. □

Exactly the same argument gives another corollary.

COROLLARY 3.2. If P is any point of $\mathscr{E}$ and $\mathscr{S}$ is any non-empty, compact subset of $\mathscr{E}$, then there is a point of $\mathscr{S}$ farthest from P.

It is instructive to see why a stronger foot property than Corollary 3.1 holds in euclidean space.

COROLLARY 3.3. Each point of $\mathscr{E}^n$ has a foot in each non-empty, closed subset of $\mathscr{E}^n$.

PROOF. Let P be a point of $\mathscr{E}^n$ and let $\mathscr{S}$ be a closed subset of $\mathscr{E}^n$ such that there exists $Q \in \mathscr{S}$. If $P = Q$, the theorem follows trivially, so suppose that $d(P, Q) = \delta > 0$, and define $\mathscr{S}_1 = \mathrm{B}(P, \delta) \cap \mathscr{S}$. Since the ball $\mathrm{B}(P, \delta)$ is obviously a closed and bounded subset of $\mathscr{E}^n$, it is compact (we could not draw this conclusion if the space were $\mathscr{E}$ rather than $\mathscr{E}^n$). Because $\mathscr{S}$ is closed, and the ball $\mathrm{B}(P, \delta)$ is closed, then $\mathscr{S}_1$ is closed. It is thus a closed subset of the compact ball $\mathrm{B}(P, \delta)$, and so $\mathscr{S}_1$ is compact. Now, by Theorem 3, the function $f(X) = d(P, X)$, $X \in \mathscr{S}_1$ is a continuous, real valued function on the compact domain $\mathscr{S}_1$. Therefore, by the min–max theorem of Sec. 1.7, f achieves a minimum at some point $X_o \in \mathscr{S}_1$. Then $d(P, X_o) \leq d(P, X)$ for all $X \in \mathscr{S}_1$. Also, $d(P, X_o) \leq d(P, Q) = \delta$. Thus if X is in $\mathscr{S}$ and not in $\mathscr{S}_1$, we have $d(P, X_o) \leq \delta < d(P, X)$. Thus $d(P, X_o) \leq d(P, X)$ for all $X \in \mathscr{S}$, so P has X_o as a foot in $\mathscr{S}$. □

Although it is not true that each point of a general metric space $\mathscr{E}$ necessarily has a foot in each closed set, a converse of this is correct.

THEOREM 4. If $\mathscr{S}$ is a subset of $\mathscr{E}$ such that each point of $\mathscr{E}$ has a foot in $\mathscr{S}$, then $\mathscr{S}$ is a closed set. (E)

Later we see that Theorem 4 is a useful tool in showing that certain geometric sets are closed.

We now turn to the generalization of another simple geometric idea. The diameter of the ball $\mathrm{B}(P, \delta)$ in $\mathscr{E}^3$ is, of course, 2δ. This is also the maximum distance between any two points of the ball. By generalizing this last property, we can define the diameter of any bounded set in a metric space.

DEFINITION (Diameter of a set). If $\mathscr{S}$ is a non-empty, bounded set in a metric space $\mathscr{E}$, the diameter of $\mathscr{S}$, denoted by $\mathrm{Dm}(\mathscr{S})$, is the number defined by

$$\mathrm{Dm}(\mathscr{S}) = \mathrm{lub}\{d(X, Y)\colon X, Y \in \mathscr{S}\}.$$

It is clear that for any non-empty, bounded set $\mathscr{S}$, the diameter of $\mathscr{S}$ exists and is a non-negative number. One would guess that in euclidean space the diameter of $\mathscr{S}$ is actually the distance between some pair of points in $\mathscr{S}$ whenever $\mathscr{S}$ is closed as well as bounded. The counterpart to this is the guess that in $\mathscr{E}$ the diameter is actually achieved at some pair in $\mathscr{S}$ whenever $\mathscr{S}$ is compact.

To investigate the questions just raised, we consider a compact set $\mathscr{S}$ in $\mathscr{E}$ and the variation of the "farthest distance from $\mathscr{S}$" of a point X, as X varies.

THEOREM 5. Let $\mathscr{S}$ be a non-empty, compact set in a metric space $\mathscr{E}$. Corresponding to each $X \in \mathscr{E}$, there is a point X^* contained in $\mathscr{S}$ and farthest from X, and the function f defined by

$$f(X) = d(X, X^*)$$

is continuous on $\mathscr{E}$.

PROOF. By Corollary 3.2, for each X in $\mathscr{E}$ there is at least one point X^* in $\mathscr{S}$ farthest from X. Since all points of $\mathscr{S}$ farthest from X are the same distance from X, the function f has a well-defined function value at each point of $\mathscr{E}$.

Now, let X_o be a point of $\mathscr{E}$. Corresponding to an arbitrary positive number ε, let $\delta = \varepsilon$, and consider $X \in \mathrm{N}(X_o, \delta)$. The continuity of f at X_o follows if we can show that necessarily $|f(X) - f(X_o)| < \varepsilon$. Let X_o^* designate a point of $\mathscr{S}$ farthest from X_o and let X^* be a point of $\mathscr{S}$ farthest from X. From the definition of f and the triangle inequality, we have

$$\begin{aligned} f(X) = d(X, X^*) &\leq d(X, X_o) + d(X_o, X^*) \leq d(X, X_o) + d(X_o, X_o^*) \\ &= d(X, X_o) + f(X_o), \end{aligned} \tag{1}$$

hence

$$f(X) - f(X_o) \leq d(X, X_o). \tag{2}$$

But the same relations (1) and (2) hold if we interchange X with X_o and X^* with X_o^*, so

$$f(X_o) - f(X) \leq d(X_o, X). \tag{3}$$

Now, (2) and (3) imply that

$$|f(X) - f(X_o)| \leq d(X, X_o) < \delta = \varepsilon, \tag{4}$$

and hence that f is continuous at X_o. □

COROLLARY 5. If $\mathscr{S}$ is a non-empty, compact subset of $\mathscr{E}$, then there exist points P, P^* in $\mathscr{S}$ such that $\mathrm{Dm}(\mathscr{S}) = d(P, P^*)$.

PROOF. Corresponding to $\mathscr{S}$, let f be the function defined in Theorem 4. Since f is continuous on $\mathscr{E}$, its restriction to $\mathscr{S}$ is continuous on $\mathscr{S}$. This restriction is therefore a continuous, real valued function on a compact domain $\mathscr{S}$. By the min–max theorem of Sec. 1.7, it follows that f achieves a maximum on $\mathscr{S}$. That is, there exists $P \in \mathscr{S}$ such that

$$f(P) \geq f(X) \text{ for all } X \in \mathscr{S}.$$

Let P^* denote any point of $\mathscr{S}$ farthest from P, so $f(P) = d(P, P^*)$. If X, Y are points of $\mathscr{S}$, then

$$d(P, P^*) = f(P) \geq f(X) = d(X, X^*) \geq d(X, Y)$$

shows that $d(P, P^*)$ is an upper bound to the set $\{d(X, Y) : X, Y \in \mathscr{S}\}$. Since $P, P^* \in \mathscr{S}$, it follows that $d(P, P^*)$ is also the least upper bound to the set, hence $d(P, P^*) = \text{Dm}(\mathscr{S})$. □

We conclude this section with a numerical measure on pairs of sets $\mathscr{R}$, $\mathscr{S}$ that is related to our intuition of how close the sets are to each other.

DEFINITION (Nearness of two sets). If $\mathscr{R}$ and $\mathscr{S}$ are non-empty sets in a metric space $\mathscr{E}$, then the *nearness* of $\mathscr{R}$ and $\mathscr{S}$ is the number, denoted by $\text{n}(\mathscr{R}, \mathscr{S})$, defined by

$$\text{n}(\mathscr{R}, \mathscr{S}) = \text{glb}\{d(X, Y) : X \in \mathscr{R}, Y \in \mathscr{S}\}.$$

In a later chapter we define an actual metric $d(\mathscr{R}, \mathscr{S})$, where the sets $\mathscr{R}$ and $\mathscr{S}$ play the role of "points" with respect to d. Clearly, $\text{n}(\mathscr{R}, \mathscr{S})$ is not a metric in this sense, since $\text{n}(\mathscr{R}, \mathscr{S}) = 0$ obviously does not imply $\mathscr{R} = \mathscr{S}$. Although it is true that intersecting sets have zero nearness, the converse is false. In fact, $\mathscr{R}$ and $\mathscr{S}$ may be separated sets and have zero nearness. However, separatedness and nearness do have the following relation.

THEOREM 6. If $\text{n}(\mathscr{R}, \mathscr{S}) > 0$, then $\mathscr{R}$ and $\mathscr{S}$ are separated sets. (E)

In the spirit of the questions we have considered in this section, it is natural to ask, when is $\text{n}(\mathscr{R}, \mathscr{S})$ achieved at some point X_0 in $\mathscr{R}$ and Y_0 in $\mathscr{S}$? To investigate this, we look at another distance variation.

THEOREM 7. If $\mathscr{S}$ is a non-empty set in a metric space $\mathscr{E}$, then the function f defined by

$$f(P) = d(P, \mathscr{S})$$

is continuous on $\mathscr{E}$.

PROOF. Corresponding to P_o in $\mathscr{E}$ and an arbitrary positive number ε, let $\delta = \varepsilon$, and consider $P \in N(P_o, \delta)$. For every point X in $\mathscr{S}$, the definition of f and the distance triangle inequality imply that

$$f(P) = d(P, \mathscr{S}) \leq d(P, X) \leq d(P, P_o) + d(P_o, X), \tag{1}$$

and hence that

$$f(P) - d(P, P_o) \leq d(P_o, X). \tag{2}$$

Since (2) holds for all X in $\mathscr{S}$, then $f(P) - d(P, P_o)$ is a lower bound to the set $\{d(P_o, X) : X \in \mathscr{S}\}$. Therefore

$$f(P) - d(P, P_o) \leq \text{glb}\{d(P_o, X) : X \in \mathscr{S}\}, \tag{3}$$

or

$$f(P) - d(P, P_o) \leq d(P_o, \mathscr{S}) = f(P_o), \tag{4}$$

hence

$$f(P) - f(P_o) \leq d(P, P_o). \tag{5}$$

But the relations (1)–(5) follow in exactly the same way if we interchange the roles of P and P_o. That is, for all X in $\mathscr{S}$

$$f(P_o) = d(P_o, \mathscr{S}) \leq d(P_o, X) \leq d(P_o, P) + d(P, X) \tag{6}$$

leads to

$$f(P_o) - f(P) \leq d(P_o, P). \tag{7}$$

Now, from (5) and (7) we have

$$|f(P) - f(P_o)| \leq d(P, P_o) < \delta = \varepsilon. \tag{8}$$

Thus f is continuous at P_o and hence at all points of $\mathscr{E}$. □

COROLLARY 7.1. If $\mathscr{R}$ and $\mathscr{S}$ are non-empty, compact sets, then there exists $P_o \in \mathscr{R}$ and $Q_o \in \mathscr{S}$ such that $d(P_o, Q_o) = n(\mathscr{R}, \mathscr{S})$.

PROOF. Corresponding to $\mathscr{S}$, let f be the function defined in Theorem 7. Since f is continuous on $\mathscr{E}$, its restriction to $\mathscr{R}$ is continuous on $\mathscr{R}$. Thus $f(P) = d(P, \mathscr{S})$, $P \in \mathscr{R}$, is a continuous, real valued function on the compact domain $\mathscr{R}$. By the min–max theorem of Sec. 1.7, f achieves a minimum on $\mathscr{R}$. Thus there exists $P_o \in \mathscr{R}$ such that

$$f(P_o) \leq f(P) \text{ for all } P \in \mathscr{R}.$$

By Corollary 3.1, P_o has a foot Q_o in $\mathscr{S}$, so $f(P_o) = d(P_o, \mathscr{S}) = d(P_o, Q_o)$. Now, if X is a point of $\mathscr{R}$ and Y is a point of $\mathscr{S}$, then

$$d(P_o, Q_o) = f(P_o) \leq f(X) = d(X, \mathscr{S}) \leq d(X, Y)$$

shows that $d(P_o, Q_o)$ is a lower bound to $\{d(X, Y): X \in \mathscr{R}, Y \in \mathscr{S}\}$. Because $P_o \in \mathscr{R}$ and $Q_o \in \mathscr{S}$, $d(P_o, Q_o)$ must be the greatest lower bound to the set, hence $d(P_o, Q_o) = \mathrm{n}(\mathscr{R}, \mathscr{S})$. □

COROLLARY 7.2. If one of two closed, non-null sets $\mathscr{R}$, $\mathscr{S}$ is compact, then $\mathscr{R} \cap \mathscr{S} = \varnothing$ implies that $\mathrm{n}(\mathscr{R}, \mathscr{S}) > 0$. (E)

Exercises – Section 1.8

1. Prove Corollary 2.
2. If P is a point in a metric space $\mathscr{E}$, then, by Theorem 3, the function $f(X) = d(P, X)$ is continuous on $\mathscr{E}$. Use this fact to prove that a ball $\mathrm{B}(P, \delta)$ and a sphere $\mathrm{S}(P, \delta)$ are closed sets in $\mathscr{E}$.
3. Using Exercise 2, prove that in $\mathscr{E}$ the balls and spheres are closed, bounded sets. Prove also that if $\mathscr{S}$ is bounded, then $\mathrm{Cl}(\mathscr{S})$ is bounded. In particular, if $\mathscr{E} = \mathscr{E}^n$, prove that the closure of a bounded set is compact and that the balls and spheres of $\mathscr{E}^n$ are compact.
4. If X_o, Y_o, X, Y are points of a metric space, and if ε is a positive number, prove that there exists a positive number δ such that $d(X, X_o) < \delta$ and $d(Y, Y_o) < \delta$ imply that $|d(X, Y) - d(X_o, Y_o)| < \varepsilon$.
5. If $\mathscr{S}$ is a bounded set in $\mathscr{E}^n$, use Exercises 3 and 4 and Corollary 5 to show that $\mathrm{Dm}(\mathscr{S}) = \mathrm{Dm}[\mathrm{Cl}(\mathscr{S})]$.
6. Let $\mathscr{S}$ and $\mathscr{T}$ be two non-empty, closed, disjoint sets in a metric space $\mathscr{E}$ and let f be the function defined on $\mathscr{E}$ by

$$f(X) = \frac{d(X, \mathscr{S})}{d(X, \mathscr{S}) + d(X, \mathscr{T})}.$$

   a. Prove that the denominator sum is never zero.
   b. Prove that f is a continuous, real valued function. (Hint: recall that if h and g are continuous, real valued functions, so is $f + g$ and also f/g if g has no zeros.)

c. Show that the range of f is contained in $[0, 1]$ and that f maps $\mathscr{S}$ onto $\{0\}$ and $\mathscr{T}$ onto $\{1\}$.
d. Explain why $\{X: f(X) < \frac{1}{2}\}$ and $\{X: f(X) > \frac{1}{2}\}$ are disjoint, open sets containing $\mathscr{S}$ and $\mathscr{T}$, respectively.

7. Prove Theorem 4.
8. If $\mathscr{S}$ is a cube n cell in $\mathscr{E}^n$ of side length δ, show that $\text{Dm}(\mathscr{S}) = \sqrt{n}\,\delta$.
9. Give examples of sets $\mathscr{R}$, $\mathscr{S}$, $\mathscr{T}$ such that

$$\text{n}(\mathscr{R}, \mathscr{S}) + \text{n}(\mathscr{S}, \mathscr{T}) < \text{n}(\mathscr{R}, \mathscr{T}).$$

10. Prove Theorem 6.
11. In $\mathscr{E}^2$, let $\mathscr{R} = \{X: x_1x_2 \geq 1\}$ and let $\mathscr{S} = \{X: x_1x_2 \leq -1\}$. Show that $\mathscr{R}$ and $\mathscr{S}$ are closed, disjoint sets with zero nearness.
12. Prove Corollary 7.2.
13. Corresponding to a bounded, non-empty set $\mathscr{S}$ and a point P in a metric space $\mathscr{E}$, let $\varphi(P, \mathscr{S})$ be defined by

$$\varphi(P, \mathscr{S}) = \text{lub}\{d(P, X): X \in \mathscr{S}\}.$$

Does $\varphi(P, \mathscr{S})$ exist for every point P in $\mathscr{E}$? Is $\varphi(P, \mathscr{S})$ a continuous function on $\mathscr{E}$?
14. In $\mathscr{E}^2$, if $\mathscr{R} = \text{Sg}[AB]$ and $\mathscr{S} = \text{Sg}[AC]$ are noncollinear, noncongruent segments, sketch and describe the set $\{X: d(X, \mathscr{R}) = d(X, \mathscr{S})\}$.
15. If F is a foot of P in set $\mathscr{S}$, must F belong to $\text{Bd}(\mathscr{S})$? If F is a farthest point of $\mathscr{S}$ from P, must F belong to $\text{Bd}(\mathscr{S})$?

2
THE STRUCTURE OF EUCLIDEAN *n*-SPACE

INTRODUCTION

Our objective in this chapter is to establish the principal characteristics of $\mathscr{E}^n$ as the framework within which geometric questions are considered. We wish to do this in a way that identifies both $\mathscr{E}^2$ and the two-dimensional subspaces of $\mathscr{E}^n$ with the plane of "plane geometry" and identifies $\mathscr{E}^3$ and the three-dimensional subspaces of $\mathscr{E}^n$ with the space of "solid geometry."

SECTION 2.1. LINEAR AND PLANAR CONCEPTS IN $\mathscr{E}^n$

In a beginning study of euclidean geometry, a point is an undefined element, and lines and planes are sets of points that satisfy certain existence and intersection axioms. Other assumptions are made about a metric and about certain congruence relations. With this basis, various geometric figures can be defined and their properties deduced. When coordinates appear, they do so as a reference system imposed on a known space. For example, when the coordinate triple of a point is identified as a vector and different vector relations are studied, what is being established is the vector character of the space.

Our point of view is radically different. A point in $\mathscr{E}^n$ is now not an undefined element but something quite explicit, namely, an ordered n-tuple of real numbers. We do not have any axioms about lines and planes. Instead, we must define what a line is and what a plane is as well as all the other basic sets such as segments, rays, and angles.

In this section, we give parametric definitions for some basic sets in $\mathscr{E}^n$ that are simply generalizations to $\mathscr{E}^n$ of similar definitions used in plane and solid analytic geometry. Later these objects appear naturally from a more synthetic point of view, and the two aspects are reconciled.

CONVENTION. By **AB** we mean the vector $b - a$.

DEFINITION (Line, ray, segment). If $A, B \in \mathscr{E}^n$, and $A \neq B$, then the collection of points

$$X = a + \lambda\mathbf{AB} = (1 - \lambda)a + \lambda b$$

is

(i) the *line* of A, B, denoted Ln(AB), for $-\infty < \lambda < \infty$
(ii) the *open ray* from A through B, denoted Ry(A, B), for $\lambda > 0$
(iii) the *closed ray* from A through B, denoted Ry[A, B), for $\lambda \geq 0$
(iv) the *open segment* of A and B, denoted Sg(AB), for $0 < \lambda < 1$
(v) the *closed segment* of A and B, denoted Sg[AB], for $0 \leq \lambda \leq 1$
(vi) the *half open segment* Sg(A, B], for $0 < \lambda \leq 1$, and the half open segment Sg[A, B) for $0 \leq \lambda < 1$.

The point A is the *origin* of the rays Ry(A, B) and Ry[A, B), and A, B are the *endpoints* of the four types of segments.

CONVENTIONS. A set is said to be a *linear* or *nonlinear* set according as it is or is not a subset of some line. Points (or sets) that are contained in a line are said to be *collinear*, otherwise they are *noncollinear*.

It should be noted that the terms "open" and "closed" are used in connection with rays and segments to indicate whether the origin or endpoints do not or do belong to the set and are *not* used to affirm that the ray or segment is an open set or a closed set. Later we consider the openess or closedness of these and many other geometric sets.

DEFINITION (Plane). If A, B, C are noncollinear points of $\mathscr{E}^n$, then the collection of points

$$X = a + \lambda_1\mathbf{AB} + \lambda_2\mathbf{AC} = (1 - \lambda_1 - \lambda_2)a + \lambda_1 b + \lambda_2 c,$$
$$-\infty < \lambda_1, \lambda_2 < \infty$$

is the *plane* of A, B, C, denoted by Pl(ABC).

There are two aspects of the definitions for line and plane that we want to point out, a linear aspect and an affine aspect, that are themes of later generalizations. The linear aspect concerns the following observations. If $A \neq B$, then $p = \mathbf{AB}$ is not the null vector. By definition, the line Ln(OP), $x = o + \lambda\mathbf{OP} = \lambda p$, $-\infty < \lambda < \infty$, is just the linear span of the vector p. The line Ln(AB) is the set $a + \lambda p$, $-\infty < \lambda < \infty$, obtained by adding the fixed vector a to each vector in the linear span of p. In other words, the line Ln(AB) is the translate of a linear span. Similarly, if A, B, C are noncollinear, then $p = \mathbf{AB}$ and $q = \mathbf{AC}$ are noncollinear vectors, and the plane Pl(OPQ) is $x = \lambda_1 p + \lambda_2 q$, $-\infty < \lambda_1, \lambda_2 < \infty$, which is just the linear span of p and q. The plane Pl(ABC) is the set $a + \lambda_1\mathbf{AB} + \lambda_2\mathbf{AC} = a + \lambda_1 p + \lambda_2 q$, $-\infty < \lambda_1, \lambda_2 < \infty$, which is obtained by adding the fixed vector a to each vector in the linear span of p and q. Thus a plane through the origin, Pl(OPQ), is the linear span of two noncollinear vectors, and a plane in general position, Pl(ABC), is the translate of such a span. There is thus a natural connection between the subspaces of $\mathscr{E}^n$ natural to geometry and the subspaces studied in linear algebra, and we want to exploit this connection.

What we called the affine aspect of our definitions for line and plane also follows from a simple observation. In the representation of a line

$$\text{Ln}(AB)\colon x = (1 - \lambda)a + \lambda b, \qquad -\infty < \lambda < \infty,$$

we have a special kind of linear combination of a and b, namely, one in which the sum of the coefficients is one. This permits an equivalent and more symmetric representation

$$\text{Ln}(AB)\colon x = \lambda_1 a + \lambda_2 b, \qquad \lambda_1 + \lambda_2 = 1, \qquad -\infty < \lambda_1, \lambda_2 < \infty.$$

Similarly, the plane

$$\text{Pl}(ABC)\colon x = (1 - \lambda_1 - \lambda_2)a + \lambda_1 b + \lambda_2 c, \qquad -\infty < \lambda_1, \lambda_2 < \infty$$

has an equivalent symmetric form

$$x = \eta_1 a + \eta_2 b + \eta_3 c, \qquad \eta_1 + \eta_2 + \eta_3 = 1, \qquad -\infty < \eta_1, \eta_2, \eta_3 < \infty.$$

A linear combination in which the sum of the coefficients is one is called an affine combination of the vectors, and the set of all such combinations is the affine span of the vectors. Thus we have in lines and planes our first indication of a modification of linear algebra that corresponds to what is known as "affine geometry."

Exercises – Section 2.1

1. Write parametric equations for Ln(AB) if

   a. $A = (1, 0), B = (0, 1)$
   b. $A = (0, 0, 0), B = (1, 2, 1)$
   c. $A = (-1, 2, 3, 4), B = (2, 1, -1, -7)$.

2. Let A, B denote distinct points in $\mathscr{E}^n$, and let a function f be defined on $\mathscr{E}^1$ by

$$f(\lambda) = a + \lambda(b - a).$$

   a. Show that $d[f(\lambda_1), f(\lambda_2)] = |\lambda_1 - \lambda_2| d(A, B)$.
   b. Show that f is a continuous one to one mapping of $\mathscr{E}^1$ onto Ln(AB).
   c. How does (b) imply that Sg[A, B] (and hence all closed segments in $\mathscr{E}^n$) are compact?
   d. If C, D, E on Ln(AB) correspond to λ_1, λ_2, λ_3, respectively, and if $\lambda_1 < \lambda_2 < \lambda_3$, show that $d(C, D) + d(D, E) = d(C, E)$. (Thus, of three points on a line, one is always between the other two.)

3. Write parametric equations of the plane Pl(ABC) if

   a. $A = (0, 0, 0), B = (0, 1, 0), C = (0, 0, 1)$
   b. $A = (-1, 2, 3, 4), B = (3, 1, 8, 1), C = (0, 0, 0, 0)$
   c. $A = (1, 0, 0, 0, 0), B = (0, 1, 0, 0, 0), C = (0, 0, 1, 0, 0)$

4. In synthetic geometry, the "flatness" of planes comes from an axiom that if two distinct points belong to a plane, then the line of the points is contained in the plane. How does the symmetric, or affine, representation of Pl(ABC) imply at once that Ln(AB), Ln(BC) and Ln(CA) are contained in Pl(ABC)?

5. If $x = a + \lambda\mathbf{AB}$ is an equation for Ln(AB), for what values of λ is

   a. X on Sg(AB)?
   b. X on Ry(A, B)?
   c. X on Cp[Ry(A, B)]?

6. If h is a positive number, for what value of λ in Exercise 5 is X on Ry(A, B) and such that $d(A, X) = h$?
7. Prove that if points A, B in a set $\mathscr{S}$ in $\mathscr{E}^n$ are such that Dm($\mathscr{S}$) $= d(A, B)$, then A and B belong to Bd($\mathscr{S}$).

SECTION 2.2. ANGLES AND THE EUCLIDEAN INNER PRODUCT

Although angles and measure of angles play a fundamental role in euclidean geometry, there is sometimes ambiguity in distinguishing between an angle and its measure and between an angle and a rotation by an angle. In this section we establish the angles of our system as a definite class of geometric figures. We then define a measure for the size of an angle that associates a particular number with each angle in the same way, for example, that a diameter associates a number with each bounded set. We define this angle measure in terms of an inner product. It is of great importance that, although the concrete nature of angles as geometric figures is intuitively and conceptually valuable, it is the identification of angle properties with inner products that provides a powerful and efficient way of making algebraic calculations with these properties.

We begin with the directional nature of rays.

DEFINITION (Ray direction). Corresponding to a ray from A through B, the set of vectors $\{k\mathbf{AB} : k > 0\}$ are the *directional vectors* of the ray, or simply the *vectors of the ray*, and both R$[A, B)$ and R(A, B) are defined to be *like directed* to each of their directional vectors. Two rays are like directed (opposite directed) if a vector of one is like directed (opposite directed) to a vector of the other. Two rays are *opposite rays* if they have a common origin and are opposite directed.

DEFINITION (Angle). The union of two closed rays with a common origin is an *angle*. It is a *regular angle* if the rays are noncollinear. It is a *zero angle* if the rays are coincident, and it is a *straight angle* if the rays are opposite rays. If A is the common origin of the rays, it is the *vertex* of the angle, and the rays R$[A, B)$ and R$[A, C)$ of the angle are the *arms* of the angle. The angle R$[A, B)$ $\cup$ R$[A, C)$ is also denoted by either $\sphericalangle BAC$ or $\sphericalangle CAB$.

Corresponding to an angle $\sphericalangle BAC$, $p = \mathbf{AB}$ and $q = \mathbf{AC}$ are directional vectors of the two arms. In $\mathscr{E}^2$ or $\mathscr{E}^3$, we know that the $\sphericalangle POQ$ determined by the vectors p and q is congruent to $\sphericalangle BAC$. Thus a measure for $\sphericalangle POQ$ should give a measure for $\sphericalangle BAC$. To pursue this, we introduce the following notion.

DEFINITION (Angle of two vectors). If p, q are nonnull vectors in $\mathscr{E}^n$, then the *angle of the vectors* p, q, denoted by $\sphericalangle(p, q)$ is $\sphericalangle POQ$.

As motivation for the measure we assign to $\sphericalangle(p, q)$, consider a triangle $\triangle POQ$ in the cartesian setting of $\mathscr{E}^2$. By the cosine law, we have

$$d(P, Q)^2 = d(O, P)^2 + d(O, Q)^2 - 2d(O, P)d(O, Q)\cos \sphericalangle POQ,$$

or

$$(p_1 - q_1)^2 + (p_2 - q_2)^2 = p_1^2 + p_2^2 + q_1^2 + q_2^2 - 2\sqrt{p_1^2 + p_2^2}\sqrt{q_1^2 + q_2^2}\cos \sphericalangle POQ,$$

which can be written as

$$\cos \sphericalangle POQ = \frac{p_1q_1 + p_2q_2}{\sqrt{p_1^2 + p_2^2}\sqrt{q_1^2 + q_2^2}}. \tag{1}$$

This formula was known in analytic geometry before the advent of vectors. With the introduction of vectors, the formula was one motivation for the inner product of vectors, since it then takes the simpler form

$$\cos \sphericalangle POQ = \frac{p \cdot q}{\sqrt{p \cdot p}\sqrt{q \cdot q}} = \frac{p \cdot q}{|p||q|}. \tag{2}$$

Since the right side of (2) is expressed entirely in vector terms that are meaningful in $\mathscr{E}^n$, we can use (2) to *define* a measure for $\sphericalangle(p, q)$ in $\mathscr{E}^n$. From Theorem 2 and its corollaries in Sec. 1.1, the number $p \cdot q/|p||q|$ lies in the closed interval $[-1, 1]$. Since we do not need negative angle measures or measures greater than π radians or 180 degrees, we adopt the following angle measure.

DEFINITION (Angle measure). If p, q are nonnull vectors in $\mathscr{E}^n$, the angle of the vectors p, q has *radian measure* $\sphericalangle(p, q)^r$ defined by

$$\sphericalangle(p, q)^r = \text{Arccos}\,\frac{p \cdot q}{|p||q|},$$

and has *degree measure* $\sphericalangle(p, q)^\circ$ defined by

$$\sphericalangle(p, q)^\circ = \frac{180}{\pi}\,\text{Arccos}\,\frac{p \cdot q}{|p||q|}.$$

The $\sphericalangle BAC$ in $\mathscr{E}^n$ has radian measure $\sphericalangle BAC^r$ defined by

$$\sphericalangle BAC^r = \text{Arccos}\frac{\mathbf{AB}\cdot\mathbf{AC}}{|\mathbf{AB}||\mathbf{AC}|}$$

and degree measure $\sphericalangle BAC^\circ$ defined by

$$\sphericalangle BAC^\circ = \frac{180}{\pi}\text{Arccos}\frac{\mathbf{AB}\cdot\mathbf{AC}}{|\mathbf{AB}||\mathbf{AC}|}.$$

CONVENTION. Unless something is said to the contrary, it is understood that when we talk about the measures of different angles, all the measures are in the same units.

The terminology of angles in plane and solid geometry carries over to angles in $\mathscr{E}^n$ without change, and it would be tedious and to no purpose to redefine a host of familiar ideas. To mention a few, an angle is *acute*, or *right*, or *obtuse* according as its measure is less than, equal to, or greater than 90 degrees ($\pi/2$ radians). Equality of angles means their identity as sets, hence that the arms of one are the arms of the other. If angles $\sphericalangle ABC$ and $\sphericalangle DEF$ have equal measures, then they are *congruent* angles, and we denote the congruence by $\sphericalangle ABC \cong \sphericalangle DEF$. One angle is smaller or larger than a second according as its measure is less than or greater than that of the second, and so on.

One notion, related to angles, is of such importance that we define it explicitly.

DEFINITION (Orthogonal vectors). Two nonnull vectors x, y are *orthogonal*, or *perpendicular*, if $x \cdot y = 0$. A null vector is orthogonal to every vector.

The usefulness of the inner product in relation to angles is apparent in the following central theorem of this section.

THEOREM 1. If x and y are nonnull vectors, then $\sphericalangle(x, y)$ is

(i) a zero angle and has zero measure iff* $x \cdot y/|x||y| = 1$
(ii) an acute angle iff $x \cdot y > 0$
(iii) a right angle iff $x \cdot y = 0$
(iv) an obtuse angle iff $-1 < x \cdot y < 0$
(v) a straight angle and has radian measure π iff $x \cdot y/|x||y| = -1$. (E)

* The notation "iff" is a common abbreviation of "if and only if."

COROLLARY 1.1. If x and y are nonnull, then $\measuredangle(x, y)$ is a zero angle if and only if x and y are like directed and is a straight angle if and only if x and y are opposite directed.

COROLLARY 2.1. If neither B or C is A, then $\measuredangle BAC$ is

(i) a zero angle iff $\mathbf{AB} \cdot \mathbf{AC}/|\mathbf{AB}||\mathbf{AC}| = 1$
(ii) an acute angle iff $\mathbf{AB} \cdot \mathbf{AC} > 0$
(iii) a right angle iff $\mathbf{AB} \cdot \mathbf{AC} = 0$
(iv) an obtuse angle iff $-1 < \mathbf{AB} \cdot \mathbf{AC} < 0$
(v) a straight angle iff $\mathbf{AB} \cdot \mathbf{AC}/|\mathbf{AB}||\mathbf{AC}| = -1$

Exercises – Section 2.2

1. Verify that if $A = (0, 0)$, $B = (2, 1)$, $C = (1, 2)$, $D = (4, 8)$, and $E = (14, 7)$, then $\measuredangle BAC \cong \measuredangle EAD$.
2. Verify that $x - (x \cdot y/|y|^2)y$ is perpendicular to y.
3. If O, X, Y are noncollinear and $(x - y) \cdot y = 0$, why is $\triangle OXY$ a right triangle? Prove the Pythagorean theorem for the lengths of the sides.
4. In the equality of the Cauchy–Schwartz inequality (Sec. 1.1), it was shown that if $x, y \neq o$ and k is any number, then

$$|x - ky|^2 = |x|^2 - 2k(x \cdot y) + k^2|y|^2. \qquad (*)$$

   a. Regarding x, y as fixed and k variable, find the value of k which minimizes $|x - ky|^2$. (Use calculus or else an algebraic argument based on "completing the square" on the right side of (*).)
   b. If Z is a general point in $\text{Ln}(OY)$, then $z = ky$, and $d(X, Z) = |x - z| = |x - ky|$. Since distances are nonnegative, the Z for which $d(X, Z)^2$ is a minimum is also the Z for which $d(X, Z)$ is a minimum. Using this fact and (a), what is the foot Z_o of X in $\text{Ln}(OY)$?
   c. Check that Z_o is the point on $\text{Ln}(OY)$ for which $\mathbf{XZ}_o$ is perpendicular to y.

5. In $\mathscr{E}^4$, is the angle of the vectors $(2, 1, 3, 3)$ and $(-1, -2, 3, 2)$ acute or or obtuse?
6. Prove Theorem 1.

SECTION 2.3. A REVIEW OF SOME BASIC LINEAR ALGEBRA

In this chapter, we are concerned with the structure of $\mathscr{E}^n$, and the first two sections dealt with lines and planes and with angles, which are plane figures. We now want to develop the character of those subspaces which are not two-dimensional. The properties we need, in particular a precise formulation of "dimension," come directly from linear algebra with which we suppose the reader is familiar. This section is thus a review of some basic linear algebra, but a review with a bias for the geometric purposes we have in mind.

In Sec. 2.1 we referred to lines and planes through the origin as linear spans. We now make that concept precise and general.

DEFINITION (Linear combination, linear span). In $\mathscr{V}^n$, a *linear combination* of vectors $a_1, a_2, \ldots, a_m$* is any vector of the form $x = k_1a_1 + k_2a_2 + \cdots + k_ma_m$, where k_i is a real number, called the *coefficient* of a_i in the combination, $i = 1, 2, \ldots, m$. If all the numbers k_i are zero, the combination is the *null combination* (and x is the null vector). The collection of all linear combinations of the vectors $a_1, a_2, \ldots, a_m$ is the *linear span* of the vectors, denoted by $\text{LS}(a_1, a_2, \ldots, a_m)$.

If O, A, and B are noncollinear, then $\text{LS}(a, b)$ is the plane $\text{Pl}(OAB)$. But if B is on the line $\text{Ln}(OA)$, that is, if b is a multiple of a, then $\text{LS}(a, b) = \text{LS}(a) = \text{Ln}(OA)$. To ensure that the span of m vectors generates the maximum size space, we need the condition that no one of the vectors belongs to the linear span of the others. This property of independence is commonly defined in the following way.

DEFINITION (Linear independence, dependence). In $\mathscr{V}^n$, a finite set of vectors is a *linearly independent* set if the only linear combination of the vectors that is the null vector is the null combination. A finite set of vectors is a *linearly dependent* set if it is not a linearly independent set.

THEOREM 1. A finite set of two or more vectors in $\mathscr{V}^n$ is a linearly independent set if and only if no one of the vectors is a linear combination of the others. (E)

DEFINITION (Linear subspace, basis, hyperplane). A set $\mathscr{S}$ is an *m-dimensional linear subspace* of $\mathscr{V}^n$ or $\mathscr{E}^n$ if it is the linear span of m linearly independent vectors. Any m linearly independent vectors whose linear span is $\mathscr{S}$ is a *basis* for $\mathscr{S}$. Corresponding to $m = 1$, $m = 2$, and $m = n - 1$, the linear subspace is a *line*, a *plane*, or a *hyperplane*.

* The vector a_i, with index i, is the n-tuple $(a_{i1}, a_{i2}, \ldots, a_{in})$.

Since the linear span of any set of vectors contains all linear combinations of them, it contains the null combination, which is the null vector. Thus the origin belongs to the linear span of every set of vectors, hence to every linear subspace.

Any finite set of vectors that contains the null vector, say $\mathscr{S} = \{o, a_2, a_3, \ldots, a_m\}$ is a linearly dependent set. This is so because $1(0) + Oa_2 + Oa_3 + \cdots + Oa_m$ is not the null combination but is the null vector. One consequence is that a singleton set $\{a\}$ is linearly independent if and only if $a \neq o$. If $a \neq o$, then LS(a) is, by definition, a one-dimensional linear subspace. On the other hand, $\text{LS}(a) = \{\lambda a: -\infty < \lambda < \infty\}$ is the line Ln(OA) by the definition in Sec. 2.2. Similarly, if $\{a, b\}$ is a linearly independent set then LS(a, b) is a two-dimensional linear subspace. But if a, b are independent, then neither is null and they are not like or opposite directed. Thus, O, A and B are noncollinear, and $\text{LS}(a, b) = \{\lambda_1 a + \lambda_2 b: \lambda_1, \lambda_2 \text{ real}\}$ is the Pl(OAB), by the definition of Sec. 2.2.

There are a number of natural questions one can ask about linear subspaces. For example, is the dimension of such a subspace uniquely determined? That is, if a set of i linearly independent vectors and a set of j linearly independent vectors have the same linear span, must $i = j$? Must every linear subspace of $\mathscr{E}^n$ have dimension equal to or less than n? Are there linear subspaces of every dimension from 1 to n? Most important to us, can we automatically apply the facts of plane and solid geometry, respectively, to the two- and three-dimensional linear subspaces of $\mathscr{E}^n$?

The uniqueness of linear subspace dimension and the range of dimensions possible for linear subspaces are determined by the following basic theorem of linear algebra which we accept without proof.*

THEOREM 2. In $\mathscr{V}^n$, if $\mathscr{S}$ is a linear subspace of dimension m, then $1 \leq m \leq n$. Every basis for $\mathscr{S}$ has exactly m linearly independent vectors, and each vector in $\mathscr{S}$ is represented by one and only one linear combination of each basis. Moreover, if $a_1, a_2, \ldots, a_m$, $m < n$, is a basis for $\mathscr{S}$, then there exist vectors $a_{m+1}, a_{m+2}, \ldots, a_n$ such that $a_1, a_2, \ldots, a_m, a_{m+1}, \ldots, a_n$ is a basis for $\mathscr{V}^n$. (A basis for $\mathscr{S}$ can be "completed" to a basis for $\mathscr{V}^n$).

We can establish the existence of subspaces of every dimension from 1 to n from some facts we need later in other contexts.

THEOREM 3. Each non-empty subset of a finite number of linearly independent vectors is also a linearly independent set. (E)

* A proof can be found in any standard text on linear algebra.

THEOREM 4. In $\mathscr{E}^n$, let e_i denote the vector whose ith component is 1 and whose other components are zero, so $e_1 = (1, 0, \ldots, 0)$, $e_2 = (0, 1, 0, \ldots, 0)$, $\ldots, e_n = (0, 0, \ldots, 1)$. Then the vectors $e_1, e_2, \ldots, e_n$ form a basis for $\mathscr{E}^n$.

PROOF. Any linear combination of the e_i vectors is an n-tuple, hence $\text{LS}(e_1, e_2, \ldots, e_n) \subset \mathscr{E}^n$. On the other hand, for any $X \in \mathscr{E}^n$, we have

$$(x_1, x_2, \ldots, x_n) = x_1(1, 0, \ldots, 0) + x_2(0, 1, 0, \ldots, 0) + \cdots + x_n(0, 0, \ldots, 1)$$
$$= x_1 e_1 + x_2 e_2 + \cdots + x_n e_n,$$

hence $\mathscr{E}^n \subset \text{LS}(e_1, e_2, \ldots, e_n)$. Thus $\mathscr{E}^n$ is the linear span of the vectors. That the e_i vectors are linearly independent follows from the fact that $k_1 e_1 + k_2 e_2 + \cdots + k_n e_n = o$ *implies* $(k_1, k_2, \ldots, k_n) = o = (0, 0, \ldots, 0)$, or $k_i = 0$, $i = 1, 2, \ldots, n$. □

DEFINITION (Standard basis). The vectors $e_1, e_2, \ldots, e_n$ of Theorem 4 form the *standard basis* for $\mathscr{E}^n$.

COROLLARY 4. The linear span of each m of the standard basis vectors is an m-dimensional subspace of $\mathscr{E}^n$, $m = 1, 2, \ldots, n$. (E)

In the basic theorem from linear algebra, our Theorem 2, we have a clear indication of the sense in which an m-dimensional linear subspace of $\mathscr{V}^n$ is a vector space $\mathscr{V}^m$. The heart of the matter is that in a subspace of dimension m, say the space $\mathscr{S} = \text{LS}(a_1, a_2, \ldots, a_m)$, each vector is a *unique* linear combination of the basis vectors. Thus for each $x \in \mathscr{S}$, there is a unique m-tuple $(u_1, u_2, \ldots, u_m)$ such that

$$x = u_1 a_1 + u_2 a_2 + \cdots + u_m a_m. \tag{1}$$

Conversely, to each $u \in \mathscr{V}^m$ there corresponds a unique vector x in $\mathscr{S}$ represented by (1).

We can regard the correspondence (1) in two different ways. One point of view is that the m-tuple u gives us a new set of coordinates for x in $\mathscr{S}$, and if $m < n$, the new coordinates are simpler. A different point of view is that (1) defines a function f whose domain is $\mathscr{V}^m$ and whose range is $\mathscr{S}$. Then

$$x = f(u) = u_1 a_1 + u_2 a_2 + \cdots + u_m a_m, \qquad u \in \mathscr{V}^m \tag{2}$$

is a one-to-one mapping of $\mathscr{V}^m$ onto $\mathscr{S}$. The inverse function is then a one-to-one mapping of $\mathscr{S}$ onto $\mathscr{V}^m$. This function f has very important special properties.

DEFINITION (Linear function, linear functional). A function f defined on a vector space with function values in a vector space is a *linear function*, or *linear transformation*, if the range of f is not $\{o\}$ and if f maps every linear combination of vectors u, v in $\mathscr{S}$ onto the same linear combination of their images, that is, if

$$f(\lambda_1 u + \lambda_2 v) = \lambda_1 f(u) + \lambda_2 f(v).$$

A linear function that is real valued is also called a *linear functional.*

THEOREM 5. The function $f : \mathscr{V}^m \to \mathscr{S}$ defined in (2) above is a linear function. (E)

Because the function f in (2) is both one to one and linear, all the vector space properties of $\mathscr{V}^m$ are transferred, via f, to $\mathscr{S}$. Because of this, $\mathscr{V}^m$ and the m-dimensional linear subspaces of $\mathscr{V}^n$ are said to be abstractly identical, or to be *isomorphic*. Since an m-dimensional linear subspace of $\mathscr{E}^n$ is also an m-dimensional linear subspace of $\mathscr{V}^n$, then, as far as vector space properties are concerned, we can regard the linear subspaces of $\mathscr{E}^n$ as lower dimensional vector spaces.

For the purposes of geometry, we want a stronger result, namely, that the linear subspaces of $\mathscr{E}^n$ are congruent to lower dimensional euclidean spaces. We can express general congruence of sets as a special case of similar sets in the following way.

DEFINITION (Similarities, isometries, congruence). A function f that is a mapping from a metric space to a metric space is a *similarity* if there exists a positive number k such that $d[f(X), f(Y)] = k\, d(X, Y)$, for all X, Y in the domain of f. Sets $\mathscr{R}$ and $\mathscr{S}$ are *similar*, denoted by $\mathscr{R} \sim \mathscr{S}$, if one is the image of the other in a similarity. A similarity in which the *ratio of similarity* k is 1 is an *isometry*, or distance preserving mapping. Sets $\mathscr{R}$ and $\mathscr{S}$ are *congruent*, denoted by $\mathscr{R} \cong \mathscr{S}$, if one is the image of the other in an isometry.

Every similarity, hence every isometry, is automatically a one-to-one mapping because $X \neq Y$ implies $d(X, Y) \neq 0$. This, with $k > 0$, implies $k\, d(X, Y) \neq 0$, hence $d[f(X), f(Y)] \neq 0$, so $f(X) \neq f(Y)$.

Since an isometry is a one-to-one mapping that preserves distance, it maps any set onto a metric copy of itself. Thus if we can show that an m-dimensional subspace of $\mathscr{E}^n$ is congruent to $\mathscr{E}^m$, we can conclude that one is a metric copy of the other. In particular, all the metric properties that we know are valid in $\mathscr{E}^2$ and $\mathscr{E}^3$ hold automatically in the two- and three-dimensional subspaces of $\mathscr{E}^n$.

We can obtain the important properties just described from a basic theorem in linear algebra. To state it, we need the following concept.

DEFINITION (Orthonormal basis). A basis $a_1, a_2, \ldots, a_m$ for a linear subspace $\mathscr{S}$ of $\mathscr{E}^n$ is an *orthonormal basis* if each of the basis vectors is a unit vector (i.e., $|a_i| = 1$) and if each two of the basis vectors are orthogonal (i.e., $a_i \cdot a_j = 0$ if $i \neq j$).

There is a method, called the Gram–Schmidt Process, for converting any basis of a linear subspace to an orthonormal basis, and this basis in turn can be extended to an orthonormal basis for the whole space $\mathscr{E}^n$. How this can be proved is discussed in the exercises, but we state it now as a fact of linear algebra.

THEOREM 6. An m-dimensional subspace $\mathscr{S}$ in $\mathscr{E}^n$ has an orthonormal basis $a_1, a_2, \ldots, a_m$ and there exist vectors $a_{m+1}, a_{m+2}, \ldots, a_n$ such that $a_1, a_2, \ldots, a_m, \ldots, a_n$ is an orthonormal basis for $\mathscr{E}^n$.

We can now establish one of the key theorems of this section.

THEOREM 7. Every m-dimensional linear subspace $\mathscr{S}$ of $\mathscr{E}^n$ is congruent to $\mathscr{E}^m$, and there is an isometric mapping of $\mathscr{E}^m$ onto $\mathscr{S}$ by a linear function.

PROOF. By Theorem 6, $\mathscr{S}$ has an orthonormal basis $b_1, b_2, \ldots, b_m$. Let f denote the function defined on $\mathscr{E}^m$ by

$$f(u) = u_1 b_1 + u_2 b_2 + \cdots + u_m b_m, \qquad u \in \mathscr{E}^m. \tag{1}$$

By Theorem 5, f is a linear function and is clearly a one-to-one mapping of $\mathscr{E}^m$ onto $\mathscr{S}$. Corresponding to u, v in $\mathscr{E}^m$, let $x = f(u)$ and $y = f(v)$. Then

$$\begin{aligned} d(X, Y)^2 &= d[f(u), f(v)]^2 = (x - y) \cdot (x - y) \\ &= \left[\sum_1^m (u_i - v_i) b_i\right] \cdot \left[\sum_1^m (u_i - v_i) b_i\right] \end{aligned} \tag{2}$$

Since the b_i basis is orthonormal, $b_i \cdot b_i = 1$ and $b_i \cdot b_j = 0$, for $i \neq j$, hence (2) reduces to

$$d(X, Y)^2 = \sum_1^m (u_i - v_i)^2 = d(U, V)^2. \tag{3}$$

Because distances are nonnegative, (3) implies

$$d(X, Y) = d(U, V), \tag{4}$$

hence that f is an isometry. □

The usefulness of Theorem 7 can be illustrated in a developmental discussion of some further properties of subspaces. Consider a fixed vector a and an independent vector x. Then O, A, and X are noncollinear and determine a plane $\mathscr{S}$, which, by Theorem 7, is a metric copy of the "plane" of plane geometry. Thus we know that X has a foot F in Ln(OA) and that $\sphericalangle OFX$ is a right angle. From plane geometry, through each point Y of Sg[OX] there is a line perpendicular to Ln(OA) at the foot of Y in the line, and the mapping of Y into

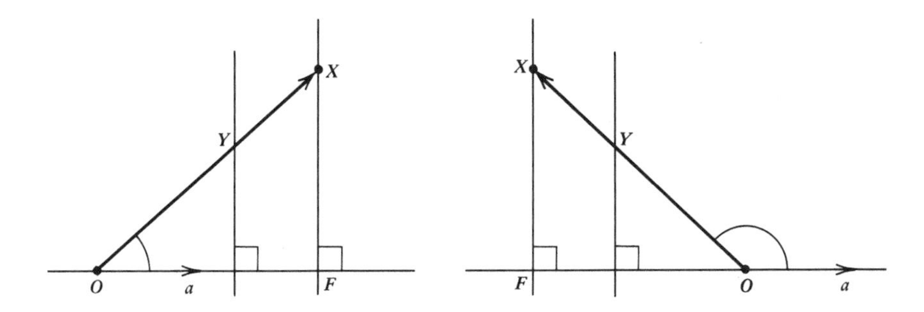

its foot is called an orthogonal projection of Y onto line Ln(OA). The segment Sg[OX] projects onto Sg[OF] and X projects onto F. By standard trigonometry, $d(O, F) = d(O, X)|\cos \sphericalangle(a, x)^\circ|$, so

$$|f| = |x|\frac{|a \cdot x|}{|a||x|}.$$

If $\sphericalangle(a, x)$ is acute, then F is on Ry(O, A), so f and a are like directed and $|a \cdot x| = a \cdot x$. Thus $f = (a/|a|)|f| = (a/|a|)\ (a \cdot x/|a||x|) = (a \cdot x/|a|^2)a$. When $\sphericalangle(a, x)$ is obtuse, F is on the ray opposite to Ry(O, A), f and a are opposite directed, and $|a \cdot x| = -(a \cdot x)$. Thus, again, $f = (-a/|a|)|f| = (a \cdot x/|a|^2)a$. In the remaining case, when $\sphericalangle(a, x)$ is a right angle, $a \cdot x = 0$ and $f = o$, so again $f = (a \cdot x/|a|^2)a$. Finally, if X is actually on Ln(OA), so $X = F$ and Sg[OX] projects onto itself, it is easily verified that the same relations holds. Thus in all cases

$$f = \frac{a \cdot x}{|a|^2}\, a.$$

We gave the previous argument to indicate the kind of relations from plane geometry that are automatically valid in a linear two space of $\mathscr{E}^n$. Also, the elementary, case-counting, angle argument provides a contrast to the analytic derivation of the same result indicated in the exercises of Sec. 2.2.

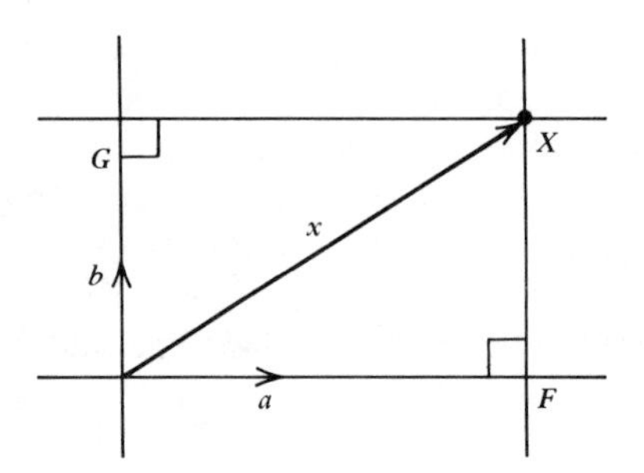

Now consider two unit, orthogonal vectors a, b and a vector x in LS(a, b). The point X has an orthogonal projection to F in Ln(OA) and to G in Ln(OB). From the relation developed in the last paragraph, we have

$$f = \frac{x \cdot a}{|a|^2} a = (x \cdot a)a, \qquad \text{and} \qquad g = \frac{x \cdot b}{|b|^2} b = (x \cdot b)b.$$

The usual "parallelogram rule" now becomes a "rectangle rule" and

$$x = f + g = (x \cdot a)a + (x \cdot b)b. \tag{*}$$

We know, of course, that if $x \in$ LS(a, b) then x has a representation $x = \lambda_1 a + \lambda_2 b$. The (*) relation gives us a simple formula for computing λ_1 and λ_2 in terms of inner products.

It is natural to wonder if the (*) relation generalizes, and it is easily seen to do so.

THEOREM 8. If $a_1, a_2, \ldots, a_m$ is an orthonormal basis for a linear subspace $\mathscr{S}$ in $\mathscr{E}^n$, and $X \in \mathscr{S}$, then

$$x = (x \cdot a_1)a_1 + (x \cdot a_2)a_2 + \cdots + (x \cdot a_m)a_m.$$

PROOF. Since $x \in$ LS($a_1, a_2, \ldots, a_m$) then x has a representation

$$x = \lambda_1 a_1 + \lambda_2 a_2 + \cdots + \lambda_m a_m. \tag{1}$$

For a fixed integer i, $1 \leq i \leq m$, the distributive property of the inner product gives

$$x \cdot a_i = \lambda_1(a_1 \cdot a_i) + \lambda_2(a_2 \cdot a_i) + \cdots + \lambda_m(a_m \cdot a_i). \tag{2}$$

Since the basis is orthonormal, $a_j \cdot a_i = 0, j \neq i$, and $a_i \cdot a_i = 1$. Thus, (2) reduces to

$$x \cdot a_i = \lambda_i. \tag{3}$$

Since (3) holds for $i = 1, 2, \ldots, m$, then (1) can be expressed as

$$x = (x \cdot a_1)a_1 + (x \cdot a_2)a_2 + \cdots + (x \cdot a_m)a_m. \quad \square$$

We can use the same sort of intuitive ideas that led to Theorem 8 to obtain a surprising and important generalization of it. Suppose that $\mathscr{S}$ is a plane through the origin in $\mathscr{E}^4$ or $\mathscr{E}^5$ and that X is a point not in $\mathscr{S}$. Now, there is no obvious direction perpendicular to $\mathscr{S}$, so it is not obvious how to find the

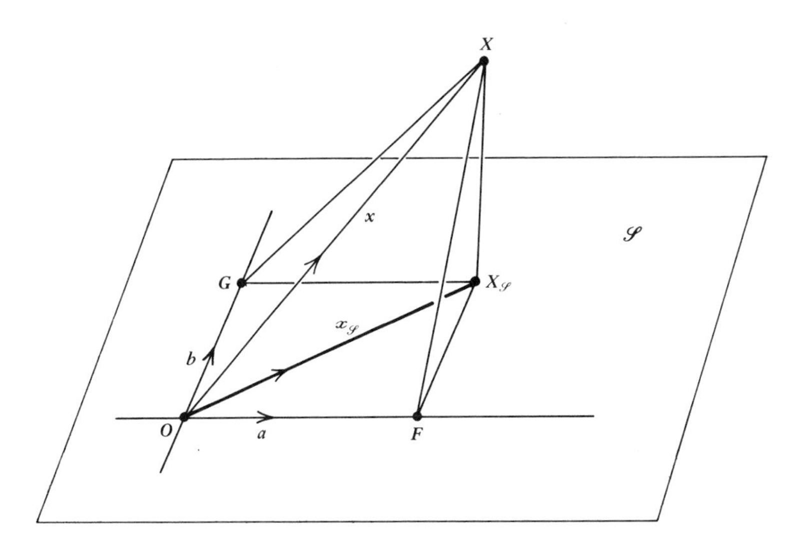

foot of X in $\mathscr{S}$ or even that such a foot exists. To investigate the problem, let us first suppose the space to be $\mathscr{E}^3$, and let $\{a, b\}$ be an orthonormal basis for $\mathscr{S}$. Now, X has a foot $X_{\mathscr{S}}$ in $\mathscr{S}$, and in this plane, $X_{\mathscr{S}}$ has an orthogonal projection to F in Ln(OA) and to G in Ln(OB). From our previous discussions, and Theorem 8, we know that

$$x_{\mathscr{S}} = (x_{\mathscr{S}} \cdot a)a + (x_{\mathscr{S}} \cdot b)b = f + g. \tag{1}$$

Because X has foot $X_{\mathscr{S}}$ in $\mathscr{S}$, and $X_{\mathscr{S}}$ has foot F in Ln(OA) $\subset \mathscr{S}$, it follows from ordinary solid geometry that X has foot F in Ln(OA). That is, in the plane Pl(OFX), F is the orthogonal projection of X into Ln(OA). Thus

$$f = (x \cdot a)a. \tag{2}$$

By the same logic, in the plane Pl(OGX), G is the orthogonal projection of X into Ln(OB), so

$$g = (x \cdot b)b \tag{3}$$

Now, from (1), (2), and (3), we have

$$x_{\mathscr{S}} = (x \cdot a)a + (x \cdot b)b. \tag{4}$$

In $\mathscr{E}^3$, (4) is a correct formula for computing the foot of X in the space LS(a, b). Happily the same formula holds in a much more general context and in fact includes Theorem 8 as a special case.

THEOREM 9. Let $\mathscr{S}$ be an m-dimensional linear subspace of $\mathscr{E}^n$ with an orthonormal basis $a_1, a_2, \ldots, a_m$. If X is any point of $\mathscr{E}^n$, then

(i) X has a unique foot $X_{\mathscr{S}}$ in $\mathscr{S}$ given explicitly by

$$x_{\mathscr{S}} = (x \cdot a_1)a_1 + (x \cdot a_2)a_2 + \cdots + (x \cdot a_m)a_m$$

(ii) $d(X, X_{\mathscr{S}})^2 = |x|^2 - \sum_1^m (x \cdot a_i)^2$

(iii) $X \in \mathscr{S}$ if and only if $X = X_{\mathscr{S}}$, hence $X \in \mathscr{S}$ implies that

$$x = \sum_1^m (x \cdot a_i)a_i$$

(iv) the vector $\mathbf{XX}_{\mathscr{S}}$ is perpendicular to every vector in $\mathscr{S}$.

PROOF. Let X be an arbitrary but fixed point of $\mathscr{E}^n$, and let Y in $\mathscr{S}$, with the representation

$$y = \sum_1^m \lambda_i a_i \tag{1}$$

be a variable point in $\mathscr{S}$. We want to show that there is a unique position of Y, that is, unique values of the coefficients λ_i, for which $d(X, Y)^2$, and hence $d(X, Y)$, is a minimum. At this position, $Y = X_{\mathscr{S}}$ will be the unique foot of X in $\mathscr{S}$.

We begin with

$$\begin{aligned} d(X, Y)^2 &= |x - y|^2 = (x - y) \cdot (x - y) \\ &= (x \cdot x) - 2(x \cdot y) + (y \cdot y). \end{aligned} \tag{2}$$

For the term $(x \cdot y)$ in (2), we have

$$x \cdot y = x \cdot \left[\sum_1^m \lambda_i a_i\right] = \sum_1^m \lambda_i (x \cdot a_i). \tag{3}$$

For the term $(y \cdot y)$ in (2), we have

$$y \cdot y = \left(\sum_1^m \lambda_i a_i\right) \cdot \left(\sum_1^m \lambda_i a_i\right). \tag{4}$$

From the distributive properties of the inner product, the right side of (4) can be multiplied like a product of polynomials with a general term of the form $(\lambda_i a_i) \cdot (\lambda_j a_j) = \lambda_i \lambda_j (a_i \cdot a_j)$. But, since the a_i basis of $\mathscr{S}$ is orthonormal, $a_i \cdot a_j = 0$, for $i \neq j$, and $a_i \cdot a_i = 1$. Thus (4) reduces to

$$y \cdot y = \sum_1^m \lambda_i^2. \tag{5}$$

Now, applying (3) and (5) in (2), we obtain

$$\begin{aligned} d(X, Y)^2 &= |x|^2 - 2 \sum_1^m \lambda_i (x \cdot a_i) + \sum_1^m \lambda_i^2 \\ &= |x|^2 + \sum_1^m [\lambda_i^2 - 2\lambda_i (x \cdot a_i)] \\ &= |x|^2 + \sum_1^m [\lambda_i^2 - 2\lambda_i (x \cdot a_i) + (x \cdot a_i)^2 - (x \cdot a_i)^2] \\ &= |x|^2 + \sum_1^m [\lambda_i - (x \cdot a_i)]^2 - \sum_1^m (x \cdot a_i)^2. \end{aligned} \tag{6}$$

Because X is fixed and the vectors a_i are fixed, the only variable part of the right side of (6) is the middle sum. Thus the total sum on the right side of (6) has a minimum if and only if the middle sum,

$$\sum_1^m [\lambda_i - (x \cdot a_i)]^2 \tag{7}$$

has a minimum. Since each number in the sum (7) is a square, it is not negative, so the sum is not negative. Thus zero is the least possible value for the sum, and this occurs if and only if each square is zero, hence if and only if

$$\lambda_i = (x \cdot a_i), \qquad i = 1, 2, \ldots, m. \tag{8}$$

Corresponding to these values,

$$Y = X_{\mathscr{S}} = \sum_1^m \lambda_i a_i = \sum_1^m (x \cdot a_i)a_i \tag{9}$$

is the unique foot of X in $\mathscr{S}$, hence (i) is established.

At $Y = X_{\mathscr{S}}$, the sum in (7) is zero, so from (6) we have

$$d(X, Y)^2 = d(X, X_{\mathscr{S}})^2 = |x|^2 - \sum_1^m (x \cdot a_i)^2, \tag{10}$$

which is part (ii) of the theorem.

If $X = X_{\mathscr{S}}$, then clearly X is in $\mathscr{S}$. On the other hand, if X is in $\mathscr{S}$, it is its own unique foot in $\mathscr{S}$, hence $X = X_{\mathscr{S}}$. Thus if X is in $\mathscr{S}$, it is $X_{\mathscr{S}}$, so, from (9),

$$x = \sum_1^m (x \cdot a_i)a_i, \tag{11}$$

which establishes (iii).

Finally, to show that $\mathbf{XX}_{\mathscr{S}}$ is orthogonal to every vector $y = \sum_1^m \lambda_i a_i$ in $\mathscr{S}$, we first observe that, because of the orthonormal properties of the a_i vectors,

$$x_{\mathscr{S}} \cdot y = \left[\sum_1^m (x \cdot a_i)a_i\right] \cdot \left[\sum_1^m \lambda_i a_i\right] \tag{12}$$

reduces to

$$\begin{aligned} x_{\mathscr{S}} \cdot y &= \sum_1^m \lambda_i (x \cdot a_i)(a_i \cdot a_i) = \sum_1^m \lambda_i (x \cdot a_i) \\ &= x \cdot \left[\sum_1^m \lambda_i a_i\right] = x \cdot y. \end{aligned} \tag{13}$$

Therefore

$$\mathbf{XX}_{\mathscr{S}} \cdot y = (x_{\mathscr{S}} - x) \cdot y = x_{\mathscr{S}} \cdot y - x \cdot y = 0. \quad \square$$

COROLLARY 9. Every linear subspace of $\mathscr{E}^n$ is a closed set.

PROOF. If $\mathscr{S}$ is a linear subspace of $\mathscr{E}^n$, then, by Theorem 9, every point of $\mathscr{E}^n$ has a foot in $\mathscr{S}$. Thus, by Theorem 4, Sec. 1.8, $\mathscr{S}$ is closed. $\square$

Exercises – Section 2.3

1. Prove Theorem 1.
2. Prove Theorem 3.
3. Show that the vectors x, y, z in $\mathscr{E}^3$ are independent if and only if the only solution to the equations

$$\lambda_1 x_1 + \lambda_2 y_1 + \lambda_3 z_1 = 0$$
$$\lambda_1 x_2 + \lambda_2 y_2 + \lambda_3 z_2 = 0$$
$$\lambda_1 x_3 + \lambda_2 y_3 + \lambda_3 z_3 = 0$$

is $\lambda_1 = \lambda_2 = \lambda_3 = 0$.

4. Show that the vectors $a = (1, 0, 2)$, $b = (-1, 1, 1)$, and $c = (-1, 3, 7)$ are linearly dependent, and express c as a linear combination of a and b.
5. Prove Corollary 4.
6. Prove Theorem 5.
7. If a in $\mathscr{E}^n$ is not null, and the real number k is not zero, show that the function g defined on $\mathscr{E}^n$ by $g(x) = a \cdot x + k$ is not a linear functional.
8. Verify that if f is a linear function on $\mathscr{E}^n$, and $\mathscr{S}$ is a linear subspace of $\mathscr{E}^n$, then the image of $\mathscr{S}$ under f, namely, the set $\{f(x): x \in \mathscr{S}\}$, is also a linear subspace.
9. Let $a_1, a_2, \ldots, a_m$ be a basis for an m-dimensional linear subspace $\mathscr{S}$. Prove that if x is orthogonal to each of the basis vectors a_i, then x is orthogonal to every vector in $\mathscr{S}$.
10. Prove that if the vectors in the set $\{a_1, a_2, \ldots, a_m\}$ are nonnull and are pairwise orthogonal, then the set is linearly independent.
11. Prove that if $p, q \in \text{LS}(a_1, a_2, \ldots, a_m)$, then

$$\text{LS}(p, q) \subset \text{LS}(a_1, a_2, \ldots, a_m).$$

12. *Gram–Schmidt Process.* Let $\{a_1, a_2, \ldots, a_m\}$ be a basis for an m-dimensional linear subspace $\mathscr{S}_m$. Define $b_1 = a_1/|a_1|$.

   a. Why is $a_1 \neq 0$, and why does $\text{LS}(b_1) = \text{LS}(a_1)$? By Theorem 3, $\{a_1, a_2\}$ is a linearly independent set, hence $a_2 \notin \text{LS}(a_1) = \text{LS}(b_1)$. Then A_2 has a distinct foot F_2 in $\text{LS}(b_1)$ given by $f_2 = (a_2 \cdot b_1)b_1$, and $b_2 = \mathbf{F_2A_2}/|\mathbf{F_2A_2}|$ is a unit vector orthogonal to b_1.
   b. Show that $b_2 = [a_2 - (a_2 \cdot b_1)b_1]/|a_2 - (a_2 \cdot b_1)b_1|$. Why are b_1 and b_2 linearly independent? Why does $\text{LS}(b_1, b_2) = \text{LS}(a_1, a_2)$? Since $\{a_1, a_2, a_3\}$ are independent, $a_3 \notin \text{LS}(a_1, a_2) = \text{LS}(b_1, b_2)$. By Theorem 9, A_3 has a distinct foot F_3 in $\text{LS}(b_1, b_2)$, given by $f_3 = (a_3 \cdot b_1)b_1 + (a_3 \cdot b_2)b_2$. Define $b_3 = \mathbf{F_3A_3}/|\mathbf{F_3A_3}|$.

c. Show that

$$b_3 = \frac{a_3 - (a_3 \cdot b_1)b_1 - (a_3 \cdot b_2)b_2}{|a_3 - (a_3 \cdot b_1)b_1 - (a_3 \cdot b_2)b_2|}.$$

Why are b_1, b_2 and b_3 mutually orthogonal? Why does $\text{LS}(b_1, b_2, b_3) = \text{LS}(a_1, a_2, a_3)$?

Continuing, it is clear that in m steps we must arrive at

$$b_m = \frac{a_m - \sum_{i=1}^{m-1} (a_m \cdot b_i)b_i}{|a_m - \sum_{i=1}^{m-1} (a_m \cdot b_i)b_i|}$$

such that $b_1, b_2, \ldots, b_m$ are m mutually orthogonal unit vectors and $\text{LS}(b_1, b_2, \ldots, b_m) = \text{LS}(a_1, a_2, \ldots, a_m) = \mathscr{S}_m$. Thus as stated in Theorem 6, every subspace has an orthonormal basis.

If $\mathscr{S}_m \neq \mathscr{E}^n$, then there exists $P \notin \mathscr{S}_m = \text{LS}(b_1, b_2, \ldots, b_m)$. By Theorem 9, P has a foot F in $\mathscr{S}_m$ and $b_{m+1} = \mathbf{FP}/|\mathbf{FP}|$ is a unit vector orthogonal to b_i, $i = 1, 2, \ldots, m$. Thus the span of $b_1, b_2, \ldots, b_{m+1}$ is an $m+1$-dimensional subspace $\mathscr{S}_{m+1}$, whose b_i basis is orthonormal. Continuing in the same way, in $n - m$ steps we can complete $b_1, b_2, \ldots, b_m$ to an orthonormal basis $b_1, b_2, \ldots, b_n$ for $\mathscr{E}^n$.

13. Let $a_1 = (3, 4, 0)$ and $a_2 = (1, 2, 2)$ be two vectors in $\mathscr{E}^3$. Using the Gram–Schmidt process in Exercise 12, find unit, orthogonal vectors b_1, b_2 such that $\text{LS}(a_1, a_2) = \text{LS}(b_1, b_2)$. Using the method of the proof for Theorem 7, give an explicit representation for an isometry of $\mathscr{E}^2$ onto $\text{LS}(b_1, b_2)$.
14. Let f denote an isometry of $\mathscr{E}^n$ into $\mathscr{E}^n$ of the form in Theorem 7. If x and y are perpendicular, show that $f(x)$ and $f(y)$ are perpendicular. Prove that, in general, $\sphericalangle(x, y) \cong \sphericalangle(f(x), f(y))$.

SECTION 2.4. AFFINE CONCEPTS

In the preceding section, we defined the linear subspaces of $\mathscr{E}^n$, namely, those which contain the origin. We now want to define the general (affine) subspaces of $\mathscr{E}^n$ that do not necessarily contain O. As we saw in Sec. 1.1, for general lines and general planes, there are two equivalent ways of defining general subspaces. We may define a general subspace to be the translate of a linear subspace, or we may define it intrinsically in terms of affine spans. From the first point of view, for example, a plane in $\mathscr{E}^n$ is seen as the translate of a two-dimensional linear subspace, whereas from the second point of view it is the affine span of three of its noncollinear points. We adopt the first of these viewpoints in making our definitions, since that permits us to apply at once the properties of linear subspaces from the preceding section. We then establish the corresponding affine properties as theorems.

We begin with a particularly simple family of isometries.

DEFINITION (Translations). Corresponding to a vector $v \in \mathscr{E}^n$, the function f defined by

$$f(x) = x + v, \qquad x \in \mathscr{E}^n,$$

is a *translation* of $\mathscr{E}^n$ by v. The image of any set $\mathscr{S}$ under f is the *v-translate* of $\mathscr{S}$. If $v = 0$, then f is the *null translation.*

THEOREM 1. A translation of space is a one-to-one isometry of $\mathscr{E}^n$ onto $\mathscr{E}^n$.

PROOF. Let f be a translation of space by v. Then $f(x) - f(y) = (x + v) - (y + v) = (x - y)$, hence $d[f(x), f(y)] = |f(x) - f(y)| = |x - y| = d(X, Y)$. Thus f is an isometry and thus it is a one-to-one mapping. Because $f(x - v) = (x - v) + v = x$, each $X \in \mathscr{E}^n$ is the f image of a point in $\mathscr{E}^n$, hence f maps $\mathscr{E}^n$ onto $\mathscr{E}^n$. □

Intuitively, a translation by v moves a set $\mathscr{S}$ a distance $|v|$ in the direction v without rotation. Thus the translate of $\mathscr{S}$ is a metric copy of $\mathscr{S}$ with the same orientation in space.

We now use translations to define the general subspaces of $\mathscr{E}^n$.

DEFINITION (Flats). A translate of an m-dimensional, linear subspace of $\mathscr{E}^n$ is an *m-dimensional flat*, or simply *m-flat* in $\mathscr{E}^n$. Flats of dimension 1, 2, and $n - 1$ are also called lines, planes, and hyperplanes, respectively.

THEOREM 2. Every m-flat in $\mathscr{E}^n$ is congruent to the euclidean space $\mathscr{E}^m$.

PROOF. Let $\mathscr{A}$ denote an m-flat in $\mathscr{E}^n$. Then, by definition, there exists an m-dimensional linear subspace $\mathscr{S}$ and a vector v such that $\mathscr{A}$ is the v-translate of $\mathscr{S}$. By Theorem 7, Sec. 2.3, $\mathscr{S}$ is congruent to $\mathscr{E}^m$. Because congruence is a transitive relation, then $\mathscr{S} \cong \mathscr{E}^n$, and $\mathscr{A} \cong \mathscr{S}$, by Theorem 2, imply $\mathscr{A} \cong \mathscr{E}^n$. □

THEOREM 3. Flats are closed sets.

PROOF. Let $\mathscr{A}$ denote a flat in $\mathscr{E}^n$ and let P be an arbitrary point of $\mathscr{E}^n$. There exists a linear subspace $\mathscr{S}$ and a vector v such that $f(x) = x + v$ maps $\mathscr{S}$ onto $\mathscr{A}$. Then f maps $q = p - v$ onto p. The point Q has a foot in

$\mathscr{S}$, by Theorem 9, Sec. 2.3, and, since f preserves distances, it maps Q to P and maps the foot of Q in $\mathscr{S}$ to the foot of P in $\mathscr{A}$. Thus every point of $\mathscr{E}^n$ has a foot in $\mathscr{A}$ (in fact a unique foot), hence, by Theorem 4, Sec. 1.8, $\mathscr{A}$ is closed. □

To obtain a direct representation of flats that does not depend on linear subspaces, we need the idea of affine combinations mentioned in Sec. 2.1.

DEFINITION (Affine combination). An *affine combination* of the vectors $a_1, a_2, \ldots, a_m$ is a linear combination of the vectors in which the sum of the coefficients is 1. Thus b is an affine combination of the vectors if

$$b = k_1a_1 + k_2a_2 + \cdots + k_ma_m, \qquad \text{and} \qquad k_1 + k_2 + \cdots + k_m = 1.$$

The set of all affine combinations of the vectors is their *affine span*, denoted by $\mathrm{AS}(a_1, a_2, \ldots, a_m)$.

As one would suppose, a set of vectors is affinely independent if no one of them is an affine combination of the others. We obtain this as a property of the following analog of linear independence.

DEFINITION (Affine independence). A finite set of vectors, $a_1, a_2, \ldots, a_m$ is an *affinely independent* set if the only zero-sum linear combination of them representing the null vector is the null combination. That is, the set is affinely independent if $k_1a_1 + k_2k_2 + \cdots + k_ma_m = o$ and $k_1 + k_2 + \cdots + k_m = 0$ imply that $k_1 = k_2 = \cdots = k_m = 0$. The set is an *affinely dependent* set if it is not an affinely independent set.

THEOREM 4. A finite set of two or more vectors is an affinely independent set if and only if no one of them is an affine combination of the others.

PROOF. Let $a_1, a_2, \ldots, a_m$ be an affinely independent set, and suppose that one of the vectors, say a_m, is an affine combination of the others. Then

$$a_m = \sum_1^{m-1} h_ia_i, \qquad \text{and} \qquad \sum_1^{m-1} h_i = 1. \tag{1}$$

Therefore

$$1a_m - \sum_1^{m-1} h_ia_i = o, \tag{2}$$

is a zero-sum linear combination representing o. But the linear combination in (2) is not the null combination, since the coefficient of a_m is 1. Thus the affine independence of the set is contradicted if one of the vectors is an affine combination of the others.

Next, let $a_1, a_2, \ldots, a_m$ be a set of vectors such that no one of them is an affine combination of the others. Assume that numbers $k_1, k_2, \ldots, k_m$ exist, not all zero such that

$$\sum_1^m k_i a_i = o \qquad \text{and} \qquad \sum_1^m k_i = 0. \tag{3}$$

Since at least one of the numbers k_i is not zero, we may suppose that $k_m \neq 0$. Then, from (3)

$$a_m = \sum_1^{m-1} -\frac{k_i}{k_m} a_i. \tag{4}$$

But $\sum_1^m k_i = 0$ and $k_m \neq 0$ imply that $\sum_1^{m-1} -k_i/k_m = 1$. Thus (4) shows that a_m is an affine combination of $a_1, a_2, \ldots, a_{m-1}$, contradicting our hypothesis. Thus if no a_i is a linear combination of the others, then the set is affinely independent. □

COROLLARY 4. Each non-empty subset of an affinely independent set is again an affinely independent set. (E)

In deciding whether to call the n-tuple $(x_1, x_2, \ldots, x_n)$ a vector x or the point X, we have been guided by traditional points of view. For example, it seems natural to talk about the direction of the vector x and unnatural to talk about the direction of the point X. There is a similar reason, as well as a convenience, in using the language of points rather than vectors in connection with affine independence and dependence.

Suppose that A, B are affinely independent points (i.e., a, b are affinely independent vectors). Then neither is an affine combination of the other, so $A \neq 1B$ and $B \neq 1A$, which is to say that $A \neq B$. Thus A and B are independent if and only if they are distinct. From Sec. 2.1, AS(A, B) = Ln(A, B), thus two (affinely) independent points determine a line.

Next, consider the affinely independent set $\{A, B, C\}$. By Corollary 4, each of the sets $\{A, B\}, \{B, C\}, \{C, A\}$ is affinely independent, and the span of each is a line. Thus the independence of $\{A, B, C\}$ is equivalent to the noncollinearity of A, B, and C. By Sec. 2.1, the affine span of A, B, C is the plane Pl(ABC).

As the preceding discussion indicates, it is helpful to have the following convention.

CONVENTION. When the terms "dependent" or "independent" are used without qualification in reference to points, they refer to affine dependence or independence. When used with reference to vectors, they refer to linear dependence or independence.

The fact that the span of three independent points is a plane, and the unique plane that contains them, is a special case of the fact that the span of $k + 1$ independent points is a k-flat and is the unique k-flat that contains all the points. The proof of this general fact can be simplified by the use of two properties that are of interest in their own right.

THEOREM 5. The $k + 1$ points $A_o, A_1, \ldots, A_k$ are affinely independent if and only if the k vectors $\mathbf{A}_o\mathbf{A}_1, \mathbf{A}_o\mathbf{A}_2, \ldots, \mathbf{A}_o\mathbf{A}_k$ are linearly independent. (E)

THEOREM 6. If the vector v belongs to LS$(a_1, a_2, \ldots, a_m)$, then the translation of $\mathscr{E}^n$ by v maps the span onto itself. (E)

THEOREM 7. The span of $k + 1$ independent points is a k-flat and is the unique k-flat that contains all $k + 1$ points.

PROOF. Let $A_o, A_1, \ldots, A_k$ denote $k + 1$ independent points. By Theorem 5, the vectors $b_i = \mathbf{A}_o\mathbf{A}_i, i = 1, 2, \ldots, k$ form a linearly independent set, hence $\mathscr{S} = \text{LS}(b_1, b_2, \ldots, b_k)$ is a k-dimensional linear subspace. The a_o-translate of $\mathscr{S}$ is a k-flat, by definition, and has a representation

$$x = a_o + \sum_1^k \lambda_i b_i, \qquad \lambda = (\lambda_1, \lambda_2, \ldots, \lambda_k) \in \mathscr{V}^k. \tag{1}$$

In (1), corresponding to $\lambda = o$, $x = a_o$. When λ in $\mathscr{V}^k$ is the standard basis vector e_i, then in (1),

$$x = a_o + b_i = a_o + (a_i - a_o) = a_i, \qquad i = 1, 2, \ldots, k.$$

Thus the k-flat, call it $\mathscr{A}$, that is defined by (1) contains the $k + 1$ points $A_o, A_1, \ldots, A_k$.

To establish that $\mathscr{A}$ is unique, let $\mathscr{B}$ denote a k-flat that contains the points $A_o, A_1, \ldots, A_k$. Then, by definition, $\mathscr{B}$ has a representation of the form

$$x = v + \sum_1^k \eta_i c_i, \qquad \eta = (\eta_1, \eta_2, \ldots, \eta_k) \in \mathscr{V}^k, \tag{2}$$

in which the vectors $c_i, i = 1, 2, \ldots, k$ are linearly independent. Let $\mathscr{R}$ denote the linear subspace $\mathrm{LS}(c_1, c_2, \ldots, c_k)$. Since each of the points A_i belongs to $\mathscr{B}$, we have

$$\begin{aligned} a_o &= v + \sum_1^k \eta_{0i} c_i \\ a_1 &= v + \sum_1^k \eta_{1i} c_i \\ &\vdots \\ a_k &= v + \sum_1^k \eta_{ki} c_i. \end{aligned} \tag{3}$$

Then $b_j = a_j - a_o = \sum_{i=1}^k (\eta_{ji} - \eta_{0i}) c_i, j = 1, 2, \ldots, k$ shows that each of the vectors b_j belongs to $\mathscr{R}$. Since there are k vectors b_j, and since they are linearly independent, they form a basis for $\mathscr{R}$, hence $\mathscr{R} = \mathscr{S}$. Thus $\mathscr{B}$ is also the v-translate of $\mathscr{S}$ and hence has a representation

$$x = v + \sum_1^k \eta_i b_i, \qquad \eta \in \mathscr{V}^k. \tag{4}$$

The first equation in (3) shows that $a_o - v$ belongs to $\mathscr{R}$, hence it also belongs to $\mathscr{S}$. By Theorem 6, then, the translation of space by $a_o - v$ maps $\mathscr{S}$ onto itself. That is,

$$\left\{\sum_1^k \eta_i b_i : \eta \in \mathscr{V}^k\right\} = \left\{a_o - v + \sum_1^k \eta_i b_i : \eta \in \mathscr{V}^k\right\}. \tag{5}$$

Substituting from (5) in (4), we obtain

$$x = v + a_o - v + \sum_1^k \eta_i b_i, \qquad \eta \in \mathscr{V}^k \tag{6}$$

as a representation of $\mathscr{B}$. But the sets represented by (6) and (1) are clearly the same, hence $\mathscr{B} = \mathscr{A}$.

Finally, if we replace b_i in (1) by $a_o - a_i$, then $\mathscr{A}$ has a representation

$$x = \left(1 - \sum_1^k \lambda_i\right) a_o + \lambda_1 a_1 + \lambda_2 a_2 + \cdots + \lambda_k a_k, \quad \lambda \in \mathscr{V}^k. \tag{7}$$

Setting $\varphi_o = 1 - \sum_1^k \lambda_i$, and $\varphi_i = \lambda_i$, $i = 1, 2, \ldots, k$, the representation (7) becomes

$$x = \sum_0^k \varphi_i a_i, \qquad \sum_0^k \varphi_i = 1, \qquad \varphi \in \mathscr{V}^{k+1}. \tag{8}$$

Thus the k-flat $\mathscr{A}$ is the affine span of $A_o, A_1, \ldots, A_k$. □

COROLLARY 7. Every k-flat contains $k + 1$ independent points. (E)

Because k-flats are represented by affine spans, they are also called the k-dimensional *affine subspaces* of $\mathscr{E}^n$. Each set of $k + 1$ independent points in the k-flat forms an *affine basis* for the flat. We also have the following important analog.

THEOREM 8. Each point of a k-flat is represented by one and only one affine combination of a given affine basis for the flat.

PROOF. Let $\{A_o, A_1, \ldots, A_k\}$ denote $k + 1$ independent points whose affine span is a k-flat $\mathscr{A}$ in $\mathscr{E}^n$. Then $P \in \mathscr{A}$ implies that P has a representation

$$p = \sum_o^k \eta_i a_i, \qquad \sum_o^k \eta_i = 1. \tag{1}$$

Now, suppose that P is also represented by

$$p = \sum_o^k \varphi_i a_i, \qquad \sum_o^k \varphi_i = 1. \tag{2}$$

Using the fact that $\eta_o = 1 - \sum_1^k \eta_i$, then from (1),

$$p = \eta_o a_o + \sum_1^k \eta_i a_i = \left(1 - \sum_1^k \eta_i\right) a_o + \sum_1^k \eta_i a_i,$$

hence

$$p - a_o = \sum_1^k \eta_i a_i - \sum_1^k \eta_i a_o = \sum_1^k \eta_i \mathbf{A}_o\mathbf{A}_i. \tag{3}$$

By the same argument applied to (2), we have

$$p - a_o = \sum_1^k \varphi_i \mathbf{A}_o\mathbf{A}_i. \tag{4}$$

The relations (3) and (4) together imply that

$$\sum_{1}^{k} (\eta_i - \varphi_i)\mathbf{A}_o\mathbf{A}_i = o. \tag{5}$$

By Theorem 5, the k vectors $\mathbf{A}_o\mathbf{A}_i$ are linearly independent, so from (5) we obtain $\eta_i - \varphi_i = 0$, or $\eta_i = \varphi_i$, $i = 1, 2, \ldots, k$. Because of the unit sum property of affine coefficients, we must also have $\eta_o = \varphi_o$. Thus the representations (1) and (2) are identical. □

When a k-flat is represented by the span of independent points, the $(k + 1)$-tuples that are the coefficients in the affine combination representing a point are the *affine coordinates* of the point. By Theorem 8, these coordinates are unique.

Given a set of k independent vectors, $\{a_1, a_2, \ldots, a_k\}$, the linear span of the vectors is a k-dimensional linear subspace, and the affine span of the vectors is a $(k - 1)$-dimensional affine subspace. If $k = 2$, the linear span is the plane $\mathrm{Pl}(OA_1A_2)$, and the affine span is the line $\mathrm{Ln}(A_1A_2)$. If $k = 3$, the linear span is a linear 3-space, and the affine span is the plane $\mathrm{Pl}(A_1A_2A_3)$.

Exercises – Section 2.4

1. Prove Corollary 4.
2. Prove Theorem 5.
3. Prove Theorem 6.
4. How does Theorem 7 imply that two distinct points determine a unique line and that three noncollinear points determine a unique plane?
5. If $a_1, a_2, \ldots, a_k$ is a basis for a k-dimensional linear subspace $\mathscr{S}$, and if a k-flat $\mathscr{A}$ is the v-translate of $\mathscr{S}$, specify $k + 1$ independent points that belong to $\mathscr{A}$, hence explain Corollary 7.
6. In $\mathscr{E}^4$, let $A_o = (0, 0, 1, 1)$, $A_1 = (1, 0, 5, 0)$, $A_2 = (1, -1, -1, 1)$ and $A_3 = (1, -1, 1, 1)$.

   a. Show that the points A_i are affinely independent.
   b. Let $\mathscr{A} = \mathrm{AS}(a_o, a_1, a_2, a_3)$, and let $\mathscr{B} = \{X: x_1 + x_2 + x_4 = 1\}$. Show that $\mathscr{A}$ is a subset of $\mathscr{B}$.
   c. If X in $\mathscr{A}$ has affine coordinates $(\bar{x}_o, \bar{x}_1, \bar{x}_2, \bar{x}_3)$ with respect to the affine basis A_o, A_1, A_2, A_3, show that

$$\bar{x}_o = x_2 + x_4, \qquad \bar{x}_1 = x_1 + x_2$$
$$\bar{x}_2 = \tfrac{5}{2}x_1 + \tfrac{5}{2}x_2 - \tfrac{1}{2}x_3 + \tfrac{1}{2}x_4$$
$$\bar{x}_3 = -\tfrac{5}{2}x_1 - \tfrac{7}{2}x_2 + \tfrac{1}{2}x_3 - \tfrac{1}{2}x_4$$

   Show that $\mathscr{B}$ is a subset of $\mathscr{A}$, hence that $\mathscr{B} = \mathscr{A}$.

7. Use Theorem 7 to prove that if two lines in $\mathscr{E}^n$ intersect, their intersection is a single point.
8. Give an example in $\mathscr{E}^4$ of two planes whose intersection is a single point.
9. Let $\mathscr{S} = \{A_1, A_2, \ldots, A_k\}$ be an affinely independent set in $\mathscr{E}^n$, where $k < n$. If P is a linear combination of the points in $\mathscr{S}$, say

$$p = \sum_{i=1}^{k} \eta_i a_i,$$

then p is an n-tuple, hence a point of $\mathscr{E}^n$. We know that if the sum of the n_i's is 1, then $P \in \text{AS}(\mathscr{S})$. Give an example to show that if the sum is not 1, this does not imply that P is not in $\text{AS}(\mathscr{S})$.

SECTION 2.5. LINEAR AND AFFINE FUNCTIONS AND INVARIANTS

In Sec. 1.3, we defined a linear function on a vector space to be one that maps a linear combination of vectors onto the same linear combination of their images. Under a linear function, linear subspaces map onto linear subspaces, and, in fact, any property of the domain that can be expressed in terms of linear combinations transforms into an analogous property of the range. A property that is preserved under any linear function is called a *linear invariant*. Thus the property of being a linear subspace is a linear invariant.

The notion of relating properties and classes of transformations that preserve the properties is a deeply important idea in nearly all parts of mathematics. This concept of *invariance* was involved when we showed earlier that compactness and connectedness are preserved by continuous functions and hence are *continuous invariants.* From the analogies we have already encountered between linear and affine concepts, it is not surprising that there should be a theory of *affine invariants* paralleling that of linear invariants.

DEFINITION (Affine function). A function f defined on a vector space $\mathscr{S}$ with function values in a vector space is an *affine function*, or affine transformation, if it maps every affine combination of vectors u, v in $\mathscr{S}$ onto the same affine combination of their images. That is, f is an affine function if

$$f(\lambda_1 u + \lambda_2 v) = \lambda_1 f(u) + \lambda_2 f(v)$$

for $u, v \in \mathscr{S}$ and all $\lambda \in \mathscr{V}^2$ such that $\lambda_1 + \lambda_2 = 1$. In particular, if f is real valued, then f is an *affine functional.*

It is clear that the property of a set being a flat is an affine invariant, since the image of a flat under an affine function is again a flat.

We are mainly interested in the application of linear and affine functionals to geometric questions. However, as a matter of perspective, we state the fundamental theorem from linear algebra that gives an explicit characterization of linear and affine functions on euclidean spaces in terms of matrices. This is the theorem that the general affine transformations of $\mathscr{E}^n$ into $\mathscr{E}^m$ are precisely the mappings of the form

$$f(x) = Ax + v, \qquad x \in \mathscr{E}^n,$$

where A is an $m \times n$ matrix, x is an $n \times 1$ matrix (or column vector), and v is a fixed $m \times 1$ matrix. The linear transformations are given by the special cases in which $v = o$.

It is not difficult to establish the fact that every affine transformation differs from a linear transformation by a translation.

THEOREM 1. If $f: \mathscr{D} \to \mathscr{R}$ is an affine function, then there exists a unique vector $v \in \mathscr{R}$ and a unique linear function g such that $f(x) = g(x) + v$, $x \in \mathscr{D}$.

PROOF. We first observe that if g and v exist, then v must be $f(o)$. This is so because g, being linear, maps o onto itself. Hence $f(o) = g(o) + v = o + v = v$. This in turn tells us that if g exists satisfying the theorem, then $f(x) = g(x) + f(o)$, so $g(x) = f(x) - f(o)$, $x \in \mathscr{D}$. Thus g is uniquely determined, and the only question is whether g, defined on $\mathscr{D}$ by $g(x) = f(x) - f(o)$, is linear.

To show that for any u, v in $\mathscr{D}$, $g(\lambda_1 u + \lambda_2 v) = \lambda_1 g(u) + \lambda_2 g(v)$, we first compute

$$\begin{aligned} g(\lambda_1 u + \lambda_2 v) &= f(\lambda_1 u + \lambda_2 v) - f(o) \\ &= f[\lambda_1 u + \lambda_2 v + (1 - \lambda_1 - \lambda_2)o] - f(o). \end{aligned}$$

Since f is affine, it maps the affine combination of u, v, o onto the same combination of their images. Hence we have

$$\begin{aligned} g(\lambda_1 u + \lambda_2 v) &= \lambda_1 f(u) + \lambda_2 f(v) + (1 - \lambda_1 - \lambda_2) f(o) - f(o) \\ &= \lambda_1[f(u) - f(o)] + \lambda_2[f(v) - f(o)] \\ &= \lambda_1 g(u) + \lambda_2 g(v). \quad \square \end{aligned}$$

It is clear from Theorem 1 that for any characterization of linear functions on a vector space, there is a corresponding characterization of affine functions on the space. We next establish such corresponding characterizations for linear and affine functionals on $\mathscr{E}^n$.

THEOREM 2. A function $f : \mathscr{E}^n \to \mathscr{E}^1$ is a linear functional if and only if there exists a unique vector a in $\mathscr{E}^n$ such that

$$f(x) = a \cdot x.$$

PROOF. Suppose there does exist a unique vector a such that $f(x) = a \cdot x$. Then, from the distributive properties of the inner product, $f(\lambda_1 u + \lambda_2 v) = a \cdot (\lambda_1 u + \lambda_2 v) = \lambda_1(a \cdot u) + \lambda_2(a \cdot v) = \lambda_1 f(u) + \lambda_2 f(v)$, hence f is a linear functional.

Next, suppose that f is a linear functional on $\mathscr{E}^n$. Then f has a real number as a function value at each of the standard basis vectors e_i, $i = 1, 2, \ldots, n$, hence $a = (f(e_1), f(e_2), \ldots, f(e_n))$ is a fixed vector in $\mathscr{E}^n$. Since any x in $\mathscr{E}^n$ has a representation $x = x_1 e_1 + x_2 e_2 + \cdots + x_n e_n$, and since f is linear, then

$$f(x) = f(x_1 e_1 + \cdots + x_n e_n) = x_1 f(e_1) + \cdots + x_n f(x_n) = a \cdot x.$$

Thus there is a fixed vector a which satisfies the theorem. To see that a is unique, let b denote a vector such that $f(x) = b \cdot x$ for all $x \in \mathscr{E}^n$. Then $f(a - b) = a \cdot (a - b) = b \cdot (a - b)$, so $a \cdot (a - b) - b \cdot (a - b) = 0$. Thus $(a - b) \cdot (a - b) = 0$, hence $|a - b|^2 = 0$, so $a = b$. □

COROLLARY 2. A function $f : \mathscr{E}^n \to \mathscr{E}^1$ is an affine functional if and only if there exists a unique vector $a \in \mathscr{E}^n$ and a unique real number k such that $f(x) = a \cdot x + k$.

Because of Theorem 2 and its corollary, we use the term "linear functional" to mean an inner product function $a \cdot x$ and the term "affine functional" to mean a function of the form $a \cdot x + k$, where k is a real number.

It is a straightforward computation to establish the following fact.

THEOREM 3. Linear and affine functionals are continuous functions. (E)

Because of Theorem 3, any continuous invariant, such as compactness or connectedness, is of course preserved by linear and affine functionals. But these functionals possess important invariants not common to all continuous functions.

THEOREM 4. If $a \neq o$, then the linear functional $f(x) = a \cdot x$ and the affine functional $g(x) = a \cdot x + k$ map bounded sets onto bounded sets, neighborhoods onto neighborhoods, balls onto balls, and open sets onto open sets.

PROOF. If we can establish the theorem for g, it will obviously hold for f as a special case of g with $k = 0$.

Let $\mathscr{S}$ be a bounded set in $\mathscr{E}^n$. Then $\mathrm{Cl}(\mathscr{S})$ is both closed and bounded and hence compact. Because g is continuous on $\mathscr{E}^n$, it is continuous on $\mathrm{Cl}(\mathscr{S})$. Thus the g-image of $\mathrm{Cl}(\mathscr{S})$, namely, the set $g[\mathrm{Cl}(\mathscr{S})]$, is compact and is therefore bounded. But $\mathscr{S} \subset \mathrm{Cl}(\mathscr{S})$ implies $g(\mathscr{S}) \subset g[\mathrm{Cl}(\mathscr{S})]$. Thus $g(\mathscr{S})$ is a subset of a bounded set and is itself a bounded set.

Next, let y be the number $g(x) = a \cdot x + k$. We contend that g maps any neighborhood $\mathrm{N}(X_o, \delta)$ onto the neighborhood $\mathrm{N}(Y_o, \delta_1)$, where $y_o = g(x_o)$ and $\delta_1 = |a|\delta$. To show this, we first observe that if $X \in \mathrm{N}(X_o, \delta)$, then $|x - x_o| < \delta$, so

$$\begin{aligned} |f(x) - f(x_o)| &= |y - y_o| = |(a \cdot x + k) - (a \cdot x_o + k)| \\ &= |a \cdot (x - x_o)| \leq |a||x - x_o| < |a|\delta = \delta_1. \end{aligned}$$

Therefore $f[\mathrm{N}(X_o, \delta)] \subset \mathrm{N}(Y_o, \delta_1)$.

Now let t be any number in $\mathrm{N}(Y_o, \delta_1)$, that is, a number in the open interval $(y_o - \delta_1, y_o + \delta_1)$, and let $s = t - y_o$. Then $|s| < \delta_1$. Define the point X in $\mathscr{E}^n$ by $x = x_o + s(a/|a|^2)$. From

$$d(X, X_o) = |x - x_o| = |s|\frac{|a|}{|a|^2} = \frac{|s|}{|a|} < \frac{\delta_1}{|a|} = \frac{|a|\delta}{|a|} = \delta,$$

it follows that $X \in \mathrm{N}(X_o, \delta)$. On the other hand,

$$g(x) = a \cdot \left(x_o + s\frac{a}{|a|^2}\right) + k = a \cdot x_o + k + s\frac{a \cdot a}{|a|^2} = y_o + s = t.$$

Therefore each point of $\mathrm{N}(Y_o, \delta_1)$ has a pre-image in $\mathrm{N}(X_o, \delta)$. This, with our previous result, implies that $g[\mathrm{N}(X_o, \delta)] = \mathrm{N}(Y_o, \delta_1)$.

The same argument shows that the ball $\mathrm{B}(X_o, \delta)$ maps onto the closed interval $[y_o - \delta_1, y_o + \delta_1]$, that is, the ball $\mathrm{B}(Y_o, \delta_1)$. The only difference is that $t \in [y_o - \delta_1, y_o + \delta_1]$ and $s = t - y_o$ imply $|s| \leq \delta_1$, which in turn implies $d(X, X_o) \leq \delta$.

Finally, let $\mathscr{S}$ be an open set in $\mathscr{E}^n$ and consider any number y_o in the image set $g(\mathscr{S})$. The number y_o has at least one pre-image X_o in $\mathscr{S}$, and, since $\mathscr{S}$ is open, there exists $\mathrm{N}(X_o, \delta) \subset \mathscr{S}$. By the previous part of the proof, $g[N(X_o, \delta)] = \mathrm{N}(Y_o, \delta_1)$, where $y_o = g(x_o)$ and $\delta_1 = |a|\delta$. Since $\mathrm{N}(X_o, \delta) \subset \mathscr{S}$ implies $g[N(X_o, \delta)] \subset g(\mathscr{S})$, it follows that $\mathrm{N}(Y_o, \delta_1) \subset g(\mathscr{S})$. Thus y_o is interior to $g(\mathscr{S})$. Since all its points are interior points, $g(\mathscr{S})$ is an open set. □

Exercises – Section 2.5

1. Prove that a nonconstant affine function maps a line onto a line, a ray onto a ray, and a segment onto a segment.
2. Prove by direct computation that a functional f defined by $f(x) = a \cdot x + k$, $a \in \mathscr{E}^n$, $k \in \mathscr{E}^1$, preserves affine combinations.
3. Prove Theorem 3.
4. In analytic geometry courses, a function of three variables of the form $f(x, y, z) = ax + by + cz$ is sometimes referred to as a "linear function." What is the relation of that term to the "linear function" of this section?
5. Prove that a nonconstant affine functional assumes a maximum and a minimum on a ball and that points at which these function values are achieved must belong to the sphere.
6. Let the real valued function g be defined on $\mathscr{E}^2$ by $g(x) = x_1x_2$.

   a. Prove that g is continuous.
   b. Prove that the set $\mathscr{S} = \{X : g(x) \geq 1\}$ is closed.
   c. Let f be the linear functional defined on $\mathscr{E}^2$ by $f(x) = (1, 0) \cdot x = x_1$. Show that $\mathscr{R} = \{X : f(x) \in \mathscr{S}\}$ is not closed.

SECTION 2.6. HYPERPLANES, HALF-SPACES, CONVEXITY

Our purpose in this chapter—to outline the structure of $\mathscr{E}^n$ in relation to its basic subsets—is nearly accomplished. We have precise meanings for the lower dimensional subspaces, as well as for segments, rays, angles, and angle measure. We have identified the subspaces with euclidean spaces, and we have representations for these spaces both in linear and affine terms. However, a basic class of sets we have not considered is the half-spaces, or opposite sides of hyperplanes. We examine them in this concluding section.

We begin by looking at a familiar situation from a new point of view. Consider the linear equation

$$a_1x_1 + a_2x_2 = 0, \qquad (a_1, a_2) \neq (0, 0). \tag{1}$$

In the ordinary cartesian plane, (1) represents a line L_o through the origin. If we rewrite (1) in the form

$$a \cdot x = 0, \qquad a \neq o, \tag{2}$$

it is clear that L_o is the line perpendicular to $\text{Ln}(OA)$ at O. A different way of

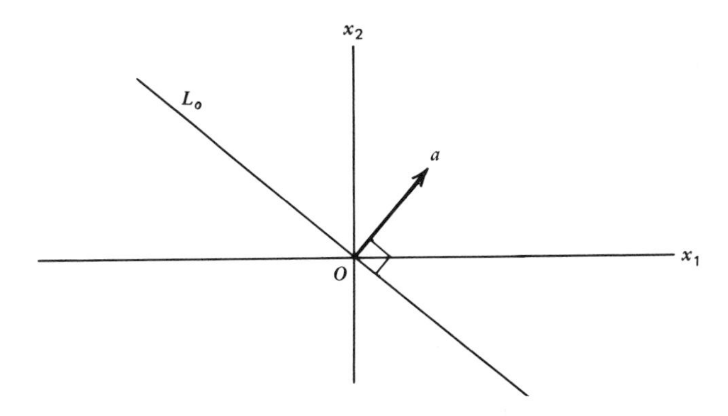

regarding L_o is to consider it in relation to the linear functional

$$f(x) = a \cdot x \tag{3}$$

defined on all $\mathscr{E}^2$. Just as we can speak of -1 and 2 as the zeros of the polynomial function $g(t) = 3(t + 1)(t - 2)$, so we can speak of the line L_o as the zero set of the linear functional f.

In the same way, it makes sense to speak of the 3-set of f, meaning the set of points X for which $f(x) = 3$. Since this is the set

$$a \cdot x = 3,$$

it is clearly a line L_3 parallel to L_o. Expressed another way, the line L_3 is the pre-image in $\mathscr{E}^2$ of the number 3 in $\mathscr{E}^1$. In general, a number k in $\mathscr{E}^1$ has a line L_k as its pre-image in $\mathscr{E}^2$. In $\mathscr{E}^1$, the number 0 separates the positive numbers from the negative numbers, and correspondingly the lines L_k for $k > 0$ lie in one side of L_o and the lines L_k for $k < 0$ lie in the opposite side of L_0. Thus the set on which the linear functional f is positive, namely, the graph of $f(x) > 0$, is one side of L_o and the graph of $f(x) < 0$ is the other side.

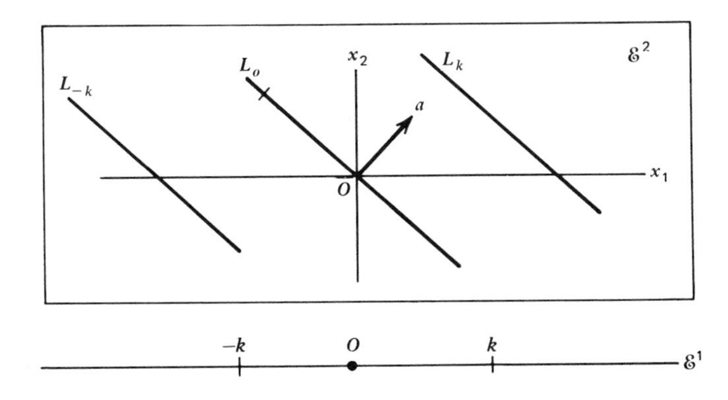

It is also true that any specific number in $\mathscr{E}^1$, say $k = 3$, separates the real numbers into those greater than 3 and those less than 3. Corresponding to 3, the line L_3 is the 3-constant set of f. As k increases from 3, the lines L_k sweep out one side of L_3, and as k decreases from 3 the lines L_k sweep out the opposite side. Thus the sides of L_3 correspond to the graphs of $f(x) > 3$ and $f(x) < 3$.

In $\mathscr{E}^2$, the lines are hyperplanes, since they are the flats of dimension $n - 1$, when $n = 2$. If the pattern we have just discussed generalizes, then a hyperplane in $\mathscr{E}^n$ should be the k-constant set of some linear functional $f(x) = a \cdot x$, and the opposite sides of this hyperplane $\mathscr{H}$ should be the graphs of $f(x) > k$ and $f(x) < k$. That is, in $\mathscr{E}^n$, as in $\mathscr{E}^2$, the separation of space by $\mathscr{H}$ should correspond to the separation of the real numbers by k. We show that this is in fact the case.

THEOREM 1. Corresponding to a hyperplane $\mathscr{H}$, there exists a non-null vector a and a real number k such that $\mathscr{H}$ is the graph of $a \cdot x = k$. The vector a is orthogonal to **PQ** for all P, Q in $\mathscr{H}$, and LS(a) is precisely the set of vectors that have this property.

PROOF. By the definition of a hyperplane $\mathscr{H}$, there exists a hyperplane $\mathscr{S}$ that contains the origin, and there is a vector v such that $\mathscr{H}$ is the v-translate of $\mathscr{S}$. As we saw in Sec. 2.3 (Theorems 6 and 9), there exists an orthonormal basis $a_1, a_2, \ldots, a_{n-1}$ for $\mathscr{S}$ and a vector a_n such that $a_1, a_2, \ldots, a_{n-1}, a_n$ is an orthonormal basis for $\mathscr{E}^n$. By Theorem 8, Sec. 2.3, each point X of $\mathscr{E}^n$ is uniquely represented in the form

$$x = \sum_1^n (x \cdot a_i)a_i. \tag{1}$$

By the same theorem, each point X in $\mathscr{S}$ is uniquely represented in the form

$$x = \sum_1^{n-1} (x \cdot a_i)a_i. \tag{2}$$

Comparing (1) and (2), we see that the points of space that belong to $\mathscr{S}$ are precisely those for which $(x \cdot a_n)a_n = o$. Since $a_n \neq o$, it follows that X is in $\mathscr{S}$ if and only if

$$a_n \cdot x = 0, \tag{3}$$

hence $\mathscr{S}$ is the graph of this equation.

Now consider $\mathscr{H}$. A point X is in $\mathscr{H}$ if and only if $x - v$ is in $\mathscr{S}$, hence if and only if

$$a_n \cdot (x - v) = 0, \tag{4}$$

hence if and only if

$$a_n \cdot x = a_n \cdot v. \tag{5}$$

Setting $a = a_n$ and $k = a_n \cdot v$, it follows that $\mathscr{H}$ is the graph of

$$a \cdot x = k. \tag{6}$$

To establish the second part of the theorem, we first observe that $a = a_n$ is an orthonormal basis for $\text{Ln}(OA) = \text{LS}(a_n)$. Thus, again by Theorem 8, Sec. 2.3, each point X of $\text{Ln}(OA)$ is uniquely represented in the form

$$x = (x \cdot a_n)a_n. \tag{7}$$

Now, comparing (1) and (7), it follows that the points of space that belong to $\text{Ln}(OA)$ are precisely those for which

$$\sum_{1}^{n-1} (x \cdot a_i)a_i = o. \tag{8}$$

But, because the vectors $a_1, a_2, \ldots, a_{n-1}$ are linearly independent, the coefficients in (8) must be zero. Hence X is in $\text{Ln}(OA)$ if and only if

$$x \cdot a_i = 0, \qquad i = 1, 2, \ldots, n - 1. \tag{9}$$

Thus $\text{Ln}(OA)$ is the graph of these equations.

If P, Q are any two points of $\mathscr{H}$, they belong to the graph of equation (6), hence $a \cdot p = k$ and $a \cdot q = k$, so

$$a \cdot (q - p) = a \cdot \mathbf{PQ} = 0. \tag{10}$$

Thus a is orthogonal to $\mathbf{PQ}$, so clearly every vector in $\text{LS}(a)$ is orthogonal to $\mathbf{PQ}$, for all P, Q in $\mathscr{H}$.

To show that $\text{LS}(a)$ is the set of all vectors with the previous property, let b denote any vector that is orthogonal to PQ for all P, Q in $\mathscr{H}$. Since the points $O, A_1, A_2, \ldots, A_{n-1}$ are in $\mathscr{S}$, their v-translates $O' = V, A_1', A_2', \ldots, A_{n-1}'$ are in $\mathscr{H}$. Thus b is orthogonal to each of the vectors $\mathbf{A}_i'\mathbf{V}$, hence

$$b \cdot (v - a_i') = b \cdot [v - (a_i + v)] = b \cdot a_i = 0, \qquad i = 1, 2, \ldots, n - 1. \tag{11}$$

Thus B satisfies the equations in (9) and so belongs to $\mathrm{Ln}(OA)$, which is to say that $b \in \mathrm{LS}(a)$. $\square$

We introduce a name for the vector a in Theorem 1.

DEFINITION (Normals to a hyperplane). A non-null vector a is *normal* to a hyperplane $\mathscr{H}$ if $a \cdot \mathbf{PQ} = 0$ for all P, Q in $\mathscr{H}$. The directions of a and $-a$ are called opposite normal directions of $\mathscr{H}$, or simply opposite directions of $\mathscr{H}$.

In Theorem 1 we showed that every hyperplane is the k-constant set of some linear functional $f(x) = a \cdot x$, or, what is equivalent, the zero set of an affine functional $g(x) = a \cdot x - k$. The converse is also correct, namely, that the k-constant set of a linear functional (or zero set of an affine functional) is a hyperplane.

THEOREM 2. The graph of $a \cdot x = k$, $a \neq o$, $k \in \mathscr{E}^1$ is a hyperplane for which a is a normal vector.

PROOF. Let $\mathscr{H} = \{X : a \cdot x = k\}$ and let $\mathscr{S} = \{X : a \cdot x = 0\}$. The linear span $\mathrm{LS}(a)$ is a one-dimensional subspace of $\mathscr{E}^n$ and $b_1 = a/|a|$ is an orthonormal basis for $\mathrm{LS}(a)$. By Theorem 6, Sec. 2.3, the basis b_1 can be completed to an orthonormal basis $b_1, b_2, \ldots, b_n$ for $\mathscr{E}^n$. By Theorem 9, Sec. 2.3, each point X in $\mathscr{E}^n$ has a unique representation $x = \sum_1^n (x \cdot b_i)b_i$ and $x \in \mathrm{LS}(b_2, b_3, \ldots, b_n)$ if and only if $x = \sum_2^n (x \cdot b_i)b_i$, that is, if and only if $x \cdot b_1 = 0$, hence if and only if $x \cdot a = 0$. Thus $\mathscr{S}$, which is the graph of $a \cdot x = 0$, is also $\mathrm{LS}(b_2, b_3, \ldots, b_n)$ and hence is a hyperplane through the origin.

Now let V be any point of $\mathscr{H}$. Then, by the definition of $\mathscr{H}$, $a \cdot v = k$. Since a general point X is in $\mathscr{H}$ if and only if $a \cdot x = k$, it follows that X is in $\mathscr{H}$ if and only if $a \cdot (x - v) = k - k = 0$. But $a \cdot (x - v) = 0$ if and only if $x - v$ is in $\mathscr{S}$. Since X is in $\mathscr{H}$ if and only if $x - v$ is in $\mathscr{S}$, then $\mathscr{H}$ is the v-translate of $\mathscr{S}$ and hence is a hyperplane. That the vector a is normal to $\mathscr{H}$ now follows from Theorem 1. $\square$

If $\lambda \neq 0$, it is clear that $a \cdot x = k$ and $(\lambda a) \cdot x = \lambda k$ have the same graphs. Thus, although it is true that $a \cdot x = k$, $a \neq 0$, represents a unique hyperplane $\mathscr{H}$, it is not true that $\mathscr{H}$ has a unique equation. The nonuniqueness is limited in the following way.

THEOREM 3. If $a \cdot x = k$ and $b \cdot x = h$, $a \neq o$, $b \neq o$, are both representations of a hyperplane $\mathscr{H}$, then there exists a real nonzero number λ such that $b = \lambda a$ and $h = \lambda k$. Moreover, $\mathscr{H}$ has an equation $c \cdot x = j$, where $|c| = 1$. (E)

CONVENTION. We speak of "the hyperplane $a \cdot x = k$" meaning that $a \neq o$ and that the hyperplane referred to is the graph of $a \cdot x = k$.

It is clear that if $h \neq k$, then the hyperplanes $a \cdot x = h$ and $a \cdot x = k$ are nonintersecting. We now introduce the usual name for this.

DEFINITION (Parallel hyperplanes). Hyperplanes $\mathscr{S}$ and $\mathscr{H}$ are parallel, denoted by $\mathscr{S} \parallel \mathscr{H}$, if they do not intersect.

THEOREM 4. Two distinct hyperplanes are parallel if and only if they have a common normal direction and hence are different constant sets of the same linear functional.

PROOF. Let $\mathscr{S}$ and $\mathscr{H}$ denote distinct hyperplanes. If they have a common normal direction a, then they have respective equations $a \cdot x = h$ and $a \cdot x = k$. Since $\mathscr{S} \neq \mathscr{H}$, it follows from Theorem 3 that $h \neq k$. Thus the the equations have no common solution, so $\mathscr{S}$ and $\mathscr{H}$ do not intersect and hence are parallel.

On the other hand, suppose that $\mathscr{S}$ and $\mathscr{H}$ have independent normal directions a and b, which we can take to be unit vectors. Then $\mathscr{S}$ and $\mathscr{H}$ have equations of the form $a \cdot x = h$ and $b \cdot x = k$, respectively, and we contend that these equations do have a common solution. More particularly, let $y = \lambda_1 a + \lambda_2 b$ be a point in LS(a, b). Then Y is in both $\mathscr{S}$ and $\mathscr{H}$ if the equations

$$a \cdot (\lambda_1 a + \lambda_2 b) = h$$

$$b \cdot (\lambda_1 a + \lambda_2 b) = k,$$

that is, the pair

$$\lambda_1 + (a \cdot b)\lambda_2 = h$$

$$(a \cdot b)\lambda_1 + \qquad \lambda_2 = k$$

have a solution in λ_1, λ_2. But the coefficient determinant of the system has the value $1 - (a \cdot b)^2$. This is zero if and only if $a \cdot b = \pm 1 = \pm |a||b|$, hence if and only if a and b are like or opposite directed (Theorem 2, Sec. 1.1). Since a and b are independent, they are not like or opposite directed, so the coefficient determinant is not zero. Thus, by Cramer's rule, the equations do have a solution in λ_1, λ_2, so $\mathscr{S}$ and $\mathscr{H}$ intersect. Thus if $\mathscr{S}$ and $\mathscr{H}$ are parallel, they cannot have different (i.e., independent) normal directions. Therefore $\mathscr{S} \parallel \mathscr{H}$ implies that $\mathscr{S}$ is some h-constant set of a linear functional $f(x) = a \cdot x$ and that $\mathscr{H}$ is some k-constant set of the same functional, where $h \neq k$, because $\mathscr{S} \neq \mathscr{H}$. $\square$

We now have in $\mathscr{E}^n$ an analog of the situation with which we began. That is, a hyperplane $\mathscr{H}$ is the k-constant set of a linear functional $f(x) = a \cdot x$, and thus can be regarded as the pre-image of k in $\mathscr{E}^1$. If we write $\mathscr{H}_k = \mathscr{H}$, and write $\mathscr{H}_t$ for the hyperplane $f(x) = t$, then as t increases from k, $\mathscr{H}_t$ moves parallel to $\mathscr{H}_k$ and sweeps out one half-space of $\mathscr{H}_k$. As t decreases from k, $\mathscr{H}_t$ sweeps out the opposite half-space. As before, the separation of space by $\mathscr{H}_k$ corresponds to the separation of the real numbers by k.

Before we introduce half-spaces in a formal way, there is another interesting and useful fact about the situation just described. Not only are the hyperplanes $\mathscr{H}_t$ and $\mathscr{H}_k$ equidistant, as one might expect, but their distance apart is directly proportional to $|t - k|$. We can, in fact, express the distance of any point P from a hyperplane $\mathscr{H} : f(x) = k$ in terms of the function values of the linear functional f at P and on $\mathscr{H}$.

THEOREM 5. Each point P in space has a unique foot P_o in a hyperplane $\mathscr{H} : f(x) = a \cdot x = k$, and

$$d(P, \mathscr{H}) = \frac{|f(p) - k|}{|a|} = \frac{|a \cdot p - k|}{|a|}.$$

PROOF. Consider the line $\mathscr{L}: x = p + \lambda a, \lambda \in \mathscr{E}^1$. The points of $\mathscr{L} \cap \mathscr{H}$ correspond to the λ-solutions of

$$a \cdot (p + \lambda a) = k. \tag{1}$$

Since (1) has the unique solution

$$\lambda_o = -\frac{a \cdot p - k}{a \cdot a},$$

$\mathscr{L}$ intersects $\mathscr{H}$ at P_o defined by

$$p_o = p + \lambda_o a = p - \frac{a \cdot p - k}{|a|^2} a. \tag{2}$$

If $P \in \mathscr{H}$, it is its own unique foot in $\mathscr{H}$. On the other hand, if $P \in \mathscr{H}$, then (2) shows that $P = P_o$, so P_o is the foot of P. If P is not in $\mathscr{H}$, then (2) shows that $\mathbf{PP}_o$ is a nonzero multiple of a and hence is a nonnull vector in LS(a). Thus, by Theorem 1, $\mathbf{PP}_o$ is orthogonal to all $\mathbf{XY}$, for X, Y in $\mathscr{H}$. If X in $\mathscr{H}$ is not X_o, then from

$$\mathbf{PX} = \mathbf{PP}_o + \mathbf{P}_o\mathbf{X} \tag{3}$$

we have

$$\mathbf{PX} \cdot \mathbf{PX} = \mathbf{PP}_o \cdot \mathbf{PP}_o + 2\mathbf{PP}_o \cdot \mathbf{P}_o\mathbf{X} + \mathbf{P}_o\mathbf{X} \cdot \mathbf{P}_o\mathbf{X}, \tag{4}$$

Because $\mathbf{PP}_o \cdot \mathbf{P}_o\mathbf{X} = 0$, (4) is just the Pythagorean theorem,

$$d(P, X)^2 = d(P, P_o)^2 + d(P_o, X)^2, \tag{5}$$

which implies that

$$d(P, X) > d(P, P_o). \tag{6}$$

Thus P_o is the unique foot of P in $\mathscr{H}$, so $d(P, \mathscr{H}) = d(P, P_o)$. So, from (2), we have

$$d(P, \mathscr{H}) = d(P, P_o) = \frac{|a \cdot p - k|}{|a|}. \quad \square$$

COROLLARY 5. If $\mathscr{H}_1 : a \cdot x = k_1$ and $\mathscr{H}_2 : a \cdot x = k_2$ are parallel hyperplanes, then

$$d(X, \mathscr{H}_1) = d(Y, \mathscr{H}_2) = \frac{|k_1 - k_2|}{|a|}$$

for all $X \in \mathscr{H}_2$ and all $Y \in \mathscr{H}_1$. (E)

DEFINITION (Distance between parallel hyperplanes). If $\mathscr{H}_1$ and $\mathscr{H}_2$ are parallel hyperplanes, then the common distance, given in Corollary 5, from each point of $\mathscr{H}_1$ to $\mathscr{H}_2$, and each point of $\mathscr{H}_2$ to $\mathscr{H}_1$, is called "the distance between $\mathscr{H}_1$ and $\mathscr{H}_2$" and denoted by $d(\mathscr{H}_1, \mathscr{H}_2)$.

We now introduce half-spaces.

DEFINITION (Half-spaces). If $a \cdot x = k$, $a \neq o$, is a representation of the hyperplane $\mathscr{H}$, then the graphs of $a \cdot x > k$ and $a \cdot x < k$ are the *opposite sides*, or *opposite open half-spaces* of $\mathscr{H}$. The graphs of $a \cdot x \geq k$ and $a \cdot x \leq k$ are the *opposite closed half-spaces* of $\mathscr{H}$. The hyperplane $\mathscr{H}$ is the *face* of the four half-spaces.

THEOREM 6. Corresponding to a hyperplane $\mathscr{H}$ there exists a unique pair of sets $\mathscr{S}_1$ and $\mathscr{S}_2$ that are the opposite sides of $\mathscr{H}$.

PROOF. By Theorem 1, $\mathscr{H}$ has a representation of the form $a \cdot x = k$, $a \neq o$. Let $\mathscr{S}_1$ and $\mathscr{S}_2$ denote the graphs of $a \cdot x > k$ and $a \cdot x < k$, respectively. Then, by definition, $\mathscr{S}_1$ and $\mathscr{S}_2$ are opposite sides of $\mathscr{H}$. If $b \cdot x = h$,

$b \neq o$, also represents $\mathscr{H}$, then $\mathscr{R}_1$ and $\mathscr{R}_2$, the respective graphs of $b \cdot x > h$ and $b \cdot x < h$, are also opposite sides of $\mathscr{H}$. By Theorem 3, there exists $\lambda \neq 0$ such that $b = \lambda a$ and $h = \lambda k$. Thus $\mathscr{R}_1$ and $\mathscr{R}_2$ are the graphs, respectively, of $\lambda(a \cdot x) > \lambda k$ and $\lambda(a \cdot x) < \lambda k$. If $\lambda > 0$, then $\lambda(a \cdot x) > \lambda k$ is equivalent to $a \cdot x > k$, hence $\mathscr{R}_1 = \mathscr{S}_1$ and similarly $\mathscr{R}_2 = \mathscr{S}_2$. If $\lambda < 0$, then $\lambda(a \cdot x) > \lambda k$ is equivalent to $a \cdot x < k$, hence $\mathscr{R}_1 = \mathscr{S}_2$ and similarly $\mathscr{R}_2 = \mathscr{S}_1$. Thus in all cases the pair of sets $\{\mathscr{R}_1, \mathscr{R}_2\}$ is the pair of sets $\{\mathscr{S}_1, \mathscr{S}_2\}$. □

THEOREM 7. Open half-spaces are open sets and closed half-spaces are closed sets.

PROOF. By Theorem 3, Sec. 2.5, the function $f(x) = a \cdot x$ is continuous on $\mathscr{E}^n$. Then, by Corollary 3.2, Sec. 1.5, since the interval (k, ∞) is an open set, so is the half-space $\{x : f(x) > k\} = \{x : f(x) \in (k, \infty)\}$. Similarly, by Corollary 4.2, Sec. 1.5, since the interval $[k, \infty)$ is a closed set, then so is the closed half-space $\{x : f(x) \geq k\} = \{x : f(x) \in [k, \infty)\}$. □

To say that $\mathscr{S}_1$ and $\mathscr{S}_2$ are opposite sides of the hyperplane $\mathscr{H}$ is to imply that in some sense $\mathscr{H}$ is between them. We can express this betweeness in a precise way in terms of the following familiar notions.

DEFINITION (Betweeness of numbers and points). In $\mathscr{E}^1$, the number p is between a and b if $a \neq b$ and p belongs to the open interval of a and b. Thus $a < p < b$ or $b < p < a$. In $\mathscr{E}^n$, the point P is between A and B if $A \neq B$ and P belongs to the open segment Sg(A, B).

Now consider a hyperplane $\mathscr{H}$ that is the k-constant set of the linear functional $f(x) = a \cdot x$. Let the sides of $\mathscr{H}$ be $\mathscr{S}_1 : f(x) > k$ and $\mathscr{S}_2 : f(x) < k$. In $\mathscr{E}^1$, let $\mathscr{R}_1 = (k, \infty) = \{x : x > k\}$ and let $\mathscr{R}_2 = (-\infty, k) = \{x : x < k\}$. Then k separates $\mathscr{R}_1$ and $\mathscr{R}_2$ in the sense that k is between every number in

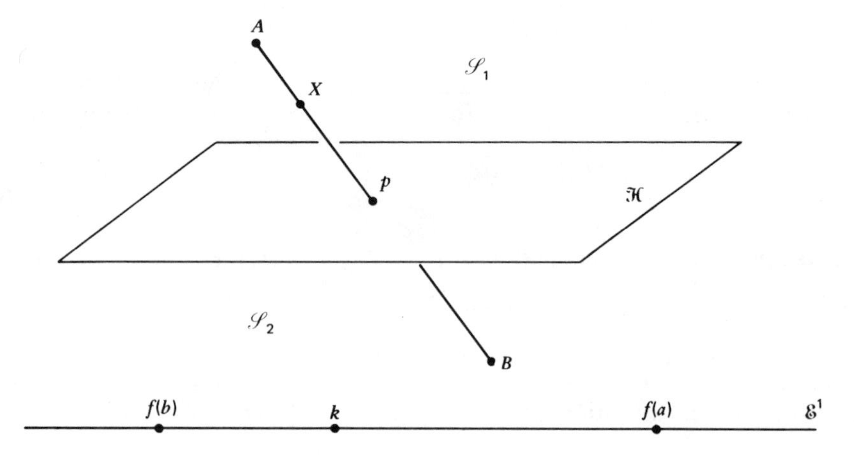

$\mathscr{R}_1$ and every number in $\mathscr{R}_2$. If A belongs to $\mathscr{S}_1$ and B belongs to $\mathscr{S}_2$, then $f(a)$ belongs to $\mathscr{R}_1$ and $f(b)$ belongs to $\mathscr{R}_2$. As a point X in $\text{Ln}(AB)$ moves from A to B, $f(x)$ decreases from $f(a)$ to $f(b)$ and hence passes through the intermediate value k. Correspondingly, X must pass through a point P in $\mathscr{H}$ since $\mathscr{H}$ is the locus of all points X for which $f(x) = k$.

We can calculate the point P explicitly. The line $\text{Ln}(AB)$ is given by

$$x = (1 - \lambda)a + \lambda b, \qquad \lambda \in \mathscr{E}^1,$$

and $\text{Sg}(A, B)$ corresponds to the values $0 < \lambda < 1$. The intersection of $\mathscr{H}$ and $\text{Ln}(AB)$ consists of the points in (1) for which

$$f[(1 - \lambda)a + \lambda b] = k. \tag{2}$$

Because f is linear, (2) is equivalent to

$$(1 - \lambda)f(a) + \lambda f(b) = k,$$

or

$$\lambda[f(a) - f(b)] = f(a) - k. \tag{3}$$

Since $f(a) > k > f(b)$, then $f(a) - f(b) > 0$, and (3) has the unique solution

$$\lambda_o = \frac{f(a) - k}{f(a) - f(b)} > 0. \tag{4}$$

Then

$$1 - \lambda_o = \frac{k - f(b)}{f(a) - f(b)} > 0. \tag{5}$$

Since λ_o, and $1 - \lambda_o$ are both positive, then $0 < \lambda_o < 1$. Thus

$$p = (1 - \lambda_o)a + \lambda_o b, \qquad \text{and} \qquad 0 < \lambda_o < 1 \tag{6}$$

shows that $P = \text{Ln}(AB) \cap \mathscr{H}$ belongs to $\text{Sg}(AB)$. Thus a hyperplane is between its sides in the following sense.

THEOREM 8. If A and B belong to opposite sides of a hyperplane $\mathscr{H}$, then there exists a unique point of $\mathscr{H}$ that is between A and B.

Just as important as Theorem 8 is the fact (which we prove later) that if A and B lie in the same side of $\mathscr{H}$, then Sg[AB] does not intersect $\mathscr{H}$. This is so because if A, B belong to one side of $\mathscr{H}$, so does the segment Sg[AB]. The property, called convexity, plays a dominant role in this study, and it is defined in $\mathscr{E}^n$ just as it is defined in elementary geometry.

DEFINITION (Convexity). A set $\mathscr{S}$ is convex if $A \in \mathscr{S}$ and $B \in \mathscr{S}$ imply that Sg(AB) $\subset \mathscr{S}$.

It follows trivially from the definition of convexity that the null set and singleton sets are convex. To establish the convexity of other sets, we first want to consider segments from a new point of view. We defined Sg[AB] as the set

$$x = a + \lambda \mathbf{AB}, \qquad 0 \leq \lambda \leq 1,$$

which is equivalent to

$$x = (1 - \lambda)a + \lambda b, \qquad 0 \leq \lambda \leq 1.$$

If we set $\eta_1 = (1 - \lambda)$ and $\eta_2 = \lambda$, then Sg[AB] is the set

$$x = \eta_1 a + \eta_2 b, \qquad \eta_1 + \eta_2 = 1, \qquad \eta_1, \eta_2 \geq 0.$$

From this form it is clear that the points of Sg[AB] are affine combinations of a, b of a special type, namely, combinations in which the coefficients are non-negative. We give these a special name.

DEFINITION (Convex combination). A *convex combination* of a finite set of vectors $a_1, a_2, \ldots, a_n$ is an affine combination of the vectors in which the coefficients are nonnegative. The set of all convex combinations of the vectors is their *convex span*, denoted by CS($a_1, a_2, \ldots, a_n$).

In terms of the language just introduced, a segment Sg[AB] is the convex span of a, b (or of A, B) and a convex set is one that contains the convex span of each pair of its points.

From the definitions for linear, affine, and convex spans, it is clear that if $\mathscr{S}$ is a finite set of vectors, then

$$\text{CS}(\mathscr{S}) \subset \text{AS}(\mathscr{S}) \subset \text{LS}(\mathscr{S}),$$

thus the convexity of flats is a special case of the following fact.

THEOREM 9. A flat contains the affine span of each pair of its points.

PROOF. Let P, Q denote two points in a k-flat $\mathscr{S}$, and let X be a point of AS(p, q). Then X has a representation

$$x = \lambda_1 p + \lambda_2 q, \qquad \lambda_1 + \lambda_2 = 1. \tag{1}$$

By Corollary 7, Sec. 2.4, $\mathscr{S}$ has an affine basis $A_1, A_2, \ldots, A_{k+1}$, so (by Theorem 8, Sec. 2.4) P and Q have representations

$$\begin{aligned} p &= \sum_1^{k+1} \eta_i a_i, \qquad \sum_1^{k+1} \eta_i = 1 \\ q &= \sum_1^{k+1} \varphi_i a_i, \qquad \sum_1^{k+1} \varphi_i = 1. \end{aligned} \tag{2}$$

Using the relations (2) in conjunction with (1), X is represented by

$$x = \lambda_1 \left(\sum_1^{k+1} \eta_i a_i \right) + \lambda_2 \left(\sum_1^{k+1} \varphi_i a_i \right)$$

or

$$x = \sum_1^{k+1} (\lambda_1 \eta_i + \lambda_2 \varphi_i) a_i.$$

Since the sum of the coefficients in (3) is

$$\begin{aligned} \sum_1^{k+1} (\lambda_1 \eta_i + \lambda_2 \varphi_i) &= \lambda_1 \left(\sum_1^{k+1} \eta_i \right) + \lambda_2 \left(\sum_1^{k+1} \varphi_i \right) \\ &= \lambda_1(1) + \lambda_2(1) = \lambda_1 + \lambda_2 = 1, \end{aligned}$$

if follows that $x \in \text{AS}(a_1, a_2, \ldots, a_k) = \mathscr{S}$. Therefore AS$(p, q) \in \mathscr{S}$. □

COROLLARY 9.1. A flat contains the affine span of each of its finite subsets. (E)

COROLLARY 9.2. Flats are convex sets.

PROOF. If P and Q are two points of a flat $\mathscr{S}$, then from Sg$[PQ]$ = CS$(p, q) \subset$ AS(p, q) and AS$(p, q) \subset \mathscr{S}$, it follows that Sg$(P, Q) \subset \mathscr{S}$, hence $\mathscr{S}$ is convex. □

THEOREM 10. Open and closed half-spaces are convex.

PROOF. Let $\mathscr{S}$ denote an open half-space whose face is the hyperplane $\mathscr{H}$. By Theorems 1 and 6, there exists an affine functional $g(x) = a \cdot x + h$, $a \neq o$, such that $\mathscr{H}$ is the graph of $g(x) = 0$ and $\mathscr{S}$ is the graph of $g(x) > 0$. If P and Q are two points of $\mathscr{S}$, then

$$g(p) > 0 \qquad \text{and} \qquad g(q) > 0. \tag{1}$$

If $X \in \mathrm{Sg}(PQ)$, then x is a convex combination of p and q of the form

$$x = \lambda_1 p + \lambda_2 q, \qquad \lambda_1 + \lambda_2 = 1, \qquad \lambda_1, \lambda_2 > 0. \tag{2}$$

Because g is an affine function, it preserves affine combinations, hence

$$g(x) = g(\lambda_1 p + \lambda_2 q) = \lambda_1 g(p) + \lambda_2 g(q). \tag{3}$$

From (1) and (2), $\lambda_1 g(p) > 0$ and $\lambda_2 g(q) > 0$. These relations, with (3), imply that $g(x) > 0$. Therefore $X \in \mathscr{S}$, which implies $\mathrm{Sg}(PQ) \subset \mathscr{S}$, hence $\mathscr{S}$ is convex. An entirely similar argument shows that the closed half-space $g(x) \geq 0$ is convex. $\square$

THEOREM 11. Open and closed segments are convex and open and closed rays are convex. (E)

Theorems 10 and 11 illustrate special cases of the following general and useful fact.

THEOREM 12. The closure of a convex set is convex.

PROOF. For an indirect proof, assume that $\mathscr{S}$ is a convex set, but that $\mathrm{Cl}(\mathscr{S})$ is not convex. Then there exist two points A, B in $\mathrm{Cl}(\mathscr{S})$ and a point P in $\mathrm{Sg}(AB)$ that is not in $\mathrm{Cl}(\mathscr{S})$. Since $P \in \mathrm{Sg}(AB)$, it has a representation

$$p = \eta_1 a + \eta_2 b, \qquad \eta_1 + \eta_2 = 1, \qquad \eta_1, \eta_2 > 0. \tag{1}$$

Because $P \notin \mathrm{Cl}(\mathscr{S})$, P is neither a point nor limit point of $\mathscr{S}$. Therefore there exists a neighborhood $\mathrm{N}(P, \delta)$ such that $\mathrm{N}(P, \delta) \cap \mathscr{S} = \varnothing$. Because $A \in \mathrm{Cl}(\mathscr{S})$, there exists a point A', which might be A, in $\mathrm{N}(A, \delta) \cap \mathscr{S}$. Similarly, $B \in \mathrm{Cl}(\mathscr{S})$ implies that there is a point B' in $\mathrm{N}(B, \delta) \cap \mathscr{S}$. Now, define P' by

$$p' = \eta_1 a' + \eta_2 b'. \tag{2}$$

Then, from (1) and (2),

$$\begin{aligned} d(P, P') &= |p - p'| = |\eta_1(a - a') + \eta_2(b - b')| \\ &\leq \eta_1|a - a'| + \eta_2|b - b'| < \eta_1\delta + \eta_2\delta = \delta. \end{aligned} \tag{3}$$

Thus $P' \in \mathrm{N}(P, \delta)$, which implies $P' \notin \mathscr{S}$. But since A' and B' are in $\mathscr{S}$, and $\mathscr{S}$ is convex, then $\mathrm{Sg}(A'B') \subset \mathscr{S}$. From (2), $P' \in \mathrm{Sg}(A'B')$, hence $P' \in \mathscr{S}$. The contradiction that P' is and is not in $\mathscr{S}$ shows that the initial assumption is false, and hence that $\mathrm{Cl}(\mathscr{S})$ must be a convex set. □

Exercises – Section 2.6

1. Prove Theorem 3.
2. Prove Corollary 5.
3. In $\mathscr{E}^2$, if $\mathscr{L}$ is the line through $(-1, 5)$ in the direction $v = (1, -2)$, give the opposite unit normal direction of $\mathscr{L}$. Are the points $(3, 2)$ and $(1, 0)$ on the same or opposite sides of $\mathscr{L}$?
4. In $\mathscr{E}^3$, give a vector representation for the plane $\mathscr{H}$ determined by $(1, 0, 1)$, $(3, 1, 0)$, and $(0, 1, 2)$. What is a normal direction of $\mathscr{H}$? If $P = (2, 2, 5)$ and $Q = (-1, 12, -2)$ what points of $\mathrm{Ln}(PQ)$ lie in the same side of $\mathscr{H}$ as the point $R = (1, 1, 2)$?
5. In $\mathscr{E}^2$, if $f(x) = (3, 1) \cdot x + 5$, and $g(x) = (3, 1) \cdot x - 4$, then the lines $\mathscr{L}_1 : f(x) = 0$ and $\mathscr{L}_2 : g(x) = 0$ are parallel. In terms of f and g, define the $\mathscr{L}_2$ side of $\mathscr{L}_1$ and the $\mathscr{L}_1$ side of $\mathscr{L}_2$.
6. In $\mathscr{E}^3$, the planes $\mathscr{H}_1 : (1, 1, 4) \cdot x = -2$ and $\mathscr{H}_2 : (1, 1, 4) \cdot x = -6$ are parallel. Describe the position of the point $(2, 0, -3)$ in relation to $\mathscr{H}_1$ and $\mathscr{H}_2$. What is the distance $d(\mathscr{H}_1, \mathscr{H}_2)$?
7. In $\mathscr{E}^3$, find the foot of $P = (10, -10, -17)$ in the plane $H : 2x_1 - 3x_2 - 3x_3 = 13$. Find the foot of $Q = (1, 0, 1)$ in the segment $\mathrm{Sg}[AB]$, if $A = (4, 1, 6)$ and $B = (12, 3, 14)$. What is the foot of Q in $\mathrm{Sg}(AB)$?
8. Prove the theorem: Corresponding to a point P and a nonnull vector a, there exists a unique hyperplane $\mathscr{H}$ that contains P and has normal direction a, and a representation for $\mathscr{H}$ is $a \cdot (x - p) = 0$.
9. If $\mathscr{H}$ is the hyperplane $a \cdot x = h$ and $\mathscr{L}$ is the line $x = p + \lambda v$, and $\mathscr{H} \cap \mathscr{L} = \varnothing$, prove that $a \cdot v = 0$. Does $a \cdot v = 0$ imply $\mathscr{H} \cap \mathscr{L} = \varnothing$?
10. Prove Corollary 9.1.
11. Prove that $\mathrm{N}(A, \delta)$ and $\mathrm{B}(A, \delta)$ are convex sets.
12. Prove Theorem 11.

3

DIMENSION AND BASIC STRUCTURE OF CONVEX BODIES AND SURFACES

INTRODUCTION

In the early development of euclidean geometry, figures were commonly defined by geometric conditions. We used this method in defining a sphere to be the set of all points at a fixed distance from a particular point. With the advent of coordinate geometry, figures could be defined as the graphs of equations or inequalities. For example, the conics Apollonius studied as the plane sections of a cone appear in analytic geometry as the graphs of quadratic equations. With the use of parametric equations, one could also define figures by mappings. Consider, for instance, the equations

$$x_1 = \cos t, \qquad x_2 = \sin t, \qquad x_3 = t, \qquad -\infty < t < \infty.$$

To each real number t there corresponds the point $(\cos t, \sin t, t)$, so the equations, or coordinate functions, define a mapping of $\mathscr{E}^1$ into $\mathscr{E}^3$, and the range of this mapping is a twisted space curve called a "circular helix."

The mapping point of view just described proved to be extremely fruitful and became a central concept in the development of classic differential geometry. During the nineteenth century, however, a considerable body of geometric knowledge was organized around a different concept. This "new" concept was, in fact, a return to the old notion of defining figures geometrically, but with the difference that convexity of the figure was made the central and primary requirement. Since this condition by itself provided no analytic representation of the object and covered such a wide range of sets, the full implications of the requirement were only gradually recognized. In time, however, it was found that convex sets, with various slight restrictions, provided classes of geometric objects with a surprising number of identifiable properties. The study of these grew into the subject of "convex bodies," which is now an extensive field with many ramifications.

In this brief chapter, we define convex bodies and surfaces of different dimensions and establish some basic facts about their structure. We also introduce a special class of projection mappings which we apply to a problem concerning the connectivity of surfaces. Although most of the properties we consider are intuitively "obvious," their proofs are not as apparent, and it is the methods in these proofs that are especially important.

SECTION 3.1. CONVEX BODIES AND SURFACES; SET DIMENSION

We have used the term "object set" to mean a set which in $\mathscr{E}^1$, $\mathscr{E}^2$, or $\mathscr{E}^3$ is a mathematical idealization of some physical object. From this intuition, it is natural to require that such a set be connected, closed, and bounded. Thus the following definition implies that a convex body is a special kind of object set.

DEFINITION. A set $\mathscr{K}$ is a *convex body* if it is compact, convex, and has a non-empty interior. A set $\mathscr{K}^0$ is a *closed, convex surface* if it is the boundary of a convex body $\mathscr{K}$. If $\mathscr{K}$ is planar and nonlinear, it is a *plane convex body* and its boundary $\mathscr{K}^0$ is a *closed, convex curve.*

The term "convex" in the names "closed, convex surface" and "closed, convex curve" is used in association with the set to which the surface or curve is a boundary. We prove, in fact, that a closed, convex surface or curve is never itself a convex set. Also, the term "closed" in these names might seem to be redundant, since boundaries are automatically closed sets. Initially, however, the term "closed" in the name "closed, convex surface" was introduced to indicate that such a surface actually encloses a region in the sense that the sphere $S(P, \delta)$ encloses $N(P, \delta)$.

The fact that convex bodies are connected is a special case of the fact that convex sets are connected. The following derivation of this property illustrates the methods of proof we now have available.

THEOREM 1. Every similarity is a continuous mapping.

PROOF. Let $f : \mathscr{D} \to \mathscr{R}$ denote a similarity of ratio $k > 0$. Corresponding to $P \in \mathscr{D}$, and an arbitrary $\varepsilon > 0$, let $\delta = \frac{1}{k}\varepsilon$. Then for $X \in \mathrm{N}(P, \delta) \cap \mathscr{D}$, we have $d[(f(P), f(X)] = kd(P, X) < k\delta = \varepsilon$, which implies the continuity of f at P. □

THEOREM 2. A closed segment is the image of a closed interval in a similarity mapping.

PROOF. By definition, the closed segment Sg$[AB]$ is the image of the closed interval $[0, 1]$ in the mapping f defined by $f(\lambda) = a + \lambda\mathbf{AB}$. Then $d[f(\lambda_1), f(\lambda_2)] = |f(\lambda_1) - f(\lambda_2)| = |a + \lambda_1\mathbf{AB} - (a + \lambda_2\mathbf{AB})| = |\mathbf{AB}|\,|\lambda_1 - \lambda_2| = |\mathbf{AB}|d(\lambda_1, \lambda_2)$, hence f is a similarity with ratio $k = |\mathbf{AB}|$. □

COROLLARY 2.1. Closed segments are connected sets.

PROOF. Since the continuous image of a connected set is connected (Theorem 5, Sec. 1.5), and since a closed interval is a connected set (Theorem 6, Sec. 1.6), it follows from Theorems 1 and 2 that closed segments are connected sets. □

COROLLARY 2.2. Every convex set is connected.

PROOF. The null set and singleton sets are obviously connected. If two points A, B belong to a convex set $\mathscr{S}$, then Sg$[AB] \subset \mathscr{S}$. Thus each two points of $\mathscr{S}$ belong to a connected subset of $\mathscr{S}$. Hence, by Theorem 5, Sec. 1.4, $\mathscr{S}$ is connected. □

COROLLARY 2.3. If set $\mathscr{S}$ is a non-empty, proper subset of $\mathscr{E}^n$, then $\mathscr{S}$ has a non-empty boundary.

PROOF. Assume that $\mathrm{Bd}(\mathscr{S}) = \varnothing$. Then $\mathscr{S} = \mathrm{In}(\mathscr{S}) \neq \varnothing$. Since $\mathrm{Cp}(\mathscr{S}) \neq 0$, because $\mathscr{S}$ is a proper subset of $\mathscr{E}^n$, and since $\mathrm{Bd}[\mathrm{Cp}(\mathscr{S})] = \mathrm{Bd}(\mathscr{S}) = \varnothing$, it follows that $\mathrm{Cp}(\mathscr{S}) = \mathrm{Ex}(\mathscr{S}) \neq \varnothing$. Therefore $\mathscr{E}^n = \mathrm{In}(\mathscr{S}) \cup \mathrm{Bd}(\mathscr{S}) \cup \mathrm{Ex}(\mathscr{S}) = \mathrm{In}(\mathscr{S}) \cup \mathrm{Ex}(\mathscr{S})$ expresses $\mathscr{E}^n$ as the union of two disjoint, open sets, which implies that $\mathscr{E}^n$ is disconnected. But $\mathscr{E}^n$ is convex, so $\mathscr{E}^n$ is connected. The contradiction shows that the assumption is false, hence $\mathrm{Bd}(\mathscr{S}) \neq \varnothing$. □

One indication of the scope of convex body theory is the fact that nearly all the figures in traditional plane geometry, that is, polygons and circles, are closed, convex curves. Similarly, the polyhedra, spheres, cylinders, and cones of solid geometry are closed, convex surfaces. Since we are interested not only in the general class of such curves and surfaces but also in higher dimensional analogs, we must distinguish sets of different dimension. To do so, we introduce the following global, or space, dimension of a set.

DEFINITION (Dimension of a set). In $\mathscr{E}^n$, the space dimension, or simply *the dimension*, of a non-empty, nonsingleton set $\mathscr{S}$ is 1 if $\mathscr{S}$ is contained in a line. The dimension is $k > 1$ if $\mathscr{S}$ is contained in a k-flat but is not contained in any flat of dimension less than k.

Comment. Single points are often defined to have dimension zero. But by our definition, "k-dimensional set" refers to a nonnull, nonsingleton set.

Our definition for the dimension of a set simply generalizes the usual sense in which we distinguish the figures of solid geometry from those of plane geometry. The next theorem gives an alternate characterization of dimension that is in some ways more useful.

THEOREM 3. Corresponding to a non-empty, nonsingleton set $\mathscr{S}$ in $\mathscr{E}^n$, there is a unique, positive integer k that is the dimension of $\mathscr{S}$, and $k + 1$ is exactly the maximal number of independent points in $\mathscr{S}$.

PROOF. By hypothesis, $\mathscr{S}$ contains at least two distinct points A_1 and A_2, and so contains two independent points. If $\mathscr{S} \subset \text{AS}(A_1, A_2)$, that is, if $\mathscr{S} \subset \text{Ln}(A_1 A_2)$, then, by definition, $\mathscr{S}$ has dimension 1. Any three points of $\mathscr{S}$ belong to $\text{AS}(A_1 A_2)$ and so are dependent. Thus 2 is the maximal number of independent points in $\mathscr{S}$.

If $\mathscr{S} \not\subset \text{AS}(A_1 A_2)$, then there exists a point A_3 in $\mathscr{S}$ that is not in $\text{AS}(A_1 A_2)$. Thus $\mathscr{S}$ has at least three independent points, A_1, A_2, A_3, and $\mathscr{S}_2 = \text{AS}(A_1, A_2, A_3)$ is the unique 2-flat that contains these points (Theorem 7, Sec. 2.4). Since no 1-flat, that is, no line, contains three independent points, no 1-flat contains $\mathscr{S}$. Hence if $\mathscr{S} \subset \mathscr{S}_2$, then $\mathscr{S}$ has dimension 2. In this case, any four points of $\mathscr{S}$ are contained in $\mathscr{S}_2$ and hence form a dependent set. Thus 3 is the maximal number of independent points in $\mathscr{S}$.

If $\mathscr{S} \not\subset \mathscr{S}_2$, then the argument repeats. That is, there must exist a point A_4 that is in $\mathscr{S}$ but not in $\text{AS}(A_1, A_2, A_3)$, and so $\mathscr{S}$ has at least four independent points A_1, A_2, A_3, A_4. Then $\mathscr{S}_3 = \text{AS}(A_1, A_2, A_3, A_4)$ is the unique 3-flat that contains these points. If $\mathscr{S} \subset \mathscr{S}_3$, then any five points of $\mathscr{S}$ are dependent. No flat of dimension less than 3 can contain four independent points, hence cannot contain $\mathscr{S}$, so $\mathscr{S}$ has dimension 3, and 4 is the maximal number of independent points in $\mathscr{S}$. If $\mathscr{S} \not\subset \mathscr{S}_3$, then there exists a point A_5 in $\mathscr{S}$ and not in $\mathscr{S}_3$, and so on. Because the affine span of $n + 1$ independent points is $\mathscr{E}^n$, which certainly contains $\mathscr{S}$, the stepwise argument must terminate in $k \leq n$ steps, giving k as the unique dimension of $\mathscr{S}$ and $k + 1$ as the exact maximal number of independent points in $\mathscr{S}$. □

COROLLARY 3. If a set $\mathscr{R}$ has dimension k and if $\mathscr{R} \subset \mathscr{S}$, then $\mathscr{S}$ has dimension equal to or greater than k. (E)

THEOREM 4. The neighborhoods and balls of $\mathscr{E}^n$ are n-dimensional sets.

PROOF. Corresponding to N(P, δ), B(P, δ), and the standard basis vectors $e_1, e_2, \ldots, e_n$, let n points $Q_1, Q_2, \ldots, Q_n$ be defined by the relations

$$q_i = p + \tfrac{1}{2}\delta e_i, \qquad i = 1, 2, \ldots, n.$$

From $d(P, Q_i) = |q_i - p| = \frac{1}{2}\delta|e_i| = \frac{1}{2}\delta$, it follows that the points Q_i all belong to N(P, δ). Because the vectors e_i are linearly independent, so are the vectors $\mathbf{PQ}_i = \frac{1}{2}\delta e_i$, $i = 1, 2, \ldots, n$. Thus the $n + 1$ points $P, Q_1, Q_2, \ldots, Q_n$ are affinely independent (Theorem 5, Sec. 2.4). Since both N(P, δ) and B(P, δ) contain $n + 1$ independent points, then both, by Theorem 3, have dimension n. □

COROLLARY 4.1. A set in $\mathscr{E}^n$ with a non-empty interior is an n-dimensional set. (E)

COROLLARY 4.2. A convex body in $\mathscr{E}^n$ is an n-dimensional set with an n-dimensional interior. (E)

As Corollary 4.2 shows, the way in which we have defined a convex body in euclidean space implies that the convex body has the dimension of the space. Thus a closed, bounded interval in $\mathscr{E}^1$ is a convex body but a closed segment is not a convex body in $\mathscr{E}^n$ if $n > 2$. However, suppose that $\mathscr{A}$ is an m-flat in $\mathscr{E}^n$, where $m < n$, and that $\mathscr{S}$ is a subset of $\mathscr{A}$. Then $\mathscr{S}$ is not a convex body. But $\mathscr{A}$ is a relative metric space with respect to $\mathscr{E}^n$ (cf. Sec. 1.3). In terms of the relative topology of $\mathscr{A}$, we can define $\mathscr{S}$ to be a *relative convex body* if it is a *nonsingleton, convex* set which is *relatively compact* and has a *non-empty relative interior.*

By Theorem 2, Sec. 2.4, if $\mathscr{A}$ is an m-flat in $\mathscr{E}^n$, $m < n$, then there is an isometric mapping f of $\mathscr{A}$ onto $\mathscr{E}^m$. Thus if $\mathscr{S}$ is a relative convex body in $\mathscr{A}$, then $f(\mathscr{S})$ is a convex body in $\mathscr{E}^m$. Conversely, if $\mathscr{K}$ is a convex body in $\mathscr{E}^m$, then $f^{-1}(\mathscr{K})$ is a relative convex body in $\mathscr{A}$. These facts allow information about convex bodies in $\mathscr{E}^m$ to be transferred automatically to the relative convex bodies in $\mathscr{A}$, and this is of great use in some parts of convex body theory.

If a set in $\mathscr{E}^n$ is both convex and n-dimensional, then its interior is automatically non-empty, and later we give a simple proof for this. The natural analog, which is also correct, is that a k-dimensional convex set has a non-empty, relative interior.

Not only does a convex body $\mathscr{K}$ in $\mathscr{E}^n$ have the dimension n of the space, but the same is true for its boundary surface $\mathscr{K}^0$. To establish this, we need a few more facts.

THEOREM 5. If A is an interior point of a convex set $\mathcal{S}$ and B is any other point of $\mathcal{S}$, then every point of the open segment Sg(AB) is an interior point of $\mathcal{S}$.

PROOF. Let C denote a point of Sg(AB). Because $\mathcal{S}$ is convex, Sg(AB) $\subset \mathcal{S}$, hence $C \in \mathcal{S}$, so C has a representation

$$c = b + \lambda(a - b), \qquad 0 < \lambda < 1. \tag{1}$$

Because $A \in \text{In}(\mathcal{S})$, there exists $\text{N}(A, \delta) \subset \mathcal{S}$. We show that $\text{N}(C, \lambda\delta) \subset \mathcal{S}$, hence that $C \in \text{In}(\mathcal{S})$.

Let P be an arbitrary point of $\text{N}(C, \lambda\delta)$, and let point Q be defined by

$$q = a + \frac{1}{\lambda}(p - c). \tag{2}$$

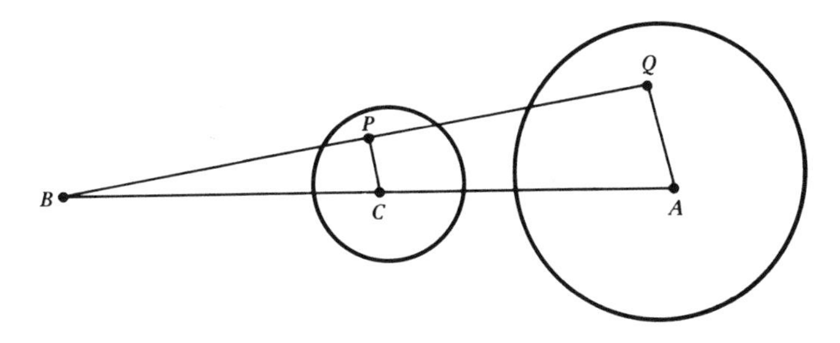

Then from

$$d(A, Q) = |q - a| = \frac{1}{\lambda}|p - c| = \frac{1}{\lambda}d(P, C) < \frac{1}{\lambda}(\lambda\delta) = \delta, \tag{3}$$

$Q \in \text{N}(A, \delta)$, so $Q \in \mathcal{S}$. Now, from (2),

$$p = c + \lambda(q - a). \tag{4}$$

Combining (1) and (4), we obtain

$$p = b + \lambda(a - b) + \lambda(q - a) = (1 - \lambda)b + \lambda q. \tag{5}$$

Since $0 < \lambda < 1$, (5) implies that $P \in \text{Sg}(BQ)$. Because B and Q belong to $\mathcal{S}$, and $\mathcal{S}$ is convex, then $\text{Sg}(BQ) \subset \mathcal{S}$, hence $P \in \mathcal{S}$. Thus $P \in \text{N}(C, \lambda\delta)$ implies $P \in \mathcal{S}$. Therefore $\text{N}(C, \lambda\delta) \subset \mathcal{S}$, hence $C \in \text{In}(\mathcal{S})$. □

COROLLARY 5. The interior of a convex set is also a convex set. (E)

We show next that a closed, convex surface $\mathscr{K}^0$ does enclose the interior of $\mathscr{K}$ in the sense that every ray from an interior point intersects $\mathscr{K}^0$. The proof uses the following facts, left as exercises.

THEOREM 7. The intersection of any number of convex sets is a convex set. (E)

THEOREM 8. A compact, convex, linear set is empty, is a single point, or else is a closed segment. (E)

THEOREM 9. If P is interior to an n-dimensional convex body $\mathscr{K}$ in $\mathscr{E}^n$, then each ray from P intersects the surface $\mathscr{K}^0$ at exactly one point.

PROOF. Let $\text{Ry}[P, X)$ be an arbitrary ray from P, and let $\mathscr{S} = \text{Ry}[P, X) \cap \mathscr{K}$. Because $\text{Ry}[P, X)$ and $\mathscr{K}$ are convex and closed, then $\mathscr{S}$ is convex and closed. Since $\mathscr{S}$ is a subset of $\mathscr{K}$, $\mathscr{S}$ is bounded, and since $\mathscr{S}$ is a subset of

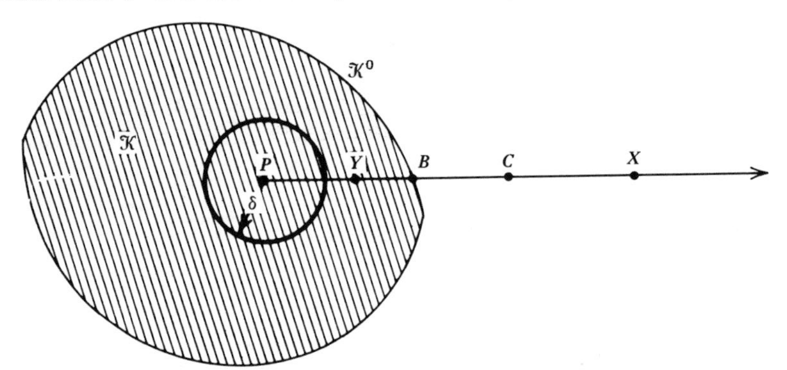

$\text{Ry}[P, X)$, $\mathscr{S}$ is linear. Thus $\mathscr{S}$ is a compact, convex, linear set. Because $P \in \mathscr{S}$, $\mathscr{S}$ is not empty, so, from Theorem 8, $\mathscr{S}$ is the single point P or else is a closed segment. But $P \in \text{In}(\mathscr{K})$ implies the existence of $\text{N}(P, \delta) \subset \mathscr{K}$, so the points of $\text{Ry}(P, X)$ at distance less than δ from P also belong to $\mathscr{S}$. Thus $\mathscr{S}$ is some closed segment $\text{Sg}[PB]$. The open ray $\text{Ry}(B, C)$ opposite to $\text{Ry}(B, P)$ is contained in the complement of $\mathscr{K}$, and B is a limit point to this ray. Thus B is a point of $\mathscr{K}$ that is a limit point to $\text{Cp}(\mathscr{K})$, hence $B \in \text{Bd}(\mathscr{K}) = \mathscr{K}^0$. By Theorem 5, every point Y between P and B is interior to $\mathscr{K}$ and so is not a point of $\mathscr{K}^0$. Because $\mathscr{K}$ contains its boundary points, and $\text{Ry}(B, C) \subset \text{Cp}(\mathscr{K})$, no point of $\text{Ry}(B, C)$ is a point of $\mathscr{K}^0$. Thus B is the unique intersection of $\text{Ry}[P, X)$ and $\mathscr{K}^0$. □

COROLLARY 9.1. Each line through P intersects $\mathscr{K}^0$ exactly twice, and P is between these intersections. (E)

COROLLARY 9.2. If $\mathscr{K}$ is an n-dimensional convex body in $\mathscr{E}^n$, then $\mathscr{K}^0$ is not a convex set. (E)

We now return to the surface dimension property mentioned earlier.

THEOREM 10. If $\mathscr{K}$ is an n-dimensional convex body in $\mathscr{E}^n$, then its surface $\mathscr{K}^0$ is also n-dimensional.

PROOF. By definition, $\text{In}(\mathscr{K}) \neq \varnothing$, so there exists $P \in \text{In}(\mathscr{K})$. Let $e_1, e_2, \ldots, e_n$ denote the standard basis for $\mathscr{E}^n$, and let $\mathscr{R}_o$ denote the ray from P in the direction $-e_1$, and let $\mathscr{R}_i$ denote the ray from P in the direction e_i, $i = 1, 2, \ldots, n$. By Theorem 9, each of the rays $\mathscr{R}_i$ intersects $\mathscr{K}^0$ at a point A_i, $i = 0, 1, 2, \ldots, n$. Thus there exist positive numbers $\lambda_o, \lambda_1, \ldots, \lambda_n$ such that

$$a_o = \lambda_o(-e_1), \qquad a_i = \lambda_i e_i, \qquad i = 1, 2, \ldots, n. \tag{1}$$

We want to show that the $n + 1$ points $A_o, A_1, \ldots, A_n$ are affinely independent. To this end, consider any zero-sum linear combination of the vectors a_i that represents the null vector. That is, consider

$$\sum_0^n \varphi_i a_i = o, \qquad \text{and} \qquad \sum_0^n \varphi_i = 0. \tag{2}$$

Using (1), the first relation in (2) becomes

$$\varphi_o(-\lambda_o e_1) + \varphi_1\lambda_1 e_1 + \sum_2^n \varphi_i\lambda_i e_i = o$$

or

$$(\varphi_1\lambda_1 - \varphi_o\lambda_o)e_1 + \sum_2^n \varphi_i\lambda_i e_i = o. \tag{3}$$

Since the vectors e_i are linearly independent, (3) implies that

$$\varphi_1\lambda_1 - \varphi_o\lambda_o = 0 \qquad \text{and} \qquad \varphi_i\lambda_i = 0, \qquad i = 2, 3, \ldots, n. \tag{4}$$

Because all the numbers λ_i are positive, (4) implies that $\varphi_i = 0$, $i = 2, 3, \ldots, n$. This, with $\sum_0^n \varphi_i = 0$, gives $\varphi_o + \varphi_1 = 0$. Thus the first equality in (4) becomes $\varphi_1(\lambda_1 + \lambda_o) = 0$, and since $\lambda_1 + \lambda_2 > 0$, then $\varphi_1 = 0$, so

$\varphi_o = -\varphi_1 = 0$. Thus all the coefficients φ_i must be zero. Since the null combination is the only zero sum linear combination of the vectors a_i that represents the null vector, the points $A_o, A_1, \ldots, A_n$ are affinely independent. Thus $\mathscr{K}^0$ contains $n + 1$ affinely independent points and hence is an n-dimensional set. □

For convex sets, certain set inclusions discussed in Chapter 1 become set equalities. These equalities are useful, and the following proofs illustrate the application of Theorem 5.

THEOREM 11. If $\mathscr{K}$ is a convex set with a non-empty interior, then $\text{Cl}(\mathscr{K}) = \text{Cl}[\text{In}(\mathscr{K})]$.

PROOF. Since $\text{In}(\mathscr{K}) \subset \mathscr{K}$, then $\text{Cl}[\text{In}(\mathscr{K})] \subset \text{Cl}(\mathscr{K})$. To obtain the opposite inclusion, consider $P \in \text{Cl}(\mathscr{K})$. If $P \in \text{In}(\mathscr{K})$, then clearly $P \in \text{Cl}[\text{In}(\mathscr{K})]$. If $P \notin \text{In}(\mathscr{K})$, then this fact, with $P \in \text{Cl}(\mathscr{K})$, implies that $P \in \text{Bd}(\mathscr{K})$. By hypothesis, there exists $Q \in \text{In}(\mathscr{K})$. Then, by Theorem 5, $\text{Sg}(PQ) \subset \text{In}(\mathscr{K})$. Since P is a limit point to $\text{Sg}(PQ)$, then $P \in \text{Lp}[\text{In}(\mathscr{K})]$. Thus $P \in \text{Cl}[\text{In}(\mathscr{K})]$. Therefore $\text{Cl}(\mathscr{K}) \subset \text{Cl}[\text{In}(\mathscr{K})]$, and this, with the opposite inclusion, implies that $\text{Cl}(\mathscr{K}) = \text{Cl}[\text{In}(\mathscr{K})]$. □

In Sec. 4.1, we show that if a convex set $\mathscr{K}$ in $\mathscr{E}^n$ has an empty interior, then it must have dimension less than n. Thus $\mathscr{K}$ must be contained in an m-flat, $m < n$. By Theorem 3, Sec. 2.4, the m-flat is closed, so $\text{Cl}(\mathscr{K})$ is contained in the flat. Thus $\text{Cl}(\mathscr{K})$ is at most m-dimensional and cannot contain a neighborhood, since the neighborhoods of $\mathscr{E}^n$, by Theorem 4, are n-dimensional sets. Thus $\text{In}[\text{Cl}(\mathscr{K})] = \varnothing$. If we anticipate this consequence of Corollary 8.2, Sec. 4.1, we can establish the following theorem in full generality.

THEOREM 12. If $\mathscr{K}$ is a convex set, then

$$\text{In}(\mathscr{K}) = \text{In}[\text{Cl}(\mathscr{K})].$$

PROOF. As indicated in the discussion preceding the theorem, if $\text{In}(\mathscr{K}) = \varnothing$, then $\text{In}[\text{Cl}(\mathscr{K})] = \varnothing$, so the sets are the same. We now consider the case where $\text{In}(\mathscr{K}) \neq \varnothing$.

From $\mathscr{K} \subset \text{Cl}(\mathscr{K})$, we have the set inclusion

$$\text{In}(\mathscr{K}) \subset \text{In}[\text{Cl}(\mathscr{K})]. \tag{1}$$

To obtain the opposite set inclusion, consider $P \in \text{In}[\text{Cl}(\mathscr{K})]$. We want to show that P must be in $\text{In}(\mathscr{K})$, and this will follow, by Theorem 5, if we can show that P lies between an interior point of $\mathscr{K}$ and any other point of $\mathscr{K}$.

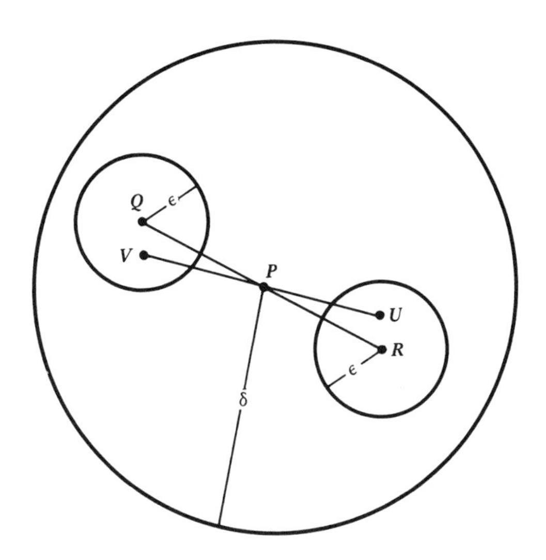

Because $P \in \text{In}[\text{Cl}(\mathscr{K})]$, there exists $\text{N}(P, \delta) \subset \text{In}[\text{Cl}(\mathscr{K})]$. By Theorem 11, $\text{Cl}(\mathscr{K}) = \text{Cl}[\text{In}(\mathscr{K})]$, so $P \in \text{Cl}[\text{In}(\mathscr{K})]$. Thus P must be a point or limit point of $\text{In}(\mathscr{K})$, and either case implies that $\text{N}(P, \delta) \cap \text{In}(\mathscr{K}) \neq \emptyset$. Thus there exists $Q \in \text{N}(P, \delta) \cap \text{In}(\mathscr{K})$. Because both $\text{N}(P, \delta)$ and $\text{In}(\mathscr{K})$ are open sets, their intersection is open, so there exists $\text{N}(Q, \varepsilon) \subset \text{N}(P, \delta) \cap \text{In}(\mathscr{K})$.

Now let R denote the point such that Q and R are symmetric with respect to P. That is, R is defined by

$$r = 2p - q. \tag{2}$$

Then, $d(P, R) = |r - p| = |p - q| = d(P, Q) < \delta$, so $R \in \text{N}(P, \delta)$. Therefore $R \in \text{In}[\text{Cl}(\mathscr{K})] \subset \text{Cl}(\mathscr{K})$. Therefore there exists a point $U \in \text{N}(R, \varepsilon) \cap \mathscr{K}$. Let V denote the point such that U, V are symmetric with respect to P, that is, define V by

$$v = 2p - u. \tag{3}$$

Then $d(Q, V) = |v - q| = |2p - u - q| = |r - u| = d(U, R) < \varepsilon$ shows that $V \in \text{N}(Q, \varepsilon)$, hence $V \in \text{In}(\mathscr{K})$. Now, from (3),

$$p = \tfrac{1}{2}u + \tfrac{1}{2}v. \tag{4}$$

Thus if $U \neq V$, then P lies between U and V. Since $V \in \text{In}(\mathscr{K})$ and $U \in \mathscr{K}$, then, by Theorem 5, $P \in \text{In}(\mathscr{K})$. If $U = V$, then (2) and (4) imply that P, Q, R, U, V are all the same point, and $P = Q$ implies $P \in \text{In}(\mathscr{K})$. Since, in all cases, $P \in \text{In}(\mathscr{K})$, then

$$\text{In}[\text{Cl}(\mathscr{K})] \subset \text{In}(\mathscr{K}). \tag{5}$$

From the opposite inclusions (1) and (5), $\text{In}[\text{Cl}(\mathscr{K})] = \text{In}(\mathscr{K})$. □

THEOREM 13. If $\mathscr{K}$ is a convex set, then

$$\text{Bd}(\mathscr{K}) = \text{Bd}[\text{Cl}(\mathscr{K})].$$

PROOF. The inclusion

$$\text{Bd}[\text{Cl}(\mathscr{K})] \subset \text{Bd}(\mathscr{K}) \tag{1}$$

holds whether or not $\mathscr{K}$ is convex (cf. Exercise 15, Sec. 1.3). Now consider $P \in \text{Bd}(\mathscr{K})$. Then P belongs to $\text{Cl}(\mathscr{K})$ but does not belong to $\text{In}(\mathscr{K})$. By Theorem 12, $\text{In}(\mathscr{K}) = \text{In}[\text{Cl}(\mathscr{K})]$. Therefore P belongs to $\text{Cl}(\mathscr{K})$ but not to $\text{In}[\text{Cl}(\mathscr{K})]$. Hence $P \in \text{Bd}[\text{Cl}(\mathscr{K})]$, so

$$\text{Bd}(\mathscr{K}) \subset \text{Bd}[\text{Cl}(\mathscr{K})]. \tag{2}$$

From (1) and (2), we have $\text{Bd}(\mathscr{K}) = \text{Bd}[\text{Cl}(\mathscr{K})]$. □

We conclude with one other property special to convex sets.

THEOREM 14. If $\mathscr{S}$ is a proper, convex subset of $\mathscr{E}^n$, then the closure of $\mathscr{S}$ is also a proper, convex subset of $\mathscr{E}^n$.

PROOF. That $\text{Cl}(\mathscr{S})$ is convex follows from Theorem 12, Sec. 2.6. Also, the theorem is trivially correct for $\mathscr{S} = \varnothing$, so we suppose that $\mathscr{S} \neq \varnothing$.

If $\text{In}(\mathscr{S}) = \varnothing$ and $n = 1$, then $\mathscr{S}$ is a point and the theorem is clearly valid. If $n > 1$, and $\text{In}(\mathscr{S}) = \varnothing$, then, by Corollary 4.1, Sec. 2.6, $\mathscr{S}$ is contained in a k-flat, $k < n$. Since this flat also contains $\text{Cl}(\mathscr{S})$, then $\text{Cl}(\mathscr{S})$ is a proper subset of $\mathscr{E}^n$.

Finally, if $\text{In}(\mathscr{S}) \neq \varnothing$, then there exists $\text{N}(P, \delta) \subset \text{In}(\mathscr{S})$. By hypothesis, there also exists $Q \in \text{Cp}(\mathscr{S})$. Let $x' = f(x) = -x + 2q$ be the reflection of space in Q. Then f maps $\text{N}(P, \delta)$ onto $\text{N}(P', \delta)$. For any Y in $\text{N}(P', \delta)$, $Y' = f(Y)$ is in $\text{N}(P, \delta)$, hence in $\mathscr{S}$, and Q is the midpoint of $\text{Sg}[YY']$. Since $\mathscr{S}$ is convex, $Y \in \mathscr{S}$ would imply $\text{Sg}[YY'] \subset \mathscr{S}$, hence $Q \in \mathscr{S}$. Therefore Y is not in $\mathscr{S}$. Since $\text{N}(P', \delta) \cap \mathscr{S} = \varnothing$, then P' is not in $\text{Cl}(\mathscr{S})$, so $\text{Cl}(\mathscr{S})$ is not $\mathscr{E}^n$ □

Exercises – Section 3.1

1. Prove Corollary 3.
2. Prove Corollary 4.1 and Corollary 4.2.
3. Prove Corollary 5.
4. Prove Theorem 7.
5. Prove Theorem 8.
6. Prove Corollary 9.1 and Corollary 9.2.

7. Prove that if A, B are two points in the boundary of a closed, convex set $\mathscr{S}$, then either $\text{Sg}(AB) \subset \text{Bd}(\mathscr{S})$ or else $\text{Sg}(AB) \subset \text{In}(\mathscr{S})$.
8. Give an example of a set $\mathscr{S}$ with a non-empty interior and such that $\text{Cl}(\mathscr{S}) \neq \text{Cl}[\text{In}(\mathscr{S})]$ (cf. Exercises, Sec. 1.3).
9. Give an example of a set $\mathscr{S}$ such that $\mathscr{S}$ is a proper subset of $\mathscr{E}^n$, but $\text{Cl}(\mathscr{S}) = \mathscr{E}^n$.

SECTION 3.2. SURFACE CONNECTIVITY, FOOT PROJECTIONS

That a convex body $\mathscr{K}$ is a connected set was established as a special case of the fact that all convex sets are connected. But this argument does not apply to the surface $\mathscr{K}^0$, which is not a convex set. To show that $\mathscr{K}^0$ is connected, we first establish the connectivity of spheres. We then show that every closed, convex surface $\mathscr{K}^0$ is the continuous image of a sphere and hence is connected.

We begin with a "coordinate-continuity" theorem.

THEOREM 1. If $f: \mathscr{D} \to \mathscr{R}$ is a mapping of $\mathscr{D} \subset \mathscr{E}^1$ into $\mathscr{R} \subset \mathscr{E}^n$, and if there exist n real valued functions $g_1, g_2, \ldots, g_n$ that are continuous on $\mathscr{D}$ and such that f has the representation

$$f(\lambda) = (g_1(\lambda), g_2(\lambda), \ldots, g_n(\lambda)), \qquad \lambda \in \mathscr{D},$$

then f is continuous on $\mathscr{D}$.

PROOF. Corresponding to $\lambda_o \in \mathscr{D}$, and an arbitrary number $\varepsilon > 0$, we must show the existence of a number $\delta > 0$ such that

$$\lambda \in \mathscr{D} \quad \text{and} \quad |\lambda - \lambda_o| < \delta \quad \text{imply} \quad |f(\lambda) - f(\lambda_o)| < \varepsilon. \quad (1)$$

Let $\varepsilon_o = \varepsilon/\sqrt{n}$. Corresponding to ε_o, the continuity of g_i at λ_o implies the existence of $\delta_i > 0$ such that

$$\lambda \in \mathscr{D} \quad \text{and} \quad |\lambda - \lambda_o| < \delta_i \Rightarrow |g_i(\lambda) - g_i(\lambda_o)| < \varepsilon_o, \quad (2)$$

for $i = 1, 2, \ldots, n$. Now, let $\delta = \min\{\delta_1, \delta_2, \ldots, \delta_n\}$. Then for $\lambda \in \mathscr{D}$ and $|\lambda - \lambda_o| < \delta$, all n inequalities in (2) are valid, so

$$\begin{aligned} |f(\lambda) - f(\lambda_o)| &= |(g_1(\lambda), g_2(\lambda), \ldots, g_n(\lambda)) - (g_1(\lambda_o), \ldots, g_n(\lambda_o))| \\ &= \sqrt{[g_1(\lambda) - g_1(\lambda_o)]^2 + \cdots + [g_n(\lambda) - g_n(\lambda_o)]^2} \\ &< \sqrt{\varepsilon_o^2 + \varepsilon_o^2 + \cdots + \varepsilon_o^2} = \sqrt{n\varepsilon_o^2} = \varepsilon. \end{aligned}$$

Thus the choice of δ does satisfy (1), hence f is continuous. □

THEOREM 2. For $k > 1$, the k-dimensional spheres in $\mathscr{E}^n$ are connected sets.

PROOF. First, consider the circles in $\mathscr{E}^2$ that are centered at the origin. If $C(O, r)$ is such a circle, then from ordinary analytic geometry, it has a rectangular coordinate equation $x_1^2 + x_2^2 = r^2$, and a parametric representation

$$x_1 = r \cos \lambda, \qquad x_2 = r \sin \lambda, \qquad 0 \leq \lambda < 2\pi.$$

The parametric representation is also a mapping

$$f(\lambda) = (g_1(\lambda), g_2(\lambda)) = (r \cos \lambda, r \sin \lambda)$$

whose domain is $[O, 2\pi)$ and whose range is the circle. Since the sine and cosine are continuous on $\mathscr{E}^1$, then, by Theorem 1, the function f is continuous. The half-open interval $[O, 2\pi)$ is connected, hence its image $C(O, r)$ is connected.

A general circle $C(A, r)$ in $\mathscr{E}^2$ is the a-translate of $C(O, r)$. Since translations are continuous mappings, and since $C(O, r)$ is connected, then $C(A, r)$ is connected.

Now, consider a circle $C(A, r)$ in $\mathscr{E}^n$. By definition, $C(A, r)$ is the set of all points in some plane $\mathscr{S}_2$ that are at distance r from the point A in $\mathscr{S}_2$. From Theorem 7, Sec. 2.3, there exists an isometric mapping f of $\mathscr{S}_2$ onto $\mathscr{E}^2$, and the image of $C(A, r)$ is a circle $C(f(A), r)$ in $\mathscr{E}^2$. Since isometries are one-to-one and continuous, and since the inverse of an isometry is an isometry, f^{-1} is a one-to-one continuous mapping of $\mathscr{E}^2$ onto $\mathscr{S}_2$. Because the circle $C(f(A), r)$ in $\mathscr{E}^2$ is connected, then its image $C(A, r)$ in $\mathscr{S}_2$ is connected. Thus the circles, namely, the 2-spheres, of $\mathscr{E}^n$ are connected.

Now, consider all the points in some k-flat $\mathscr{S}_k$, $k > 2$, that are at distance r from a point A in $\mathscr{S}_k$. Let $S_k(A, r)$ denote this k-sphere, and let P, Q denote any two points of $S_k(A, r)$. Then there is a plane $\mathscr{S}_2$ that contains A and P, Q. If A and P, Q are noncollinear, then $\mathscr{S}_2$ is their affine span. If A is collinear with P and Q, then $\mathscr{S}_2$ may be taken to be any plane through P, Q and any third point of the sphere. The intersection of $S_k(A, r)$ and the plane $\mathscr{S}_2$ is a circle $C(A, r)$, which is a connected set that contains P and Q. Thus each two points of the k-sphere belong to a connected subset of the sphere, hence the sphere is connected (Theorem 5, Sec. 1.4). □

We now want to show that a closed, convex surface is the continuous image of a sphere. The mapping we employ for this purpose, a so-called "central projection," belongs to a family of mappings called "orthogonal projections." Because these projections play an important role in geometry and are needed in later work, we introduce them here.

DEFINITION (Orthogonal projection into a hyperplane). Corresponding to a hyperplane $\mathscr{H}$ and a set $\mathscr{S}$ (which could be $\mathscr{E}^n$), for each $X \in \mathscr{S}$ there exists a unique line $\mathscr{L}_X$ that contains X and is perpendicular to $\mathscr{H}$ at a point

X'. The mapping f defined on $\mathscr{S}$ by $f(X) = X'$, $\{X'\} = \mathscr{L}_X \cap \mathscr{H}$, is the *orthogonal projection* of $\mathscr{S}$ into $\mathscr{H}$. The image set $\mathscr{S}' = f(\mathscr{S})$ is said to be "the projection of $\mathscr{S}$," and $\mathscr{S}$ is said to "project onto $\mathscr{S}'$."

DEFINITION (Orthogonal projection into a line). Corresponding to a line $\mathscr{L}$ and a set $\mathscr{S}$ (which could be $\mathscr{E}^n$), for each $X \in \mathscr{S}$ there is a unique hyperplane $\mathscr{H}_X$ that contains X and is perpendicular to $\mathscr{L}$ at a point X'. The mapping f defined by $f(X) = X'$, $\{X'\} = \mathscr{L} \cap \mathscr{H}_X$, is the *orthogonal projection* of $\mathscr{S}$ into $\mathscr{L}$.

DEFINITION (Central projection into a sphere). Corresponding to a sphere $\mathrm{S}(P, r)$ and a set $\mathscr{S}$ that does not contain P (and which could be $\mathrm{Cp}[\{P\}]$), for each $X \in \mathscr{S}$ there is a unique ray $\mathrm{Ry}[P, X)$ that intersects $\mathrm{S}(P, r)$ at a point X'. The mapping f, defined by

$$f(X) = X', \{X'\} = \mathrm{Ry}[P, X) \cap \mathrm{S}(P, r),$$

is the *central projection* of $\mathscr{S}$ into $\mathrm{S}(P, r)$ from the center P.

In all three of the mappings just defined, it may be observed that the image of a point X in the set $\mathscr{S}$ is its foot in the line, hyperplane, and sphere, respectively. This suggests a more general mapping which can be defined as follows.

DEFINITION (Foot projection of $\mathscr{S}$ into $\mathscr{R}$). Corresponding to a closed set $\mathscr{R}$ and to a set $\mathscr{S}$ (which might be $\mathscr{E}^n$), if each point X in $\mathscr{S}$ has a unique foot X' in $\mathscr{R}$, then the function f defined on $\mathscr{S}$ by $f(X) = X'$ is the *foot projection* of $\mathscr{S}$ into $\mathscr{R}$.

THEOREM 3. Foot projections are continuous mappings.

PROOF. Let f denote the foot projection of $\mathscr{S}$ into the closed set $\mathscr{R}$, where $Y = f(X)$ in $\mathscr{R}$ is the unique foot of X in $\mathscr{S}$. To show the continuity of f at $X_o \in \mathscr{S}$, let $\mathrm{N}(Y_o, \varepsilon) = \mathrm{N}(f(X_o), \varepsilon)$ be an arbitrary neighborhood of Y_o. Setting $r_o = d(X_o, Y_o)$, we consider the infinite sequence of sets $\{\mathscr{R}_n\}$ defined by

$$\mathscr{R}_n = \mathscr{R} \cap \mathrm{B}\left(X_o, r_o + \frac{1}{n}\right), \qquad n = 1, 2, \ldots. \tag{1}$$

Because the sequence of balls $\{\mathrm{B}(X_o, r_o + 1/n)\}$ is decreasingly nested, the sequence $\{\mathscr{R}_n\}$ is decreasingly nested. Since the balls are compact, and since $\mathscr{R}$ is closed, then $\mathscr{R}_n$ is compact (Theorem 11, Sec. 1.4). Because $Y_o \in \mathscr{R}$ and $d(X_o, Y_o) = r_o < r_o + 1/n$, the point Y_o belongs to the intersection set of the sequence $\{\mathscr{R}_n\}$. Moreover, it is the only point in the intersection set. To see this, let Z be any point of $\mathscr{R}$ other than Y_o. Because Y_o is the unique foot in $\mathscr{R}$

of X_o, it follows that $d(X_o, Z) > d(X_o, Y_o) = r_o$. Then, corresponding to a positive integer k such that $d(X_o, Z) - r_o > 1/k$, it follows that Z is outside $B(X_o, r_o + 1/k)$, and $Z \notin \mathscr{R}_k$ implies that Z is not in the intersection set of $\{\mathscr{R}_n\}$. Thus

$$\bigcap_1^\infty \mathscr{R}_i = \{Y_o\}. \tag{2}$$

From relation (2) and Theorem 13, Sec. 1.4, it follows that every neighborhood of Y_o contains all the sets $\mathscr{R}_n$ with at most a finite number of exceptions. Thus there exists a positive integer m_o such that

$$\mathscr{R}_n \subset N(Y_o, \varepsilon), \quad \text{for all } n \geq m_o. \tag{3}$$

We now choose δ to be a number such that

$$0 < \delta < \frac{1}{2m_o}, \tag{4}$$

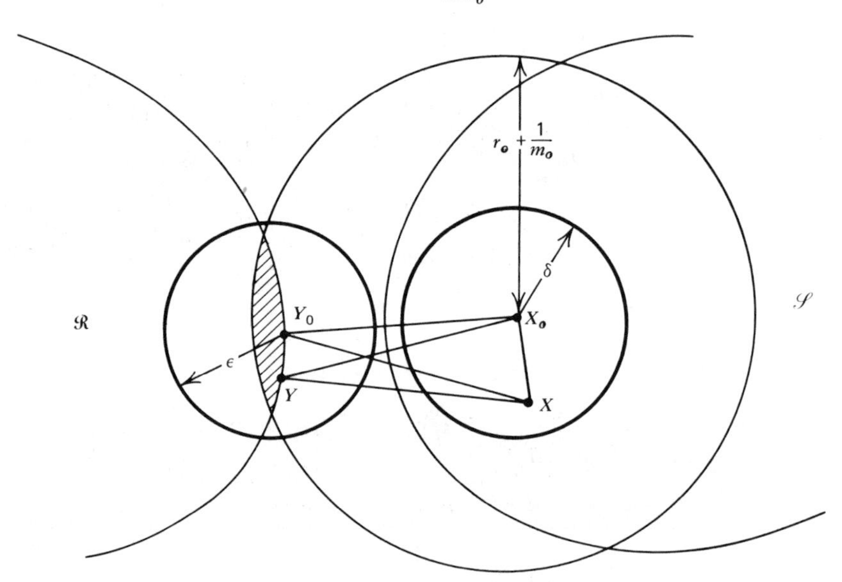

and consider $X \in \mathscr{S} \cap N(X_o, \delta)$. The point X has a unique foot Y in $\mathscr{R}$. From the triangle inequality,

$$d(X_o, Y) \leq d(X, Y) + d(X, X_o) < d(X, Y) + \delta. \tag{5}$$

Because Y in $\mathscr{R}$ is the foot of X, $d(X, Y) \leq d(X, Y_o)$, and this with (5) implies

$$d(X_o, Y) \leq d(X, Y_o) + \delta. \tag{6}$$

With the triangle inequality and (4), we obtain from (6)

$$d(X_o, Y) \leq d(X_o, Y_o) + d(X_o, X) + \delta$$

$$< r_o + 2\delta < r_o + \frac{1}{m_o}. \tag{7}$$

Since (7) shows that $Y \in \mathscr{R}_{m_o}$ and since, from (3), $\mathscr{R}_{m_o} \subset \mathrm{N}(Y_o, \varepsilon)$, it follows that $Y \in \mathrm{N}(Y_o, \varepsilon)$. Thus $X \in \mathrm{N}(X_o, \delta) \cap \mathscr{S}$ implies $f(X) \in \mathrm{N}[f(X_o), \varepsilon]$, hence f is continuous at X_o. □

COROLLARY 3. The orthogonal projection of $\mathscr{E}^n$ into a line or a hyperplane is a continuous mapping, as is the central projection from P of $\mathrm{Cp}[\{P\}]$ into the sphere $\mathrm{S}(P, r)$.

We can now complete the plan we outlined at the beginning of the section.

THEOREM 4. A closed, convex surface is the continuous image of a sphere and hence is a connected set.

PROOF. Let $\mathscr{K}^0$ denote a closed, convex surface that is the boundary of an n-dimensional convex body in $\mathscr{E}^n$. Corresponding to any point $P \in \mathrm{In}(\mathscr{K})$ and any number $r > 0$, let f denote the central projection from P of $\mathscr{K}^0$ into the sphere $\mathrm{S}(P, r)$. Since P is not in $\mathscr{K}^0$, the mapping is well defined and, by

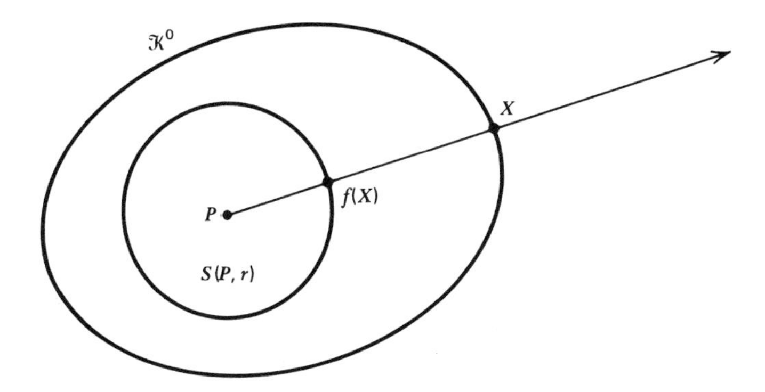

Corollary 3, is continuous. Each $X \in \mathscr{K}^0$ determines a unique ray $\mathrm{Ry}[P, X)$ because each ray from P intersects $\mathscr{K}^0$ exactly once, and $\mathrm{Ry}[P, X)$ intersects $\mathrm{S}(P, r)$ at $f(X)$. The same ray is uniquely determined by $f(X)$ on $\mathrm{S}(P, r)$, and, since each ray from P intersects both $\mathscr{K}^0$ and $\mathrm{S}(P, r)$ exactly once, f is a one-to-one mapping of $\mathscr{K}^0$ onto $\mathrm{S}(P, r)$. But, since $\mathscr{K}^0$ is closed and bounded, hence compact, it follows from Theorem 8, Sec. 1.5, that f^{-1} is also continuous. Thus f^{-1} is a continuous mapping of $\mathrm{S}(P, r)$ onto $\mathscr{K}^0$. Because $\mathrm{S}(P, r)$ is connected, then $\mathscr{K}^0$ is connected. □

The converse of Theorem 4 is obviously false, that is, a one-to-one continuous image of a sphere is not necessarily a closed, convex surface. Such an object is a *simple, closed surface*, and in $\mathscr{E}^2$ is a *simple, closed curve*. A famous problem about such curves was settled by the Jordan curve theorem, and we consider it briefly, since it relates to closed, convex curves and since the nature of the problem is sometimes misunderstood.

In intuitive and physical terms, a simple, closed curve $\mathscr{C}^0$ is obtained by placing a pen at point A on a piece of paper and then moving it in constant contact with the paper to trace a curve that does not occupy any position twice and finally returns to point A. The curve is "simple" in the sense that it does not cross itself or form loops, and it is "closed" in the sense of returning to the starting point A. The Jordan theorem is sometimes described as the property that a continuous curve $\mathscr{C}'$ that contains a point I interior to $\mathscr{C}^0$ and a point E outside $\mathscr{C}^0$ must intersect $\mathscr{C}^0$.

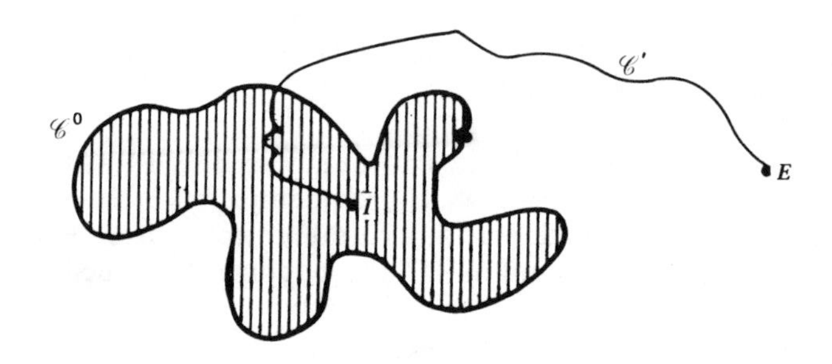

The Jordan property, as just described, seems to make the intersection of $\mathscr{C}^0$ and $\mathscr{C}'$ the heart of the matter. To see that this is not the case, consider any two open, non-empty, disjoint sets $\mathscr{S}_1$ and $\mathscr{S}_2$. A point P of $\mathscr{S}_1$ is automatically interior to $\mathscr{S}_1$, hence there exists $\mathrm{N}(P, \delta) \subset \mathscr{S}_1$. Since $\mathscr{S}_1 \cap \mathscr{S}_2 = \varnothing$, then $\mathrm{N}(P, \delta) \cap \mathscr{S}_2 = \varnothing$, so P is not a boundary point to $\mathscr{S}_2$. Because P is not in $\mathscr{S}_2$ or $\mathrm{Bd}(\mathscr{S}_2)$, then P is not in $\mathrm{Cl}(\mathscr{S}_2)$. That is, $\mathscr{S}_1 \cap \mathrm{Cl}(\mathscr{S}_2) = \varnothing$. By a symmetric argument, $\mathscr{S}_2 \cap \mathrm{Cl}(\mathscr{S}_1) = \varnothing$, so we have the following fact.

THEOREM 5. Any two non-empty, disjoint, open sets are separated.

If $\mathscr{S}_1$ and $\mathscr{S}_2$ are disjoint, open sets and if a connected set $\mathscr{C}'$ intersects both $\mathscr{S}_1$ and $\mathscr{S}_2$, then $\mathscr{C}'$ is not contained in either $\mathscr{S}_1$ or $\mathscr{S}_2$. Then, from Theorem 5, and from Theorem 4, Sec. 1.4, $\mathscr{C}'$ is not contained in the union of $\mathscr{S}_1$ and $\mathscr{S}_2$. Thus we have the following consequence of Theorem 5.

COROLLARY 5.1. A connected set that intersects each of two disjoint, open sets also intersects the complement of their union.

Now, if $\mathscr{S}$ is any set whatever, $\text{In}(\mathscr{S})$ and $\text{Ex}(\mathscr{S})$ are disjoint open sets and the complement of their union is $\text{Bd}(\mathscr{S})$. Thus Corollary 5.1 implies the following general fact.

COROLLARY 5.2. A connected set that intersects both the interior and exterior of a set $\mathscr{S}$ also intersects the boundary of $\mathscr{S}$.

As Corollary 5.2 indicates, the central difficulty in the Jordan curve problem is not the intersection question, but the existence of a region interior to $\mathscr{C}^0$, that is, the existence of a non-empty, bounded, open set having $\mathscr{C}^0$ as its boundary. With the existence of such a set $\mathscr{S}_1$ established, then $\mathscr{S}_2 = \text{Cp}[\text{Cl}(\mathscr{S}_1)]$ is a non-empty, open set that is disjoint with $\mathscr{S}_1$, and $\mathscr{C}^0$ is the complement of $\mathscr{S}_1 \cup \mathscr{S}_2$. Then any continuous curve that intersects $\mathscr{S}_1$ and $\mathscr{S}_2$ is a connected set that intersects $\mathscr{S}_1$ and $\mathscr{S}_2$ and so, by Corollary 5.2, must intersect $\mathscr{C}^0$.

For reference, we list the following special case of Corollary 5.2.

COROLLARY 5.3. A connected set that intersects the interior and exterior of a closed convex body $\mathscr{K}$ of dimension n in $\mathscr{E}^n$ also intersects the surface $\mathscr{K}^0$.

For our further work with convex bodies in the remaining chapters, it will be convenient to make the following agreement.

CONVENTION. The term "convex body," without any qualification, is understood to refer to an n-dimensional convex body, where $n \geq 2$, that is, in a space $\mathscr{E}^n$ of the same dimension. The same agreement is used for the unqualified term "convex surface."

Exercises – Section 3.2

1. Modify the parametric form for the circle in the proof of Theorem 2 to show that the ellipse $x_1^2/a_1^2 + x_2^2/a_2^2 = 1$ is connected.
2. Use a mapping argument to show the connectivity of the parabola $x_2^2 = ax_1, a > 0$.
3. The interior and exterior of the ball $\text{B}(O, r)$ are the respective graphs of $x \cdot x < r^2$ and $x \cdot x > r^2$. Give an argument based on these representations to show that a connected set which intersects the interior and exterior of $\text{B}(P, r)$ must intersect $\text{S}(P, r)$.
4. Explain why the foot projection of any set onto itself exists and is continuous.
5. If space is the union of two disjoint open sets, prove that one of these is empty.
6. Prove that if $Q \neq P$, then Q has a unique foot in the sphere $\text{S}(P, r)$ and this foot F is the intersection of $\text{Ry}(P, Q)$ and $\text{S}(P, r)$.

4
THE GENERAL GEOMETRY OF CONVEX BODIES

INTRODUCTION

There is a pattern behind much of the geometry of convex bodies that can be indicated by an example. In elementary geometry, a diameter segment of a sphere is a chord that contains the center. In this formulation, diameter segments are special to spheres and balls. One can observe, however, that among all the segments determined by pairs of points in the sphere, a diameter segment is one of maximal length. If the "diameter segment of a set" is defined by this property, then all compact sets that are not empty or singleton sets have diameter segments. Moreover, when such a set is a sphere, then the diameter segments given by the general definition coincide with the former ones defined in a more special way.

The diameter-segment example is just one case of a theme that runs through convex body theory and that consists of finding analogs to elementary concepts that can be applied to general convex bodies and surfaces and sometimes to completely arbitrary sets. In this chapter we take up a number of such analogs that are a standard part of convex body theory.

SECTION 4.1. SIMPLICES; BARYCENTRIC COORDINATES

A k-dimensional set in $\mathscr{E}^n$ must contain $k + 1$ independent points. If the set is also convex, one would suppose that it must also contain the convex span of the independent points. If so, then such spans must be basic subsets of all convex sets. We show that this is so and in the process find a natural generalization of the polygons and polyhedra of elementary geometry.

We proved in Theorem 9, Sec. 2.6, that the affine span of a finite set of points is convex. A nearly identical proof gives the following property.

THEOREM 1. The convex span of a finite set of points is a convex set. (E)

The next theorem gives a basic property of convex spans.

THEOREM 2. The convex span of m points, $A_1, A_2, \ldots, A_m$, $m \geq 2$, is the union of all the points in the segments joining A_m to the convex span of the remaining points $A_1, A_2, \ldots, A_{m-1}$.

PROOF. Let $\mathscr{R}$ denote the union of all the points belonging to segments $\text{Sg}[A_m X]$, where $X \in \text{CS}(A_1, A_2, \ldots, A_{m-1})$. Consider any point P in $\text{CS}(A_1, A_2, \ldots, A_m)$. By definition, P has a representation

$$p = \sum_1^m \lambda_i a_i, \qquad \sum_1^m \lambda_i = 1, \qquad \lambda_i \geq 0, \qquad i = 1, 2, \ldots, m. \tag{1}$$

If $\lambda_m = 1$, then $P = A_m$ and $P \in \mathscr{R}$, since obviously $A_m \in \mathscr{R}$. If $\lambda_m \neq 1$, then $1 - \lambda_m > 0$, so (1) can be put in the form

$$p = (1 - \lambda_m)\left(\sum_1^{m-1} \frac{\lambda_i a_i}{1 - \lambda_m}\right) + \lambda_m a_m, \tag{2}$$

that is,

$$p = (1 - \lambda_m)q + \lambda_m a_m, \tag{3}$$

where

$$q = \sum_1^{m-1} \frac{\lambda_i a_i}{1 - \lambda_m}. \tag{4}$$

Because, from (1), $1 - \lambda_m = \sum_1^{m-1} \lambda_i$, the sum of the $m - 1$ coefficients in (4) is 1. Moreover, since all the λ_i are nonnegative, and $1 - \lambda_m > 0$, the coefficients in (4) are nonnegative. Thus $Q \in \text{CS}(A_1, A_2, \ldots, A_{m-1})$. From (3), $P \in \text{Sg}[A_m Q]$, hence $P \in \mathscr{R}$. Therefore $\text{CS}(A_1, A_2, \ldots, A_m) \subset \mathscr{R}$.

For the converse, suppose that P is an arbitrary point of $\mathscr{R}$. Then P belongs to some segment $\text{Sg}[A_m Q]$, where Q is a point of $\text{CS}(A_1, A_2, \ldots, A_{m-1})$. Thus P has a representation

$$p = \lambda q + (1 - \lambda)a_m, \qquad 0 \leq \lambda \leq 1, \tag{5}$$

and

$$q = \sum_{1}^{m-1} \eta_i a_i, \qquad \sum_{1}^{m-1} \eta_i = 1, \qquad \eta_i \geq 0, \qquad i = 1, 2, \ldots, n-1. \quad (6)$$

Substituting from (6) in (5), we obtain

$$p = \left(\sum_{1}^{m-1} \lambda \eta_i a_i\right) + (1 - \lambda) a_m. \quad (7)$$

Because λ, $1 - \lambda$, and all the numbers η_i are nonnegative, all the coefficients in (7) are nonnegative. Moreover, their sum is 1 because

$$\left(\sum_{1}^{m-1} \lambda \eta_i\right) + (1 - \lambda) = \lambda\left(\sum_{1}^{m-1} \eta_i\right) + 1 - \lambda = \lambda + 1 - \lambda = 1.$$

The right side of (7) is therefore a convex combination of the vectors a_i, so $P \in \mathrm{CS}(A_1, A_2, \ldots, A_m)$. Consequently $\mathscr{R} \subset \mathrm{CS}(A_1, A_2, \ldots, A_m)$. This, with our previous result, implies that $\mathscr{R} = \mathrm{CS}(A_1, A_2, \ldots, A_m)$. □

The application of Theorem 2 in the next proof indicates how it can be used to construct the convex span of a finite set.

THEOREM 3. A convex set contains the convex span of each of its finite subsets.

PROOF. Let $\{A_1, A_2, \ldots, A_m\}$ denote a finite subset of a convex set $\mathscr{S}$. The convex span of A_1 is the singleton set $\{A_1\}$, which is contained in $\mathscr{S}$ by hypothesis. The convex span of A_1, A_2 is $\mathrm{Sg}[A_1 A_2]$, which is contained in $\mathscr{S}$ because $\mathscr{S}$ is convex. Then $\mathrm{CS}(A_1, A_2, A_3)$, by Theorem 2, is the union of all the points in segments $\mathrm{Sg}[A_3 A]$, $X \in \mathrm{CS}(A_1, A_2)$. Since $A_3 \in \mathscr{S}$, and each such $X \in \mathscr{S}$, the convexity of $\mathscr{S}$ implies that $\mathrm{Sg}[A_3 X] \subset \mathscr{S}$, and therefore that $\mathrm{CS}(A_1, A_2, A_3) \subset \mathscr{S}$. Continuing in this stepwise way, it follows that each of the successive spans $\mathrm{CS}(A_1)$, $\mathrm{CS}(A_1, A_2)$, $\mathrm{CS}(A_1, A_2, A_3), \ldots$, $\mathrm{CS}(A_1, A_2, \ldots, A_m)$ is contained in $\mathscr{S}$. □

COROLLARY 3. If point P belongs to $\mathrm{CS}(A_1, A_2, \ldots, A_m)$, then P is one of the points A_i or else P lies between one of the points A_i and some other point in the span. (E)

The properties established in Theorems 1, 2, and 3 are valid whether the span is the convex span of a dependent or independent set of points. However, when the points are independent, then the convex span is a basic geometric figure that has a special name.

DEFINITION (Simplex). The convex span of $k + 1$ independent points $A_1, A_2, \ldots, A_{k+1}$ is a *k-dimensional simplex*. The points A_i are the *vertices* of the simplex. The convex span of k of the points is a *face* of the simplex and is said to be the face *opposite to* the remaining vertex. The convex span of each two of the vertices is an *edge* of the simplex.

The stepwise process in the proof of Theorem 2 gives a way of constructing simplices. Any two distinct points A_1, A_2 are independent, and their convex span is the segment $\mathrm{Sg}[A_1A_2]$, which is a 1-simplex. Since all points dependent on A_1 and A_2 belong to the $\mathrm{AS}(A_1, A_2)$, we can obtain an independent triple A_1, A_2, A_3 by taking A_3 to be any nonaffine combination of A_1 and

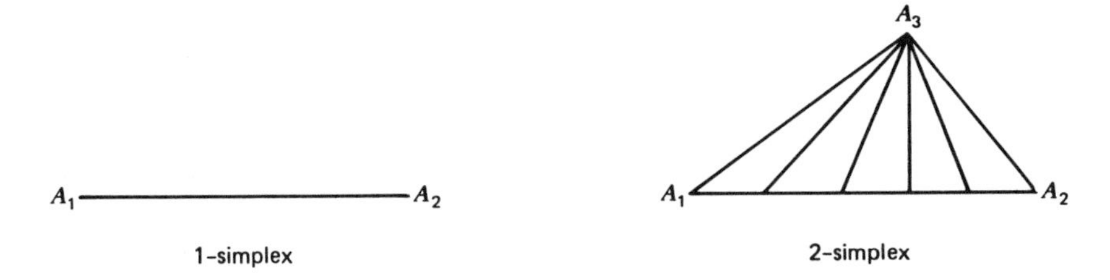

1-simplex 2-simplex

A_2, for example, $a_3 = a_1 + a_2$. Then $\mathrm{CS}(A_1, A_2, A_3)$ is a 2-simplex and is the union of the segments joining A_3 to points of $\mathrm{Sg}[A_1A_2]$. The edges of the 2-simplex are also its faces and are the sides of the triangle $\triangle A_1A_2A_3$. Since the 2-simplex also contains the points inside the triangle, it can be thought of as a *full triangle*, which we denote by $\triangle^+A_1A_2A_3$. If A_4 is not an affine combination of A_1, A_2, A_3, then $\mathrm{CS}(A_1, A_2, A_3, A_4)$ is a 3-simplex and is the union of segments joining A_4 to points of the full triangle $\triangle^+A_1A_2A_3$. It is thus a solid tetrahedron. The next stage, of course, produces a four-dimensional analog of these familiar figures. The faces of the 4-simplex are solid tetrahedra.

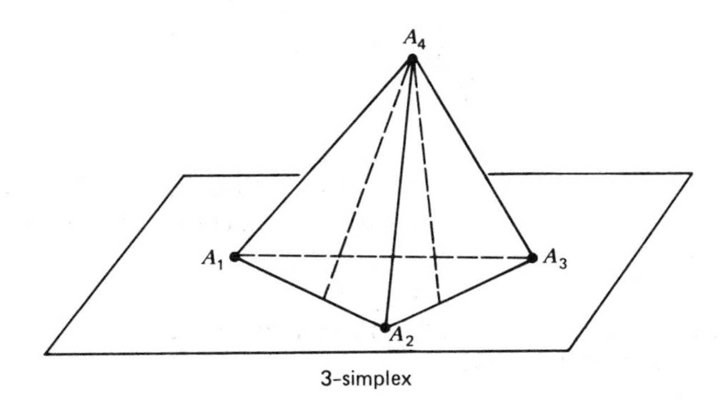

3-simplex

Because a k-dimensional set must contain $k + 1$ independent points, and since a convex set contains the convex spans of its finite subsets, the following result is immediate.

THEOREM 4. Every k-dimensional convex set contains a k-simplex.

Because the neighborhoods of $\mathscr{E}^n$ are n-dimensional sets, a k-simplex is not a convex body when $k < n$. We want to show, however, that an n-simplex is a convex body. We know that it is convex, so we must show that it is compact and has a non-empty interior. We obtain the compactness by showing that it is closed and bounded, and the boundedness is a special case of the following fact.

THEOREM 5. The convex span of a finite set of points is a bounded set. (E)

Now consider a 2-simplex in $\mathscr{E}^2$, namely, a full triangle $\triangle^+ A_1 A_2 A_3$. Each of the lines $\text{Ln}(A_i A_j)$ is the edge of a closed half-plane that contains the third vertex A_k, and the intersection of these three half-planes is the closed, triangular region. The interior of the simplex is the intersection of the corresponding open half-planes. These are the facts we want to establish, not just

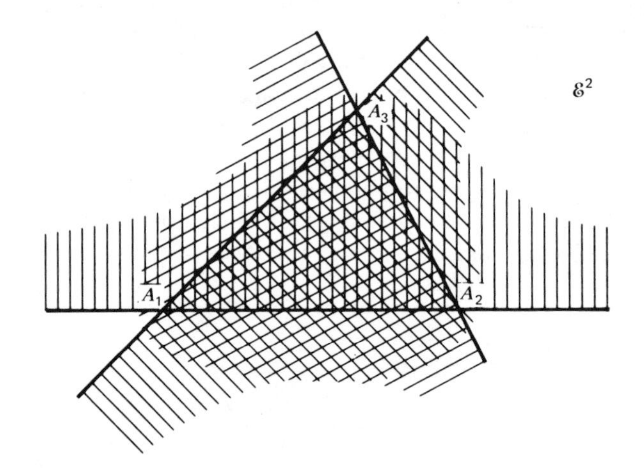

for a 2-simplex in $\mathscr{E}^2$, but the analogs for an n-simplex in $\mathscr{E}^n$. In doing so, we make use of the obvious fact that the interior of a closed half-space is its open half-space. We also need the following fact.

THEOREM 6. The interior of the intersection of a finite number of sets is the intersection of their interiors. (E)

Finally, before we turn to our main theorem, there is one new idea that is useful. A closed segment Sg$[AB]$ is a 1-simplex and consists of the points $\eta_1 a + \eta_2 b$, $\eta_1 + \eta_2 = 1$, $0 \leq \eta_1, \eta_2 \leq 1$. In $\mathscr{E}^1$, this 1-simplex is a linear convex body and its interior is just the open interval (a, b). But (a, b) consists of those elements in CS(a, b) for which $0 < \eta_1, \eta_2 < 1$. Thus the interior of the 1-simplex $[a, b]$ consists of the *positive* convex combinations of a, b. Since we will show that this property generalizes, the following definition is useful.

DEFINITION (Positive convex combinations and spans). A *positive convex combination* of a finite set of points $A_1, A_2, \ldots, A_m$ is a convex combination of the points in which all the coefficients are positive. The set of all positive convex combinations of the points is their *positive, convex span*, denoted by $\mathrm{CS}^+(A_1, A_2, \ldots, A_m)$.

The basic structure of an n-simplex can now be described in the following way.

THEOREM 7. If $A_1, A_2, \ldots, A_{n+1}$ are the vertices of an n-simplex in $\mathscr{E}^n$, and if $\mathscr{R}_i$ and $\mathscr{R}_i^*$ are the closed and open half-spaces that contain A_i and whose faces contain the simplex face opposite to A_i, $i = 1, 2, \ldots, n + 1$, then the following relations hold.

(i) The simplex is the intersection of the $n + 1$ closed, half-spaces $\mathscr{R}_i$.
(ii) The interior of the simplex is the intersection of the $n + 1$ open half-spaces $\mathscr{R}_i^*$.
(iii) The interior of the simplex is the positive convex span of the vertices.

PROOF. The simplex face opposite to A_i is the convex span of the other n vertices, and the affine span of these other n vertices is a hyperplane $\mathscr{H}_i$, $i = 1, 2, \ldots, n + 1$. If b_i is a unit vector normal to $\mathscr{H}_i$, then $\mathscr{H}_i$ is the graph of an equation of the form $b_i \cdot x + h_i = 0$. Because A_i is not in $\mathscr{H}_i$, $b_i \cdot a_i + h_i$ is either positive or negative. If it is negative, then A_i is in the negative side of $\mathscr{H}_i$. But in this case, we can, if we choose, take $(-b_i) \cdot x + (-h_i) = 0$ as a representation of $\mathscr{H}_i$, and, in terms of this representation, A_i is in the positive side of $\mathscr{H}_i$. Thus, with appropriate choices for b_i and h_i, there exist $n + 1$ affine functionals f_i with the following properties.

$$\begin{aligned}
&\text{(a)} \quad f_i(x) = b_i \cdot x + h_i, \ |b_i| = 1 \\
&\text{(b)} \quad \mathscr{H}_i \text{ is the graph of } f_i(x) = 0 \\
&\text{(c)} \quad \mathscr{R}_i \text{ is the graph of } f_i(x) \geq 0 \\
&\text{(d)} \quad \mathscr{R}_i^* \text{ is the graph of } f_i(x) > 0, \quad i = 1, 2, \ldots, n + 1
\end{aligned} \tag{1}$$

Because $\mathscr{H}_i$ contains every vertex except A_i, it follows that

$$f_i(a_j) = 0, \qquad i \neq j, \qquad i, j = 1, 2, \ldots, n + 1. \tag{2}$$

From the formula for the distance from a point to a hyperplane, Theorem 5, Sec. 2.6, we also have

$$d(A_i, \mathscr{H}_i) = \frac{|b_i \cdot a_i + h_i|}{|b_i|} = f_i(a_i) > 0, \qquad i = 1, 2, \ldots, n+1. \tag{3}$$

Now consider any point P in $\mathscr{E}^n$. Because the $n+1$ vertices A_i are independent, they form an affine basis for $\mathscr{E}^n$. Hence P has a unique representation as

$$p = \sum_1^{n+1} \lambda_j a_j, \qquad \sum_1^{n+1} \lambda_j = 1. \tag{4}$$

Because the functions f_i are affine,

$$f_i(p) = f_i\left(\sum_{j=1}^{n+1} \lambda_j a_j\right) = \sum_{j=1}^{n+1} \lambda_j f_i(a_j). \tag{5}$$

Then, from (2) and (5) it follows that

$$f_i(p) = \lambda_i f_i(a_i), \qquad i = 1, 2, \ldots, n+1. \tag{6}$$

If the point P in $\mathscr{E}^n$ is in $\mathrm{CS}(A_1, A_2, \ldots, A_{n+1})$, then the λ coefficients in (4) are all nonnegative. From (3), $f_i(a_i)$ is positive. Thus, from (6), we have

$$f_i(p) \geq 0, \qquad i = 1, 2, \ldots, n+1. \tag{7}$$

This relation, with (c) of (1), implies that P belongs to all the closed half-spaces $\mathscr{R}_i$. Thus

$$\mathrm{CS}(A_1, A_2, \ldots, A_{n+1}) \subset \bigcap_{i=1}^{n+1} \mathscr{R}_i. \tag{8}$$

If $P \in \mathrm{CS}^+(A_1, A_2, \ldots, A_{n+1})$, then the λ coefficients in (4) are all positive. This, with (3) and (6), implies that all the function values $f_i(p)$ are positive, hence that P belongs to all the open half-spaces $\mathscr{R}_i^*$. Hence

$$\mathrm{CS}^+(A_1, A_2, \ldots, A_{n+1}) \subset \bigcap_{i=1}^{n+1} \mathscr{R}_i^*. \tag{9}$$

For the converse, suppose that $P \in \bigcap_1^{n+1} \mathscr{R}_i$. Then

$$f_i(p) \geq 0, \qquad i = 1, 2, \ldots, n+1. \tag{10}$$

From (10) and (6), it follows that $\lambda_i f_i(a_i) \geq 0$, and, since $f_i(a_i) > 0$, we must have

$$\lambda_i \geq 0, \qquad i = 1, 2, \ldots, n + 1, \tag{11}$$

and therefore $P \in \mathrm{CS}(A_1, A_2, \ldots, A_{n+1})$. Thus

$$\bigcap_1^{n+1} \mathscr{R}_i \subset \mathrm{CS}(A_1, A_2, \ldots, A_{n+1}). \tag{12}$$

Similarly, $P \in \bigcap_1^{n+1} \mathscr{R}_i^*$ implies that $f_i(p) > 0$, for each i, hence that $\lambda_i f_i(a_i) > 0$, for each i, and hence that $\lambda_i > 0$, for each i, so

$$P \in \mathrm{CS}^+(A_1, A_2, \ldots, A_{n+1}).$$

Therefore,

$$\bigcap_1^{n+1} \mathscr{R}_i^* \subset \mathrm{CS}^+(A_1, A_2, \ldots, A_{n+1}). \tag{13}$$

From (8) and (12), we have

$$\mathrm{CS}(A_1, A_2, \ldots, A_{n+1}) = \bigcap_1^{n+1} \mathscr{R}_i, \tag{14}$$

which is part (i) of the theorem. From (9) and (13), we also have

$$\mathrm{CS}^+(A_1, A_2, \ldots, A_{n+1}) = \bigcap_1^{n+1} \mathscr{R}_i^*. \tag{15}$$

Now, using the fact that $\mathrm{In}(\mathscr{R}_i) = \mathscr{R}_i^*$ and that the interior of a finite intersection is the intersection of the interiors, we have

$$\mathrm{In}[\mathrm{CS}(A_1, A_2, \ldots, A_{n+1})] = \mathrm{In}\left(\bigcap_1^{n+1} \mathscr{R}_i\right) = \bigcap_1^{n+1} \mathrm{In}(\mathscr{R}_i) = \bigcap_1^{n+1} \mathscr{R}_i^*, \tag{16}$$

which is part (ii) of the theorem. From (15) and (16) it follows that

$$\mathrm{In}[\mathrm{CS}(A_1, A_2, \ldots, A_{n+1}) = \mathrm{CS}^+(A_1, A_2, \ldots, A_{n+1}), \tag{17}$$

which is part (iii) of the theorem. □

Collecting the results of several separate theorems, we can now establish the following fact.

THEOREM 8. An n-simplex in $\mathscr{E}^n$ is a convex body.

PROOF. If $\mathscr{S}$ is an n-simplex in $\mathscr{E}^n$, then, from Theorem 1, $\mathscr{S}$ is convex. Since $\mathscr{S}$ is the intersection of $n + 1$ closed half-spaces (Theorem 7), it is a closed set. From Theorem 5, $\mathscr{S}$ is a bounded set. Since $\mathscr{S}$ is closed and bounded, it is compact. From Theorem 7, the interior of $\mathscr{S}$ is the positive convex span of the vertices, hence the interior is not empty. In particular, the convex combination of the vertices, in which all the coefficients have the value $1/(n + 1)$ is a point in the interior of $\mathscr{S}$. □

COROLLARY 8.1. Every n-dimensional convex set in $\mathscr{E}^n$ has an n-dimensional interior. (E)

COROLLARY 8.2. A non-empty, convex set $\mathscr{S}$ in $\mathscr{E}^n$ with an empty interior is a single point or a set of dimension less than n. (E)

Although we do not give the proof, a k-simplex in $\mathscr{E}^n$ is congruent to a k-simplex in $\mathscr{E}^k$, and the simplex in $\mathscr{E}^n$ has a k-dimensional relative interior.

We can now describe the natural analogs of a familiar class of figures in elementary geometry. The convex span of a two-dimensional, finite set is a closed, polygonal region, and its boundary is a convex polygon. In $\mathscr{E}^3$, the convex span of a three-dimensional, finite set is a solid polyhedron, and its boundary is a convex polyhedron. In $\mathscr{E}^n$, the convex span of an n-dimensional, finite set is a convex polytope.

The relations and conventions appearing in the proof of Theorem 7 can be used to gain new insight about the nature of affine coordinates in general. To see this, suppose that $A_1, A_2, \ldots, A_{n+1}$ is any affine basis for $\mathscr{E}^n$. Then a point P has a unique representation of the form

$$p = \sum_1^{n+1} \lambda_i a_i, \qquad \sum_1^{n+1} \lambda_i = 1, \tag{1}$$

and the affine coordinates $(\lambda_1, \lambda_2, \ldots, \lambda_{n+1})$ depend solely on the points A_i.

Because the points A_i in the basis are independent, they are the vertices of an n-simplex which has face hyperplanes $\mathscr{H}_i$, and these are the zero sets of affine functionals $f_i(x) = b_i \cdot x + h_i$, $|b_i| = 1$. Then, as shown previously,

$$f_i(p) = \lambda_i f_i(a_i), \qquad i = 1, 2, \ldots, n + 1. \tag{2}$$

Since A_i is not in $\mathscr{H}_i$, $f_i(a_i) \neq 0$, so (2) implies

$$\lambda_i = \frac{f_i(p)}{f_i(a_i)}, \qquad i = 1, 2, \ldots, n + 1. \tag{3}$$

From (3) it follows that λ_i is positive if and only if $f_i(p)$ and $f_i(a_i)$ have the same sign and negative if and only if they have opposite signs. Thus P and A_i lie in the same side of $\mathscr{H}_i$ if and only if λ_i is positive and in opposite sides if and only if λ_i is negative. If $\lambda_i = 0$, then, from (2), $f_i(p) = 0$ and P is in $\mathscr{H}_i$.

The absolute magnitude of the affine coordinate λ_i gives a different kind of information about P. We showed before that

$$d(A_i, \mathscr{H}_i) = f_i(a_i), \qquad i = 1, 2, \ldots, n + 1. \tag{4}$$

The same derivation applied to P gives

$$d(P, \mathscr{H}_i) = \frac{|b_i \cdot p + h_i|}{|b_i|} = |f_i(p)|, \qquad i = 1, 2, \ldots, n + 1. \tag{5}$$

Then, from (3),

$$|\lambda_i| = \frac{|f_i(p)|}{f_i(a_i)} = \frac{d(P, \mathscr{H}_i)}{d(A_i, \mathscr{H}_i)}, \qquad i = 1, 2, \ldots, n + 1. \tag{6}$$

Thus P is $|\lambda_i|$ as far from $\mathscr{H}_i$ as is the vertex A_i.

To summarize, we have established the following facts.

THEOREM 9. If point P has affine coordinates $(\lambda_1, \lambda_2, \ldots, \lambda_{n+1})$ with respect to an affine basis $A_1, A_2, \ldots, A_{n+1}$ in $\mathscr{E}^n$, then

$$|\lambda_i| = \frac{d(P, \mathscr{H}_i)}{d(A_i, \mathscr{H}_i)}, \qquad i = 1, 2, \ldots, n + 1,$$

where $\mathscr{H}_i$ is the face-hyperplane opposite to A_i in the simplex determined by the basis. The ith coordinate λ_i is positive, zero, or negative, according as P is in the A_i side of $\mathscr{H}_i$, is on $\mathscr{H}_i$, or is in the non-A_i side of $\mathscr{H}_i$.

The metric interpretation we have just given for affine coordinates can be used to define coordinates directly in relation to a set of independent points. For example, suppose that A_1, A_2, and A_3 are three noncollinear points in a plane. Let t_i denote the line of the side of $\triangle A_1A_2A_3$ opposite to A_i, $i = 1, 2, 3$. Then coordinates $(\bar{p}_1, \bar{p}_2, \bar{p}_3)$ for a point P can be defined by

$$(\bar{p}_1, \bar{p}_2, \bar{p}_3) = \left(\pm \frac{d(P, t_1)}{d(A_1, t_1)}, \pm \frac{d(P, t_2)}{d(A_2, t_2)}, \pm \frac{d(P, t_3)}{d(A_3, t_3)}\right),$$

where the sign of $\bar{p}_i$ is taken positive if P and A_i are in the same side of t_i and negative if they are in opposite sides of t_i. The three coordinates are not independent, since $\bar{p}_1 + \bar{p}_2 + \bar{p}_3 = 1$. Coordinates defined in this way were

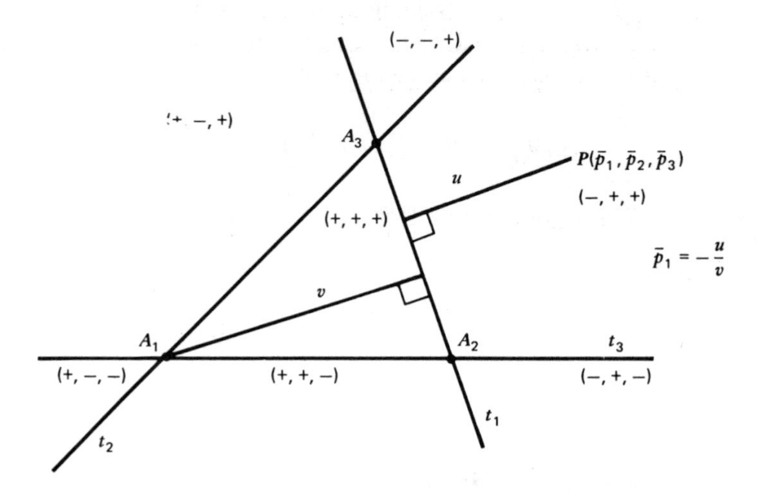

introduced in 1827 by Mobius, who interpreted $\bar{p}_i$ as a moment of force about line t_i. Because of this interpretation, the coordinates are often called "barycentric coordinates." This is simply a different name for affine coordinates.

Exercises – Section 4.1

1. Prove Theorem 1.
2. Prove Corollary 3.
3. Prove Theorem 5.
4. Prove Theorem 6 (cf. Theorem 3, Sec. 1.2).
5. Prove Corollaries 8.1 and 8.2.
6. If three noncollinear points A_1, A_2, A_3 are taken in that order as an affine basis for $\mathscr{E}^2$, then each point $X = (x_1, x_2)$ in $\mathscr{E}^2$ has affine coordinates $(\bar{x}_1, \bar{x}_2, \bar{x}_3)$. The vectors $\mathbf{A_1A_2}$ and $\mathbf{A_1A_3}$ are independent and hence form a linear basis for $\mathscr{E}^2$. Thus if P is any point, $\mathbf{A_1P}$ has a representation $\mathbf{A_1P} = \eta_1\mathbf{A_1A_2} + \eta_2\mathbf{A_1A_3}$.

   a. Show that $(\bar{p}_1, \bar{p}_2, \bar{p}_3) = (1 - \eta_1 - \eta_2, \eta_1, \eta_2)$.
   b. If $A_1 = (1, 0)$, $A_2 = (0, 3)$, $A_3 = (4, 2)$, what are the affine coordinates of $P = (-13, 6)$?
   c. What is the point Q if its affine coordinates are $(-3, 2, 2)$?

7. In $\mathscr{E}^3$, $A_1 = (2, 0, 1)$, $A_2 = (1, 1, 3)$, and $A_3 = (1, -1, 2)$.

   a. Show that $b = \mathbf{A_1A_2}$ and $c = \mathbf{A_1A_3}$ are independent vectors, and hence that A_1, A_2, A_3 are independent points.
   b. Show that $u = (3, -1, 2)$ satisfies $u \cdot b = 0$ and $u \cdot c = 0$ and hence is normal to $\mathrm{Pl}(A_1, A_2, A_3)$.
   c. From $u \cdot \mathbf{A_1X} = 0$, obtain $3x_1 - x_2 + 2x_3 - 8 = 0$ as an equation of $\mathrm{Pl}(A_1, A_2, A_3)$.
   d. Show that any affine combination of A_1, A_2, A_3 satisfies the equation in (c).
   e. Why is $a_4 = a_1 + a_2 + a_3 = (4, 0, 6)$ independent of A_1, A_2, A_3?
   f. Is $p = 3a_1 + a_2 - a_3 + a_4$ dependent on A_1, A_2, A_3, A_4?
   g. What are the affine coordinates (p_1, p_2, p_3, p_4) with respect to the affine basis A_1, A_2, A_3, A_4?
   h. Verify numerically that $d[P, \mathrm{Pl}(A_1, A_2, A_3)]$ is equal to

$$\bar{p}_4 d[A_4, \mathrm{Pl}(A_1, A_2, A_3)].$$

8. If A_1, A_2, A_3 form an affine basis for a plane $\mathscr{S}$ in $\mathscr{E}^n$, then equations and inequalities in the affine coordinates $(\bar{x}_1, \bar{x}_2, \bar{x}_3)$ of points X in $\mathscr{S}$ determine subsets of $\mathscr{S}$. For example, $\bar{x}_2 = 0$ is an affine coordinate description of the line $\mathrm{Ln}(A_1A_3)$, since

$$x = \bar{x}_1 a_1 + \bar{x}_3 a_3, \qquad \bar{x}_1 + \bar{x}_3 = 1$$

   are the points of $\mathscr{E}^n$ on this line. Give descriptions for the sets determined by the following conditions.

   a. $\bar{x}_3 > 0$
   b. $\bar{x}_1 \leq 0$
   c. $\bar{x}_1 = 0, \bar{x}_2 = 0$
   d. $\bar{x}_3 = 1$
   e. $\bar{x}_2 \leq 1$

   What affine coordinate equations or inequalities determine the following sets?

   f. $\mathrm{Ry}[A_3, A_2)$
   g. $\sphericalangle A_2 A_3 A_1$
   h. $\mathrm{In}[\sphericalangle A_2 A_3 A_1]$

9. The lines of a triangle divide a plane into seven regions. Why is it reasonable to suppose that the hyperplanes of an n-simplex divide $\mathscr{E}^n$ into $2^{n+1} - 1$ regions?

10. The *centroid* of a finite set of points $A_1, A_2, \ldots, A_m$ is defined to be the point G given by

$$g = \frac{1}{m}\sum_{1}^{m} a_i$$

and G is also defined to be the centroid of $CS(A_1, A_2, \ldots, A_m)$. In an n-simplex in $\mathscr{E}^n$, the segment joining a vertex to the centroid of the opposite face is a *median* of the simplex. In a 2-simplex, each face is a segment whose centroid is its midpoint. Thus the medians of a 2-simplex, $CS(A_1, A_2, A_3)$, are just the ordinary plane geometry medians of the triangle $\triangle A_1A_2A_3$. A standard fact of plane geometry is that the centroid G is the point of

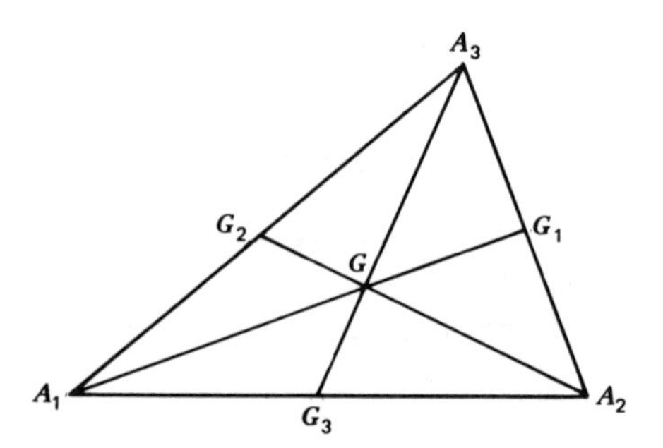

concurrency of the three medians and that $d(A_i, G) = \frac{2}{3}d(A_i, G_i)$, $i = 1, 2, 3$. Prove the theorem "The $n + 1$ medians $Sg[A_i, G_i]$ of an n-simplex $CS(A_1, A_2, \ldots, A_{n+1})$ are concurrent at G, the centroid of the simplex, and $d(A_i, G) = (n/n + 1)\, d(A_i, G_i)$, $i = 1, 2, \ldots, n + 1$."

11. Why is the centroid of an n-simplex always an interior point?

12. If $A_1, A_2, \ldots, A_n$ are the vertices of an n-simplex in $\mathscr{E}^n$, and if F_i is the foot of A_i in the face-hyperplane opposite to A_i, then line $Ln(A_iF_i)$ is an *altitude line*, $Sg[A_iF_i]$ is an *altitude segment*, and $d(A_i, F_i)$ is an altitude of the simplex, $i = 1, 2, \ldots, n + 1$. It is a theorem of plane geometry that the altitude lines of a triangle are concurrent, hence this is true in a 2-simplex. Show with an example that the altitude lines of a 3-simplex need not be concurrent.

13. In a *regular simplex*, all the edges have the same length. Thus a regular 2-simplex is a full equilateral triangle. A regular 3-simplex is a solid tetrahedron whose faces are all full equilateral triangles. Give a specific example of a regular 4-simplex in $\mathscr{E}^4$.

14. Prove McKai's theorem: If the m points of the set

$$\mathscr{S} = \{A_1, A_2, \ldots, A_m\}$$

lie on a sphere $S(P, r)$, and if P is the centroid of the set, then the sum of the squares of the distances determined by pairs of points in $\mathcal{S}$ is

$$\left(\frac{1}{2}\right)\sum_{i,j=1}^{m} d(A_i, A_j)^2 = m^2r^2.$$

In particular, then, the sum of the squares of the edges and diagonals of a regular n-gon inscribed in a unit circle is n^2.

15. The points $A_1 = (0, 0, 0)$, $A_2 = (4, 0, 0)$, $A_3 = (2, 4, 0)$, and $A_4 = (2, 2, 6)$ are the vertices of a 3-simplex in $\mathscr{E}^3$. Give an affine basis for each plane that is equidistant from all four vertices.

SECTION 4.2. TANGENT AND SUPPORTING HYPERPLANES OF A CONVEX SURFACE; REGULAR POINTS, CORNER POINTS; HYPERPLANE SEPARATION OF SETS

As mentioned in the introduction to this chapter, many of the concepts in convex body theory are generalizations of ideas encountered in elementary geometry. In particular, the notion of a tangent line to a circle in $\mathscr{E}^2$, or a tangent plane to a sphere in $\mathscr{E}^3$, generalizes to one of the most basic convex body concepts, that of a supporting hyperplane.

A line t in $\mathscr{E}^2$ is tangent to a circle $C(P, r)$ if it intersects the circle in just one point. However, we can observe that a closed half-plane of t contains $C(P, r)$. Thus if we define a "supporting line to a set $\mathcal{S}$" to be a line that intersects $\mathcal{S}$

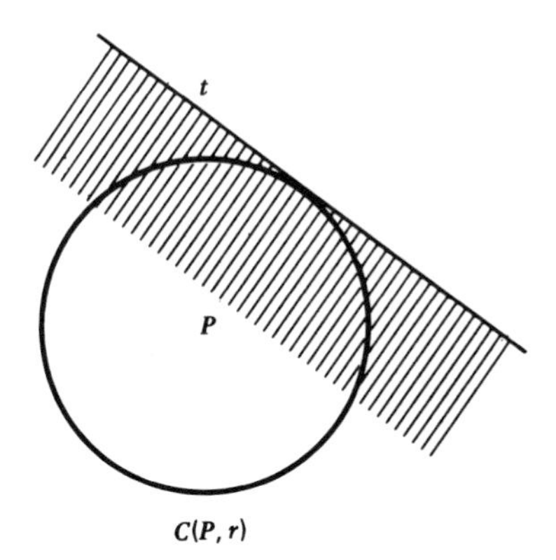

C(P, r)

and has a closed side containing $\mathcal{S}$, then the supporting lines to $C(P, r)$ are exactly the tangent lines. But the new concept has meaning for any set in $\mathscr{E}^2$. Of course, not all sets have supporting lines. In particular, if a line t intersects an open set $\mathcal{S}$, then neither closed side of t contains $\mathcal{S}$. We can extend the class of sets having supporting lines to include certain open sets if we merely require that a supporting line to $\mathscr{S}$ intersect the closure of $\mathscr{S}$ and have a closed side containing $\mathscr{S}$.

We now introduce a natural extension of the previous ideas to sets in $\mathscr{E}^n$.

DEFINITION (Supporting hyperplane). A hyperplane $\mathscr{H}$ in $\mathscr{E}^n$ is a *supporting hyperplane* to a set $\mathscr{S}$ if $\mathscr{H}$ intersects the closure of $\mathscr{S}$ and a closed side of $\mathscr{H}$ contains $\mathscr{S}$. Points in the set $\mathscr{H} \cap \mathrm{Cl}(\mathscr{S})$ are said to be the *contact points of* $\mathscr{H}$ (with $\mathrm{Cl}(\mathscr{S})$), and $\mathscr{H}$ is said to support $\mathscr{S}$ *at* each contact point.

From the definition just given, it follows that a supporting hyperplane to any one of the sets $\mathscr{S}$, $\mathrm{In}(\mathscr{S})$, $\mathrm{Cl}(\mathscr{S})$, $\mathrm{Bd}(\mathscr{S})$ is a supporting hyperplane to all of them that are non-empty. In $\mathscr{E}^3$, for example, the tangent planes to a sphere $\mathrm{S}(P, r)$ are supporting planes to $\mathrm{S}(P, r)$ and also to $\mathrm{N}(P, r)$ and $\mathrm{B}(P, r)$. In $\mathscr{E}^2$, the lines of support to a full triangle $\triangle^+ ABC$ are lines that contain a vertex but no interior point, and these lines also support $\triangle ABC$ and $\mathrm{In}(\triangle^+ ABC)$.

Supporting hyperplanes can also be described in terms of the following notion.

DEFINITION (Hyperplane separation). Two sets, or two points, are *separated by* a hyperplane $\mathscr{H}$ if they are contained in opposite sides of $\mathscr{H}$.

In terms of the type of separation just defined, a hyperplane supports a set $\mathscr{S}$ if it intersects $\mathrm{Cl}(\mathscr{S})$ but does not separate any two points of $\mathscr{S}$.

We can now distinguish tangent hyperplanes as special supporting hyperplanes.

DEFINITION (Tangent hyperplanes). A hyperplane $\mathscr{H}$ is tangent to a closed, convex surface $\mathscr{K}^0$ at point P in $\mathscr{K}^0$ if $\mathscr{H}$ is the unique hyperplane that supports $\mathscr{K}^0$ at P.

To see that the tangent hyperplanes to a sphere in $\mathscr{E}^n$ are exactly what we expect them to be, consider a point F on the sphere $\mathrm{S}(P, r)$, and let $\mathscr{H}$ be the

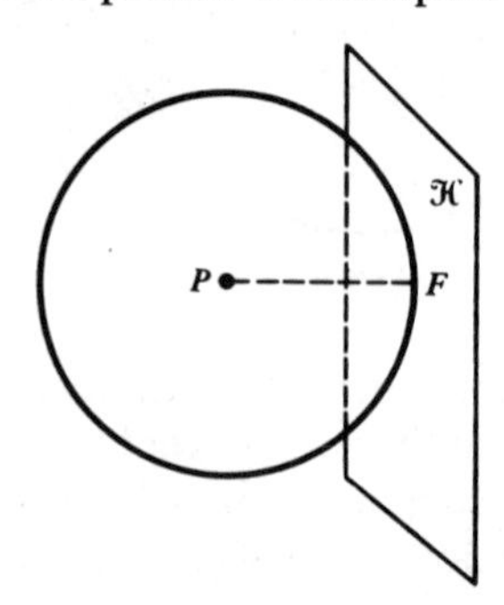

hyperplane that is perpendicular to the radial segment Sg[PF] at F. Then $\mathscr{H}$ has a representation $q(x) = 0$, where q is the affine functional defined on $\mathscr{E}^n$ by

$$q(x) = (p - f) \cdot (x - f). \tag{1}$$

Since $q(p) = |p - f|^2 > 0$, the open P side of $\mathscr{H}$ is the graph of $q(x) > 0$.

Now, let X be any point of the ball B(P, r) other than F. From $d(P, X) \leq r = d(P, F)$, we have

$$|x - p|^2 \leq |p - f|^2, \tag{2}$$

so

$$|(x - f) - (p - f)|^2 \leq |p - f|^2. \tag{3}$$

An inner product form for (3) is

$$[(x - f) - (p - f)] \cdot [(x - f) - (p - f)] \leq |p - f|^2, \tag{4}$$

and, by the distributivity of the inner product, (4) is equivalent to

$$|x - f|^2 - 2(x - f) \cdot (p - f) + |p - f|^2 \leq |p - f|^2, \tag{5}$$

which reduces to

$$|x - f|^2 - 2q(x) \leq 0, \tag{6}$$

or

$$q(x) \geq \tfrac{1}{2}|x - f|^2. \tag{7}$$

Because $X \neq F$, $|x - f| > 0$, hence (7) implies that $q(x) > 0$. Thus every point of B(P, r) except F lies in the P side of $\mathscr{H}$, so $\mathscr{H}$ supports the sphere and intersects it only at F.

To show that $\mathscr{H}$ is the only hyperplane through F that supports S(P, r), let $\mathscr{H}'$ be any other hyperplane that contains F. Since $\mathscr{H}'$ intersects the P side of $\mathscr{H}$, there exists A in $\mathscr{H}'$ and in the P side of $\mathscr{H}$. Because $\mathscr{H}'$ is convex, Sg(F, A] $\subset \mathscr{H}'$, and we contend that Sg(F, A] contains an inner point of B(P, r). If so, this inner point belongs to $\mathscr{H}'$, so $\mathscr{H}'$ does not support B(P, r) and therefore does not support S(P, r).

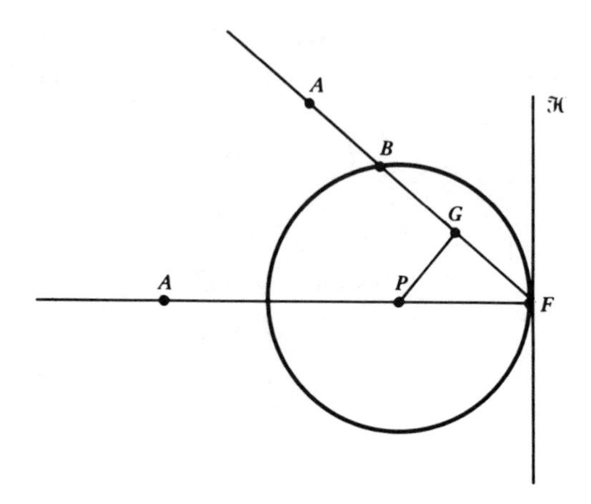

If A is on the ray Ry(F, P) or if A itself is interior to B(P, r), then Sg(F, A] obviously intersects N(P, r), so we consider P, F, and A noncollinear and A on or outside S(P, r). From the definition of angle measure, we have

$$\cos \measuredangle PFA^0 = \frac{\mathbf{FP} \cdot \mathbf{FA}}{|\mathbf{FP}||\mathbf{FA}|} = \frac{q(a)}{d(F, P)\, d(F, A)} > 0$$

because $q(a) > 0$. Thus the $\measuredangle PFA$ is acute and therefore G, the foot of P in Ln(FA), must lie on the open ray Ry(F, A). Since $d(P, G) < d(P, F) = r$, $G \in$ N(P, r). Thus, by Corollary 9.1, Sec. 3.1, the line Ln(FA) through G intersects S(P, r) at F and a second point B, and G is between F and B. Since A is B, or is on Ry(F, B) and outside the sphere, G is between A and F. In all cases, then, Sg(FA] in the hyperplane $\mathscr{H}'$ intersects the interior of B(P, r), hence $\mathscr{H}'$ does not support S(P, r). Thus $\mathscr{H}$ is the tangent hyperplane at F.

We now summarize what we have established.

THEOREM 1. If F is any point in the sphere S(P, r), the hyperplane $\mathscr{H}$ perpendicular to Sg[PF] at F is tangent to the sphere, and, except for F, all points of B(P, r) lie in the P side of $\mathscr{H}$. For every point A in the P side of $\mathscr{H}$, Sg[AF) intersects the interior of B(P, r).

The next theorem restates Exercise 6 in Sec. 3.2.

THEOREM 2. If Q is any point other than the center P of the sphere S(P, r), then the intersection of the ray Ry(P, Q) and S(P, r) is the unique foot of Q in the sphere.

From Theorems 1 and 2 together it follows that if Q is any point outside the ball B(P, r), then Q has a unique foot F in the ball and the hyperplane perpendicular to Ln(QF) at F is tangent to S(P, r). We now generalize this property.

THEOREM 3. Each point Q of space has a unique foot F in a non-empty, closed, convex set $\mathscr{S}$, and if Q is not in $\mathscr{S}$, then the hyperplane perpendicular to $\text{Ln}(QF)$ at F supports the set $\mathscr{S}$.

PROOF. That Q has a foot F in $\mathscr{S}$ follows from the fact that $\mathscr{S}$ is non-empty and closed (Corollary 3.3, Sec. 1.8). If $Q \in \mathscr{S}$, then Q is its own unique foot, so consider $Q \notin \mathscr{S}$ and hence $Q \neq F$. Let $\mathscr{H}$ be the hyperplane perpendicular to $\text{Ln}(QF)$ at F, and assume that there is some point A in $\mathscr{S}$ and in the Q side of $\mathscr{H}$. If $r = d(Q, F)$, it follows from Theorem 1 that $\mathscr{H}$ is tangent at F to $\text{S}(Q, r)$ and that $\text{Sg}[A, F)$ contains some point Y interior to $\text{B}(Q, r)$.

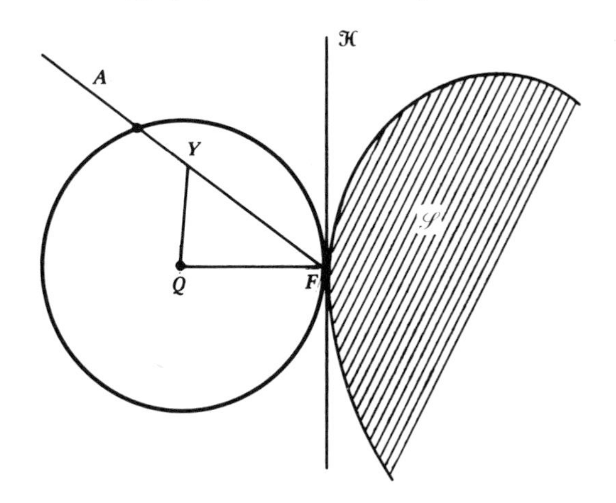

Because $\mathscr{S}$ is convex, $\text{Sg}[A, F) \subset \mathscr{S}$, hence $Y \in \mathscr{S}$. But since $Y \in \text{In}[\text{B}(Q, r)]$, $d(Q, Y) < r = d(Q, F)$, which contradicts the fact that F is the foot of Q. Hence there is no point of $\mathscr{S}$ in the Q side of $\mathscr{H}$, so $\mathscr{H}$ is a supporting hyperplane to $\mathscr{S}$.

If Q had a second foot G in $\mathscr{S}$, then $d(Q, F) = d(Q, G)$ would imply $G \in \text{S}(Q, r)$. Since, by Theorem 1, all points of $\text{B}(Q, r)$ except F lie in the Q side of $\mathscr{H}$, G would have to be in the Q side of $\mathscr{H}$. Since we have shown that no point of $\mathscr{S}$ is in the Q side of $\mathscr{H}$, it follows that no such second foot G can exist. Hence F is the unique foot in $\mathscr{S}$ of the point Q. □

COROLLARY 3. If point Q has foot F in a convex set $\mathscr{S}$ and $Q \neq F$, then every point of $\text{Ry}[F, Q)$ has F as its unique foot in $\mathscr{S}$. (E)

In our next theorem, the existence of supporting hyperplanes is based on the compactness rather than the convexity of the set.

THEOREM 4. Corresponding to a non-empty, compact set $\mathscr{S}$ in $\mathscr{E}^n$, and any point Q, there is a point F of $\mathscr{S}$ farthest from Q, and the hyperplane perpendicular to $\text{Ln}(QF)$ at F supports the set $\mathscr{S}$. (E)

We now give two useful facts that are similar in spirit to Theorems 3 and 4.

THEOREM 5. If P has foot F in set $\mathscr{S}$, and Q is between P and F, then Q has F as its unique foot in $\mathscr{S}$. If F is a farthest point of $\mathscr{S}$ from P, and P is between F and Q, then F is the unique point of $\mathscr{S}$ farthest from Q. (E)

We turn now to one of the most important and basic of all support theorems.

THEOREM 6. At each boundary point to a closed, convex set $\mathscr{S}$ there exists a supporting hyperplane to $\mathscr{S}$.

PROOF. Let B denote a boundary point to $\mathscr{S}$. By Theorem 7, Sec. 1.8, the function $g(X) = d(X, \mathscr{S})$ is continuous on $\mathscr{E}^n$, hence its restriction to the unit sphere $\mathrm{S}(B, 1)$ is continuous. Because the unit sphere is compact, it follows from the min–max theorem that g achieves a maximum at some point P in $\mathrm{S}(B, 1)$. That is,

$$d(P, \mathscr{S}) \geq d(X, \mathscr{S}), \qquad \text{for all } X \in \mathrm{S}(B, 1). \tag{1}$$

We wish to show that B is the foot in set $\mathscr{S}$ of the point P.

First, since $d(P, B) = 1$, and B belongs to $\mathscr{S}$, it follows from $d(P, \mathscr{S}) \leq d(P, B)$ that

$$d(P, \mathscr{S}) \leq 1. \tag{2}$$

Now, consider any number δ such that $0 < \delta < 1$. Because B is a boundary point to $\mathscr{S}$, there is a point Q in the neighborhood $\mathrm{N}(B, \frac{1}{2}\delta)$ that is not in $\mathscr{S}$.

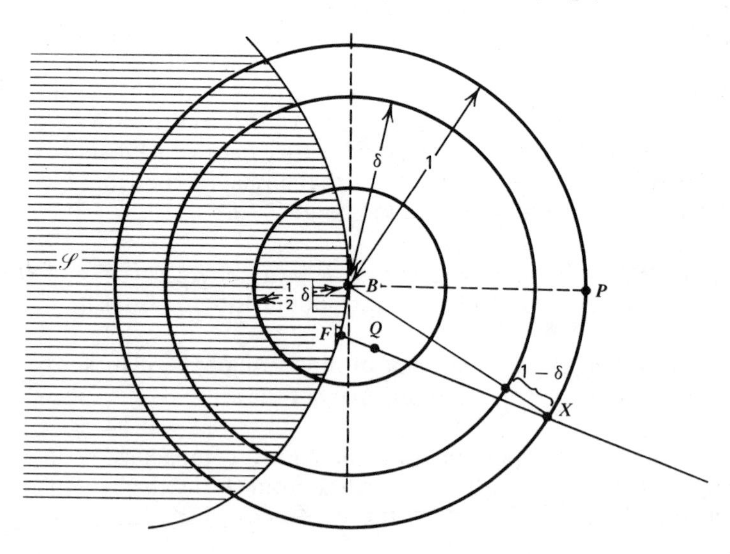

Since $\mathscr{S}$ is closed and convex, Q has a unique foot F in set $\mathscr{S}$. From $d(Q, F) \leq d(Q, B) < \frac{1}{2}\delta$, the triangle inequality gives

$$d(B, F) \leq d(F, Q) + d(Q, B) < \tfrac{1}{2}\delta + \tfrac{1}{2}\delta = \delta. \tag{3}$$

Thus F is interior to $\mathrm{B}(B, \delta)$ and hence to $\mathrm{B}(B, 1)$. Therefore the ray $\mathrm{Ry}[F, Q)$ intersects $\mathrm{S}(B, 1)$ at a point X. By Corollary 3, F is also the foot in $\mathscr{S}$ of point X, hence

$$d(X, \mathscr{S}) = d(X, F). \tag{4}$$

The distance of X from the ball $\mathrm{B}(B, \delta)$ is clearly

$$d[X, \mathrm{B}(B, \delta)] = 1 - \delta, \tag{5}$$

and, since F is interior to $\mathrm{B}(B, \delta)$,

$$d[X, \mathrm{B}(B, \delta)] > d(X, F). \tag{6}$$

From (4), (5), and (6), it follows that

$$d(X, \mathscr{S}) > 1 - \delta. \tag{7}$$

Since for every δ such that $0 < \delta < 1$ there is some point X of the sphere $\mathrm{S}(B, 1)$ such that $d(X, \mathscr{S}) > 1 - \delta$, it follows that

$$\max\{d(X, \mathscr{S}) \colon X \in \mathrm{S}(B, 1)\} \geq 1, \tag{8}$$

hence that

$$d(P, \mathscr{S}) \geq 1. \tag{9}$$

Now, from (2), (9), and $P \in \mathrm{S}(B, 1)$, we have

$$d(P, \mathscr{S}) = 1 = d(P, B), \tag{10}$$

and therefore B is the foot in $\mathscr{S}$ of the point P. Thus, by Theorem 3, the hyperplane perpendicular to $\mathrm{Ln}(PB)$ at B is a supporting hyperplane to $\mathscr{S}$. □

COROLLARY 6. At each boundary point to a convex set $\mathscr{S}$ there is a supporting hyperplane to $\mathscr{S}$.

PROOF. Each boundary point B to $\mathscr{S}$ is also a boundary point to $\mathrm{Cl}(\mathscr{S})$, and $\mathrm{Cl}(\mathscr{S})$ is a closed, convex set. By Theorem 6, there is a hyperplane of support to $\mathrm{Cl}(\mathscr{S})$ at B, and this hyperplane also supports $\mathscr{S}$. □

Since hyperplanes occur so frequently in the study of convex bodies, we sometimes use the shorter name "h-plane" for "hyperplane." It is also useful to have names that distinguish points of a surface at which tangent h-planes do and do not exist.

DEFINITION (Regular point, corner point). A point P is a *regular point* of a closed, convex surface $\mathscr{K}^0$ if there is a tangent h-plane at P. The point P is a *corner point* of $\mathscr{K}^0$ if more than one h-plane supports $\mathscr{K}^0$ at P.

The next theorem is a good example of a surprisingly general property of closed, convex surfaces that can be established by elementary arguments.

THEOREM 7. If point F is the foot in a closed, convex surface $\mathscr{K}^0$ of a point P interior to $\mathscr{K}$, then F is a regular point of $\mathscr{K}^0$, and the h-plane perpendicular to $\text{Ln}(PF)$ at F is tangent to $\mathscr{K}^0$ at F.

PROOF. Let $r = d(P, F)$ and let $\mathscr{H}$ denote the h-plane perpendicular to $\text{Ln}(PF)$ at F. Then $\mathscr{H}$ is tangent to $\text{S}(P, r)$ at F (Theorem 1). Next, let $\mathscr{H}'$ denote an arbitrary h-plane that is distinct from $\mathscr{H}$ and that also contains F. Since $\mathscr{H}'$ intersects the P side of $\mathscr{H}$, it intersects the interior of $\text{B}(P, r)$ (Theorem 1), so there exists $Y \in \mathscr{H}' \cap \text{N}(P, r)$. Then the line $\text{Ln}(FY)$, which is contained in $\mathscr{H}'$, intersects $\text{S}(P, r)$ at a second point B. Because P is interior to $\mathscr{K}$, the

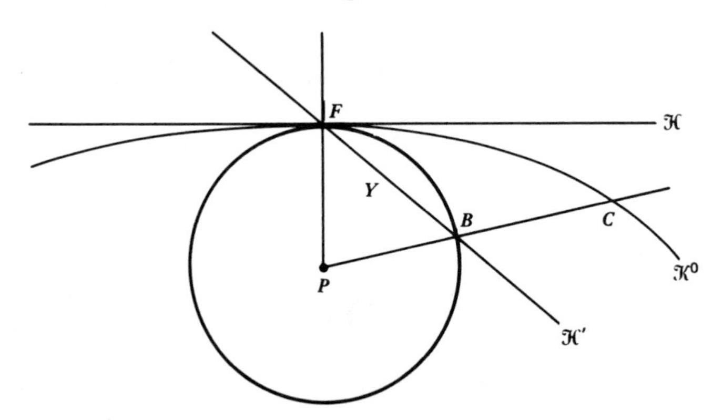

ray $\text{Ry}[P, B)$ intersects $\mathscr{K}^0$ at a point C. Since F is a foot of P in $\mathscr{K}^0$, $d(P, C) \geq d(P, F)$. This, with $d(P, F) = d(P, B)$, implies that $d(P, C) \geq d(P, B)$, so either B is between P and C, or else $C = B$. In the former case, B is interior to $\mathscr{K}$, since it is between an interior point and a boundary point (Theorem 5, Sec. 3.1). Hence $\mathscr{H}'$ intersects the interior of $\mathscr{K}$ and so does not support $\mathscr{K}$. If $C = B$, then from $F, B \in \mathscr{K}$, and $\mathscr{K}$ convex, $\text{Sg}[FB] \subset \mathscr{K}$, hence $Y \in \mathscr{K}$. Because $Y \in \text{N}(P, r)$, $d(P, Y) < r$ implies $d(P, Y) < d(P, F)$, hence $Y \notin \mathscr{K}^0$. Therefore $Y \in \text{In}(\mathscr{K})$. Thus again $\mathscr{H}'$ intersects the interior of $\mathscr{K}$ and so does not support $\mathscr{K}$.

From Theorem 6, there exists an h-plane that supports $\mathscr{K}$ at F. Because every h-plane through F that is distinct from $\mathscr{H}$ fails to support $\mathscr{K}$, it follows that $\mathscr{H}$ must be the unique supporting h-plane at F. Hence $\mathscr{H}$ is tangent to $\mathscr{K}^0$ at F. □

We conclude this section with a fundamental theorem on hyperplane separation of sets that we can obtain with the use of Theorem 3. The theorem is an answer to the question "When does there exist an h-plane separating two sets?", and the answer is not only useful in geometry, but has important implications in other parts of mathematics.

Any two closed, disjoint sets $\mathscr{R}$ and $\mathscr{S}$ that are non-empty are separated sets. It might seem that if they are also convex, then they must be separated by some hyperplane. However, this expectation is wrong. In the space $\mathscr{E}^2$, if $\mathscr{R}$ is the set $\{(x_1, x_2): x_2 \geq 1/x_1, x_1 > 0\}$, and $\mathscr{S} = \{(x_1, x_2): x_2 \leq 0\}$,

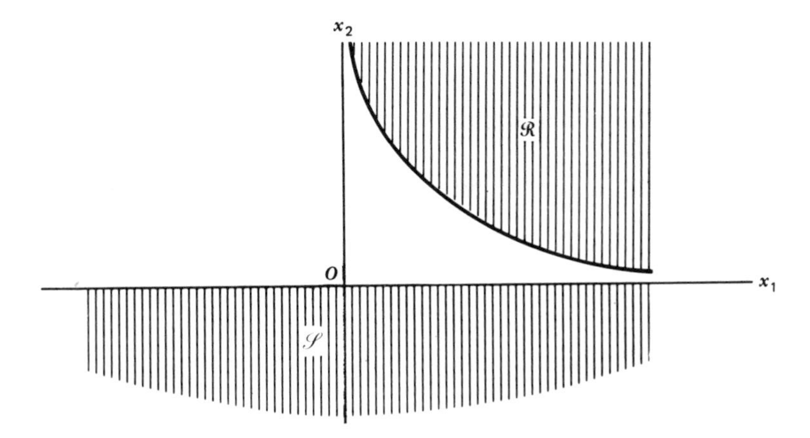

then the sets $\mathscr{R}$ and $\mathscr{S}$ are disjoint, closed, and convex. But there is no line that separates them. However, hyperplane separation is implied if we add the condition that one of the sets is bounded. In these circumstances, we can in fact establish the following stronger result.

THEOREM 8. If $\mathscr{R}$ and $\mathscr{S}$ are non-empty, disjoint sets that are closed and convex, and if one of them is bounded, then there exist two parallel hyperplanes $\mathscr{H}_1$ and $\mathscr{H}_2$ that support $\mathscr{R}$ and $\mathscr{S}$, respectively, in such a way that every hyperplane parallel to $\mathscr{H}_1$ and $\mathscr{H}_2$ and between them separates $\mathscr{R}$ and $\mathscr{S}$.

PROOF. Let f denote the function defined on the bounded set, say set $\mathscr{R}$, by $f(X) = d(X, \mathscr{S})$. By Theorem 7, Sec. 1.8, f is continuous, and since the domain $\mathscr{R}$ is compact, then by the min–max theorem it follows that f achieves a minimum function value at some point X_o in $\mathscr{R}$. Since $\mathscr{S}$ is closed and convex, X_o has a unique foot Y_o in $\mathscr{S}$, and $X_o \neq Y_o$ because $\mathscr{R}$ and $\mathscr{S}$ are disjoint.

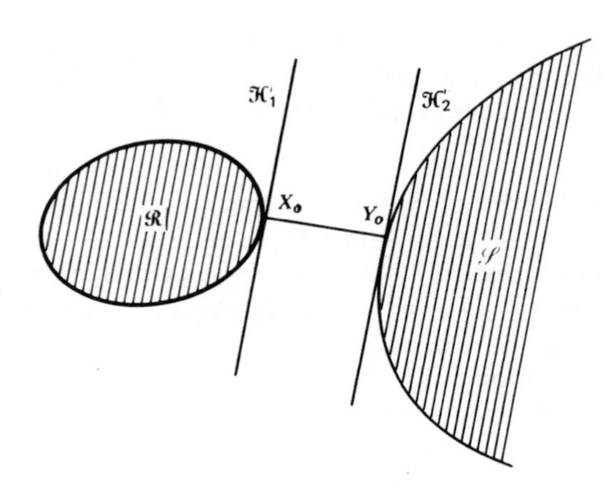

For any X in $\mathcal{R}$, $d(X, Y_o) \geq d(X, \mathcal{S}) = f(X) \geq f(X_o) = d(X_o, Y_o)$ shows that X_o is the foot in $\mathcal{R}$ of Y_o. Now, by Theorem 3, the hyperplane $\mathcal{H}_1$ perpendicular to $\text{Ln}(X_o Y_o)$ at X_o supports $\mathcal{R}$ and the hyperplane $\mathcal{H}_2$ perpendicular to $\text{Ln}(X_o Y_o)$ at Y_o supports $\mathcal{S}$. Because $\mathcal{R}$ lies in the closed, non-$\mathcal{H}_2$ side of $\mathcal{H}_1$ and $\mathcal{S}$ lies in the non-$\mathcal{H}_1$ side of $\mathcal{H}_2$, any hyperplane between $\mathcal{H}_1$ and $\mathcal{H}_2$ and parallel to them separates $\mathcal{R}$ and $\mathcal{S}$. □

Exercises – Section 4.2

1. If P is interior to a set $\mathcal{S}$, then there exists $\text{N}(P, r) \subset \mathcal{S}$. Let $\mathcal{H} : a \cdot x + h = 0$ be an arbitrary h-plane that contains P. Then $x = p + \eta a, \eta \in \mathcal{E}^1$, is a line perpendicular to $\mathcal{H}$ at P. Show that there exist points B, C on this line that belong to $\mathcal{S}$ and are separated by $\mathcal{H}$, and hence that $\mathcal{H}$ does not support $\mathcal{S}$.
2. Prove Corollary 3.
3. Prove Theorem 4. (cf. Corollary 3.2, Sec. 1.8)
4. Prove Theorem 5.
5. Let A_1, A_2, A_3 denote the vertices of a triangle in $\mathcal{E}^2$, with t_i the line of the side opposite to A_i, $i = 1, 2, 3$. Let $h_i = d(A_i, t_i)$, $i = 1, 2, 3$ and let $(\bar{x}_1, \bar{x}_2, \bar{x}_3)$ be affine coordinates of $X = (x_1, x_2)$ with respect to the affine basis A_1, A_2, A_3. An *incircle* to $\triangle A_1 A_2 A_3$ is defined to be a circle $\text{C}(P, r)$ that is tangent to the three line t_i and with P interior to $\triangle^+ A_1 A_2 A_3$. Thus if such a circle exists, P must satisfy $P \in \text{CS}^+(A_1, A_2, A_3)$ and $d(P, t_i) = r$, $i = 1, 2, 3$. Why do these conditions imply that (p_1, p_2, p_3) must be a solution to

$$h_1 \bar{x}_1 = h_2 \bar{x}_2 = h_3 \bar{x}_3, \qquad \bar{x}_1, \bar{x}_2, \bar{x}_3 > 0? \tag{*}$$

Show that the (*) conditions have a unique solution, hence that a unique incircle exists. Find the affine coordinates of P, and the value of r, in terms of the altitudes h_i.

6. Generalize Exercise 5 to show that an n-simplex $\mathrm{CS}(A_1, A_2, \ldots, A_{n+1})$ in $\mathscr{E}^n$ has a unique *insphere* tangent to the $n + 1$ face h-planes and with P interior to the simplex.
7. An *excircle* to a triangle is tangent to the lines of the sides, but the center of the circle is outside the triangle. Assuming the usual formula for the area of a triangle (half the base times the height), modify Exercise 5 to prove that a triangle has exactly three excircles.
8. Prove that the locus of points X in $\mathscr{E}^n$ that are equidistant from two points A, B is the h-plane that is the perpendicular bisector of $\mathrm{Sg}[AB]$ and has an equation

$$(x - [a + b]/2) \cdot \mathbf{AB} = 0.$$

9. Let $O, A_1, A_2, \ldots, A_n$ be the vertices of an n-simplex in $\mathscr{E}^n$. A sphere that contains all $n + 1$ vertices, if it exists, is a *circumsphere* of the simplex. By Exercise 8, the h-planes that are the perpendicular bisectors of the edges $\mathrm{Sg}[OA_i]$ are

$$\mathscr{H}_i : (x - a_i/2) \cdot a_i = 0, \qquad i = 1, 2, \ldots, n.$$

Explain why these n equations have a unique solution $(p_1, p_2, \ldots, p_n)$ and why this implies that $\mathrm{S}(P, r)$, $r = d(P, O)$, is the unique circumsphere of the simplex.

10. We can define $\mathrm{S}(P, r)$ to be a *generalized insphere* of set $\mathscr{S}$ if $\mathrm{B}(P, r) \subset \mathscr{S}$ and no ball of greater radius is contained in $\mathscr{S}$. If $\mathscr{K}$ is a convex body, make use of the function $f(X) = d(X, \mathscr{K}^0)$, $X \in \mathscr{K}$, to prove that $\mathscr{K}$ has a generalized insphere (cf. Theorem 7, Sec. 1.8).
11. We can define $\mathrm{S}(P, r)$ to be a *generalized circumsphere* of a set $\mathscr{S}$ if $\mathscr{S} \subset \mathrm{B}(P, r)$ and is not contained in any ball of smaller radius than r. Prove that if $\mathscr{S}$ is a non-empty, nonsingleton compact set, then a generalized circumsphere to $\mathscr{S}$ exists and is unique.
12. If $\mathscr{K}_1^0$ and $\mathscr{K}_2^0$ are closed, convex surfaces in $\mathscr{E}^n$, prove that if $\mathscr{K}_1^0$ intersects the interior and exterior of $\mathscr{K}_2$, then $\mathscr{K}_2^0$ intersects the interior and exterior of $\mathscr{K}_1$.

SECTION 4.3. WIDTHS OF A SET, SUPPORT AND WIDTH FUNCTIONS, CONVEX HULLS

It is intuitively clear that the support lines of a plane convex body occur in parallel pairs. For suppose that t is any line through an interior point of the set, and let u, v denote opposite normal directions of t. Because the figure $\mathscr{K}$

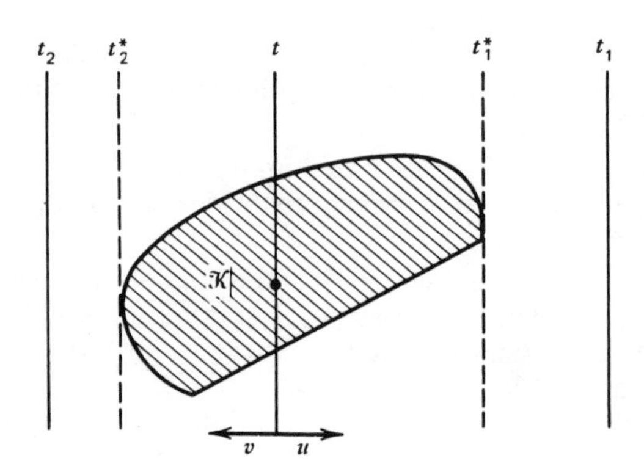

is bounded, as t moves parallel to itself in the direction u it will reach some position t_1 of noncontact with $\mathscr{K}$. In moving from t to t_1, it will pass through a position t_1^* of last contact with $\mathscr{K}$, and t_1^* supports $\mathscr{K}$. A similar motion in the direction v will produce a noncontact line t_2 and a last contact line t_2^* that supports $\mathscr{K}$. Among all the closed strips whose parallel edges have the normals directions u, v, and which contain $\mathscr{K}$, the one with edges t_1^*, t_2^* is clearly the one with the smallest width. It is natural to call this width, namely $d(t_1^*, t_2^*)$, *the width of* $\mathscr{K}$ *in the directions* u, v.

The intuitive notions just described can easily be made formal and generalized in such a way that the "widths of a set" make sense for any non-empty set in a space of dimension greater than 1. To do this, we first introduce some new terms.

DEFINITION (Supporting half-spaces). An open or closed *half-space supports* a set $\mathscr{S}$ if it contains $\mathscr{S}$ and its face supports $\mathscr{S}$.

DEFINITION (Open and closed slabs). The *open slab* between two parallel hyperplanes $\mathscr{H}_1$ and $\mathscr{H}_2$, denoted by $\text{Sl}(\mathscr{H}_1, \mathscr{H}_2)$, is the intersection of the $\mathscr{H}_1$ side of $\mathscr{H}_2$ with the $\mathscr{H}_2$ side of $\mathscr{H}_1$. The *closed slab* between $\mathscr{H}_1$ and $\mathscr{H}_2$, denoted by $\text{Sl}[\mathscr{H}_1, \mathscr{H}_2]$, is the intersection of the corresponding closed half-spaces. The hyperplanes $\mathscr{H}_1$ and $\mathscr{H}_2$ are the *faces* of both slabs, and the distance $d(\mathscr{H}_1, \mathscr{H}_2)$ between these faces is the width of both slabs. A direction of a slab is a direction normal to both faces. In $\mathscr{E}^2$, a slab is a *strip* between parallel lines that are its *edges*.

DEFINITION (Supporting slabs). An open or closed slab *supports* a non-empty set $\mathscr{S}$ if it contains $\mathscr{S}$ and both its faces support $\mathscr{S}$.

DEFINITION (Widths of a set). A non-empty set $\mathscr{S}$ in $\mathscr{E}^n$, $n > 1$, has *width zero in the direction u* if $\mathscr{S}$ is contained in a hyperplane with direction u. The set has *width h in the direction u* if there exists a slab of width h and direction u that supports $\mathscr{S}$. The width of $\mathscr{S}$ in the direction u is denoted by $W(u)$.

We can now show that the only way in which a non-empty set in $\mathscr{E}^n$, $n > 1$, can fail to have a width in some direction is by being unbounded.

THEOREM 1. A non-empty, bounded set $\mathscr{S}$ in $\mathscr{E}^n$, $n > 1$, has a width in every direction.

PROOF. Corresponding to an arbitrary direction u, let $\mathscr{L} = \text{Ln}(PQ)$ be a line in the direction u, and let f denote the function with domain $\text{Cl}(\mathscr{S})$ that is the orthogonal projection of $\text{Cl}(\mathscr{S})$ into $\mathscr{L}$. Because $\text{Cl}(\mathscr{S})$ is both closed and bounded, it is a compact set. The function f is continuous (Corollary 3, Sec. 3.2), so its range $f[\text{Cl}(\mathscr{S})]$ is a compact subset of $\mathscr{L}$. For each point $Y = f(X)$ in $\mathscr{L}$ there is a unique real number $\eta(Y)$ such that $y = p + \eta(Y)u$. The correspondence $Y \to \eta(Y)$ is a one-to-one continuous mapping (cf. Theorem 2, Sec. 3.1) of the range of f into the reals. The image of the range of f in this mapping is therefore a compact set of numbers with a minimum η_1 and a maximum η_2. Let $a = p + \eta_1 u$, $b = p + \eta_2 u$, and let $\mathscr{H}_A$ and $\mathscr{H}_B$ denote the hyperplanes perpendicular to $\mathscr{L}$ at A and B, respectively. If

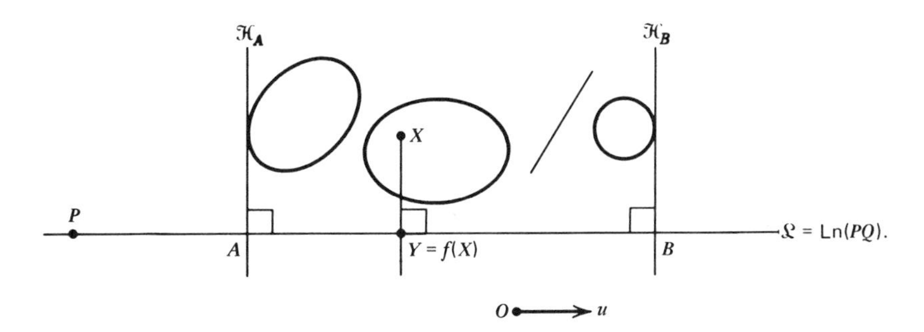

$\eta_1 = \eta_2$, then the range of f is the single point $A = B$. In this case, both $\mathscr{S}$ and $\text{Cl}(\mathscr{S})$ are contained in the hyperplane $\mathscr{H}_A = \mathscr{H}_B$, with normal direction u, so the width of $\mathscr{S}$ is zero. If $\eta_1 \neq \eta_2$, $A \neq B$, and $\eta_1 \leq \eta(Y) \leq \eta_2$ for every $Y \in f[\text{Cl}(\mathscr{S})]$ implies that $f[\text{Cl}(\mathscr{S})] \subset \text{Sg}[AB]$. Because A is in the range of f, $\mathscr{H}_A$ intersects $\text{Cl}(\mathscr{S})$. But the open ray opposite to $\text{Ry}[A, B)$ contains no point in the range of f, hence the non-B side of $\mathscr{H}_A$ contains no point in $\text{Cl}(\mathscr{S})$. Thus the closed B side of $\mathscr{H}_A$ supports $\mathscr{S}$. By a symmetric argument, the closed A side of $\mathscr{H}_B$ supports $\mathscr{S}$. The intersection of these two closed half-spaces is the closed slab $\text{Sl}[\mathscr{H}_A, \mathscr{H}_B]$ with direction u. Since $\mathscr{S}$ is contained in this slab, and each face supports $\mathscr{S}$, the slab supports $\mathscr{S}$, hence the width of $\mathscr{S}$ in the direction u is $d(\mathscr{H}_A, \mathscr{H}_B)$. □

COROLLARY 1. If a set $\mathscr{S}$ is not contained in a hyperplane with direction u, and $\mathscr{S}$ is bounded, then there are exactly two hyperplanes with direction u that support $\mathscr{S}$. The closed slab between these hyperplanes supports $\mathscr{S}$, and its width is the non-zero width of $\mathscr{S}$ in the direction u.

By following up the orthogonal projection idea used in the proof of Theorem 1, we can obtain a very useful form for the width function of a set. First, it may be observed that if $\mathscr{L}$ is a line through the origin, then the orthogonal projection $Y = f(X)$ of a set $\mathscr{R}$ into $\mathscr{L}$ has a simple explicit representation. Let u be a unit vector in the direction of $\mathscr{L}$, so

$$x = \eta u, \qquad \eta \in \mathscr{E}^1 \tag{1}$$

represents $\mathscr{L}$. If P is a point of $\mathscr{R}$, then the h-plane through P and perpendicular to $\mathscr{L}$ has an equation

$$\mathscr{H} : (x - p) \cdot u = 0. \tag{2}$$

The intersection of $\mathscr{H}$ with $\mathscr{L}$ corresponds to the η-solution of

$$(\eta u - p) \cdot u = 0,$$

hence to

$$\eta = u \cdot p. \tag{3}$$

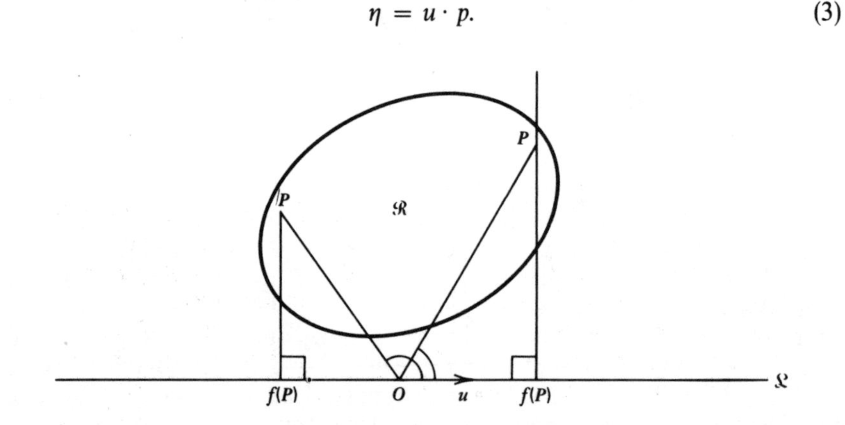

Thus $f(P) = (u \cdot p)u$, and in general f is the mapping

$$f(X) = (u \cdot x)u, \qquad X \in \mathscr{R}. \tag{4}$$

Now consider the case in which two h-planes support $\mathscr{R}$ and are perpendicular to $\mathscr{L}$ at A and B, with O between A and B. Let $\mathscr{H}_A$ and $\mathscr{H}_B$ denote these h-planes, and let P be a contact point in $\mathscr{H}_A$ and Q be a contact point in

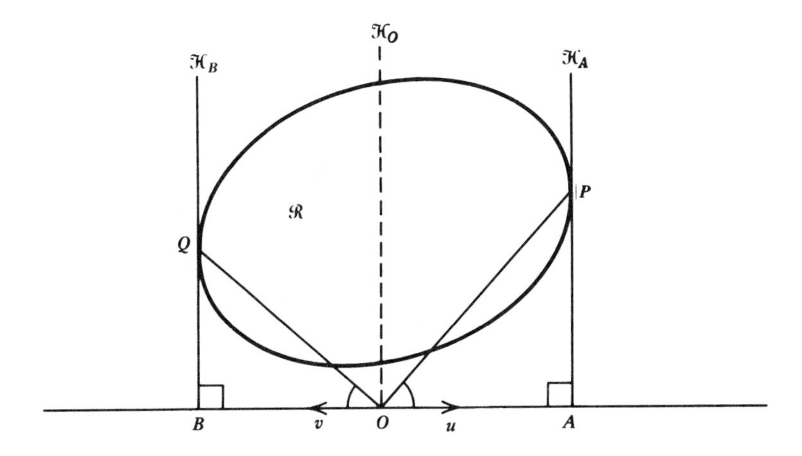

$\mathcal{H}_B$. Finally, let u and v be opposite unit vectors in $\mathcal{L}$ with the directions **OA** and **OB**, respectively. The width of $\mathcal{R}$ in the directions u, v is

$$d(\mathcal{H}_A, \mathcal{H}_B) = d(A, B) = d(O, A) + d(O, B). \tag{5}$$

From (4), $A = f(P) = (u \cdot p)u$, and since $u \cdot p > 0$,

$$d(O, A) = |(u \cdot p)u| = u \cdot p. \tag{6}$$

By precisely the same logic,

$$d(O, B) = |(v \cdot q)v| = v \cdot q. \tag{7}$$

From (5), (6), and (7), we have

$$d(\mathcal{H}_A, \mathcal{H}_B) = u \cdot p + v \cdot q. \tag{8}$$

If $\mathcal{R}$ is closed, then $P, Q \in \mathcal{R}$, and we can describe the terms $u \cdot p$ and $v \cdot q$ in (8) in a more intrinsic way. Let $\mathcal{H}_O$ denote the h-plane perpendicular to $\mathcal{L}$ at O. If $Y = f(X)$ is the orthogonal projection of $\mathcal{R}$ into $\mathcal{L}$, then for each X in $\mathcal{R}$, $u \cdot x = \pm d(O, Y)$, and $u \cdot x$ is positive, zero, or negative according as X is in the A side of $\mathcal{H}_O$, is in $\mathcal{H}_O$, or is in the B side of $\mathcal{H}_O$. Thus

$$u \cdot p = \max\{u \cdot x : X \in \mathcal{R}\}. \tag{9}$$

By the same logic,

$$v \cdot q = \max\{v \cdot x : X \in \mathscr{R}\}. \tag{10}$$

The relations (9) and (10) are the motivation for the following concept.

DEFINITION (Support function of a set). If $\mathscr{R}$ is a non-empty, compact set, the *support function* of $\mathscr{R}$ is the function H defined at each unit vector u by

$$H(u) = \max\{u \cdot x : X \in \mathscr{R}\}.$$

For a fixed vector u, the function $u \cdot x$ is continuous at all X and hence achieves a maximum on any compact set $\mathscr{R}$. Thus the support function H of such a set has a real function value at each point U of the unit sphere $S(O, 1)$. The results in (8), (9), and (10), and using $v = -u$, can now be put in the following form.

THEOREM 2. If two parallel supporting h-planes to a compact set $\mathscr{R}$ have unit normal direction u, and are such that O is between them, then the width of $\mathscr{R}$ in the direction u is given by

$$W(u) = H(u) + H(-u).$$

We can generalize Theorem 2 and remove the artificial condition about the position of the origin by making use of the following facts.

THEOREM 3. If $\mathscr{R}$ is a non-empty, compact set and $\mathscr{R}'$ is the a-translate of $\mathscr{R}$, then the width functions W and W' of the sets are the same, and the support functions H and H' are related by $H'(u) = H(u) + a \cdot u$.

PROOF. Since a translation is an isometry, and hence a one-to-one continuous mapping, $\mathscr{R}'$ is compact. Every isometry preserves intersections and containment and maps hyperplanes onto hyperplanes and half-spaces onto half-spaces. Under any isometry, a slab supporting $\mathscr{R}$ would map to a slab of the same width supporting $\mathscr{R}'$. Under a translation, which leaves the normal directions of a hyperplane invariant, the slabs have the same direction, hence $W'(u) = W(u)$. Similarly, if $\mathscr{R}$ is contained in a hyperplane $\mathscr{H}$ with direction u, then so is $\mathscr{R}'$, so both $W(u)$ and $W'(u)$ are zero.

Next, corresponding to a fixed unit vector u, there exists $X_o \in \mathscr{R}$ such that

$$H(u) = \max\{u \cdot x : X \in \mathscr{R}\} = u \cdot x_o. \tag{1}$$

From $u \cdot x_o \geq u \cdot x$, for all $X \in \mathscr{R}$, it follows that

$$u \cdot (x_o + a) \geq u \cdot (x + a) \text{ for all } X \in \mathscr{R}. \tag{2}$$

Then because $x' = x + a$ is a one-to-one mapping of $\mathscr{R}$ onto $\mathscr{R}'$,

$$\begin{aligned} H'(u) &= \max\{u \cdot x' : X' \in \mathscr{R}'\} \\ &= \max\{u \cdot (x + a) : X \in \mathscr{R}\} = u \cdot (x_o + a) \\ &= u \cdot x_o + u \cdot a = H(u) + u \cdot a. \qquad \square \end{aligned}$$

THEOREM 4. If $\mathscr{R}$ is a non-empty, compact set with support function H and width function W defined on $S(O, 1)$, then

$$W(u) = H(u) + H(-u),$$

PROOF. First, if $W(u) = 0$, then there exists a hyperplane $\mathscr{H}$ that contains $\mathscr{R}$ and is perpendicular to $\text{Ln}(OU)$ at some point F, and F is the orthogonal projection of $\mathscr{R}$ into this line. If F is the origin, then $u \cdot x = 0$ for all $X \in \mathscr{R}$, so $H(u) + H(-u) = 0 + 0 = 0$. If F is not the origin, then $H(U) = \pm d(O, F)$ and $H(-u) = \mp d(O, F)$, so again $H(u) + H(-u) = 0$.

If $W(u) \neq 0$, then there exists a slab with direction u that supports $\mathscr{R}$ and has face hyperplanes $\mathscr{H}_1$ and $\mathscr{H}_2$. If B is any point between $\mathscr{H}_1$ and $\mathscr{H}_2$, the translation that maps B to the origin maps $\mathscr{R}$ to a set $\mathscr{R}'$. If W' and H' are the width and support functions of $\mathscr{R}'$, then because O is between the translates of $\mathscr{H}_1$ and $\mathscr{H}_2$, and these support $\mathscr{R}'$, Theorems 2 and 3 imply that

$$\begin{aligned} W(u) &= W'(u) = H'(u) + H'(-u) \\ &= H(u) + u \cdot (-b) + H(-u) + (-u) \cdot (-b) \\ &= H(u) + H(-u). \quad \square \end{aligned}$$

Just as we think of *the* direction of a ray, in contrast to the two directions of a line, it is often useful to have the following agreements about the direction of a half-space.

DEFINITION (Half-space direction). The *normal direction*, or simply *the* direction, of a closed half-space $\mathscr{S}$ with face H is that of a non-null vector u such that $u \cdot \mathbf{PX} = 0$ for $P \in \mathscr{H}$ and $X \in \mathscr{S}$. The opposite direction, $v = -u$, is the *outer normal direction* of $\mathscr{S}$. An open half-space has the direction and outer normal direction of the corresponding closed half-space.

We can use the language just introduced to describe a support function in a slightly different way. If $\mathscr{R}$ is a bounded set and u is a non-null vector, there exists a supporting closed half-space $\mathscr{S}$, with face $\mathscr{H}$ and outer normal direction u. The absolute value of $H(u)$ is the distance $d(O, \mathscr{H})$. The number $H(u)$ is positive, negative, or zero according as O is in the interior, the exterior, or the boundary of the half-space $\mathscr{S}$.

Both the support and width functions of a non-empty, compact set are continuous on the sphere $S(O, 1)$, and to show this we first consider a function closely related to the support function.

THEOREM 5. If $\mathscr{S}$ is a non-empty, compact set, the function f defined at each point P of $\mathscr{E}^n$ by

$$f(P) = \max\{p \cdot x : X \in \mathscr{S}\}$$

is continuous at each point of $\mathscr{E}^n$.

PROOF. Since $\mathscr{S}$ is compact, it is bounded, hence there exists a positive number m such that

$$|x| < m \quad \text{for all} \quad X \in \mathscr{S}. \tag{1}$$

Corresponding to a point P, there is at least one point X_o in $\mathscr{S}$ such that $f(p) = p \cdot x_o$. Using (1), we obtain

$$f(p) = p \cdot x_o \leq |p \cdot x_o| \leq |p||x_o| \leq m|p|. \tag{2}$$

Now, consider any two points P, Q. Let X_o and X_1 in $\mathscr{S}$ denote points such that $f(p) = p \cdot x_o$ and $f(q) = q \cdot x_1$. From the definition of f it follows that for all X in $\mathscr{S}$,

$$f(p) = p \cdot x_o \geq p \cdot x \qquad \text{and} \qquad f(q) = q \cdot x_1 \geq q \cdot x. \tag{3}$$

Therefore, for all X in $\mathscr{S}$,

$$\begin{aligned}(p + q) \cdot x = p \cdot x + q \cdot x \leq p \cdot x_o + q \cdot x_1 \\ = f(p) + f(q).\end{aligned} \tag{4}$$

Thus

$$f(p + q) = \max\{(p + q) \cdot x : X \in \mathscr{S}) \leq f(p) + f(q), \tag{5}$$

which is the defining property for a function f to be a *subadditive function.*

To show that f is continuous at an arbitrary point P_o, let $N(f(P_o), \varepsilon)$ be an arbitrary neighborhood of $f(P_o)$. Corresponding to ε, let δ be defined as

$$\varepsilon = \delta/m \tag{6}$$

and consider $P \in N(P_o, \delta)$. From the subadditivity of f,

$$f(p_o) = f(p + p_o - p) \leq f(p) + f(p_o - p). \tag{7}$$

Now, from (7) and (2),

$$f(p_o) - f(p) \leq f(p_o - p) \leq m|p_o - p|. \tag{8}$$

Again, using subadditivity,

$$f(p) = f(p_o + p - p_o) \leq f(p_o) + f(p - p_o), \tag{9}$$

and this, with (2), gives

$$f(p) - f(p_o) \leq f(p - p_o) \leq m|p - p_o|. \tag{10}$$

Now, from (8), (10), and (6), it follows that

$$|f(p_o) - f(p)| \leq m|p_o - p| < m\delta = \varepsilon. \tag{11}$$

Thus $P \in N(P_o, \delta) \Rightarrow f(P) \in N(f(P_o), \varepsilon)$, hence f is continuous at P_o. □

COROLLARY 5. The support function H for a non-empty, compact set $\mathscr{S}$ is continuous on the unit sphere.

PROOF. The support function H is just the restriction to $S(O, 1)$ of the function f in Theorem 5. Since f is continuous on $\mathscr{E}^n$, its restriction H is continuous on $S(O, 1)$. □

THEOREM 6. The width function W for a non-empty, compact set $\mathscr{S}$ is continuous on $S(O, 1)$.

PROOF. Let U_o denote an arbitrary point of $S(O, 1)$, and let ε be any given positive number. By Corollary 5, the support function H is continuous at u_o and $-u_o$. Hence, corresponding to ε, there exist positive numbers δ_1 and δ_2 such that

$$|u| = 1 \quad \text{and} \quad |u - u_o| < \delta_1 \Rightarrow |H(u) - H(u_o)| < \tfrac{1}{2}\varepsilon, \tag{1}$$

and

$$|-u| = 1$$

and

$$|-u - (-u_o)| = |u - u_o| < \delta_2 \Rightarrow |H(-u) - H(-u_o)| < \tfrac{1}{2}\varepsilon. \quad (2)$$

Now, let $\delta = \min\{\delta_1, \delta_2\}$. Then for $|u| = 1$ and $|u - u_o| < \delta$, it follows that

$$\begin{aligned} |W(u) - W(u_o)| &= |H(u) + H(-u) - [H(u_o) - H(-u_o)]| \\ &\leq |H(u) - H(u_o)| + |H(-u) - H(-u_o)| \quad (3) \\ &< \tfrac{1}{2}\varepsilon + \tfrac{1}{2}\varepsilon = \varepsilon. \end{aligned}$$

Therefore W is continuous at U_o and hence at every point of the unit sphere. □

COROLLARY 6. If $\mathcal{S}$ is a non-empty, compact set in $\mathscr{E}^n$, $n > 1$, then its width function W assumes a minimal value in some direction u_1 and a maximal value in some direction u_2. If $W(u_1) < W(u_2)$, then W assumes every value intermediate to $W(u_1)$ and $W(u_2)$. If $n > 2$, W assumes every intermediate value infinitely often and at least once in the plane of u_1 and u_2.

PROOF. Because the domain of W is the unit sphere $S(O, 1)$, which is both compact and connected, the range of W is either a single number, and W is a constant function, or the range is a closed interval $[a, b]$. In the latter case, there exist points U_1 and U_2 on $S(O, 1)$ such that $W(u_1) = a$ and $W(u_2) = b$ are the minimal and maximal values of W. For every number x between a and b there is a pre-image u such that $W(u) = x$.

Because $W(-u_1) = W(u_1) \neq W(u_2)$, the points O, U_1 and U_2 determine a plane. If $n > 2$, this plane intersects $S(O, 1)$ in two circular arcs joining U_1 and U_2. Each of these arcs is a connected set and hence each is mapped by W onto $[a, b]$. Hence W assumes all its values in the plane of u_1 and u_2. Because there are infinitely many pairwise disjoint connected sets on the sphere joining U_1 and U_2, W must take on every value intermediate to a and b infinitely many times. □

A circle in $\mathscr{E}^2$ and a sphere in $\mathscr{E}^3$ are familiar examples of a closed convex curve and a closed convex surface that have the same widths in all directions and hence are sets of *constant width.* Since one's first impression is likely to be that the constant width property characterizes circles and spheres, it is an interesting fact that there are many other such sets. The subject of curves and surfaces of constant width is in fact an appealing special topic within convex body theory. A few examples of such curves are given in the exercises.

For a triangle in $\mathscr{E}^2$, it is clear that the minimal width is the shortest altitude and that the maximal width is the longest side. It is also clear that the length of the longest side is the diameter of the triangle. Moreover, the lines perpendicular to the longest side at its endpoints are parallel supporting lines. If one

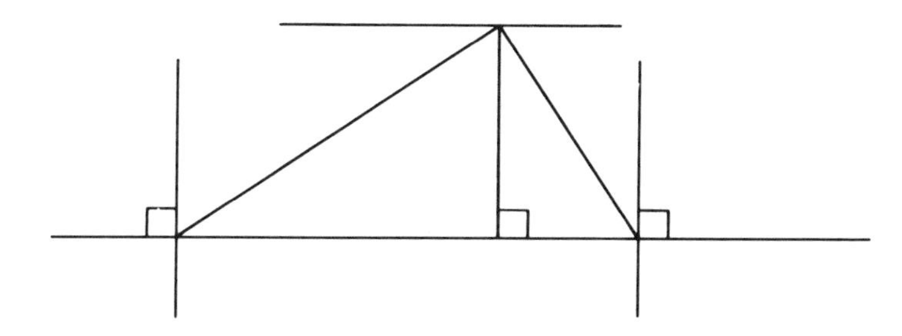

takes a different example, for instance, an ellipse, the diameter is the length of the major axis, and again the lines perpendicular to the axis at its endpoints are supporting lines that define the maximal width. We can generalize this relation of the maximal width and the diameter, and in doing so the following term is useful.

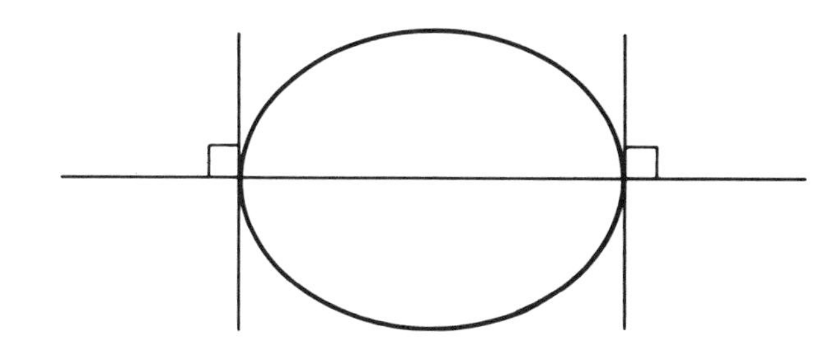

DEFINITION (Contact chord). If a slab with faces $\mathscr{H}_1$ and $\mathscr{H}_2$ supports a closed, convex surface $\mathscr{K}^0$, then a *contact chord* is a segment that joins a contact point of $\mathscr{H}_1$ and $\mathscr{K}^0$ with a contact point of $\mathscr{H}_2$ and $\mathscr{K}^0$.

The property of the triangle and the ellipse noted previously can now be generalized.

THEOREM 7. The diameter and the maximal width of a closed convex surface $\mathscr{K}^0$ are equal. The h-planes perpendicular to a diameter segment at its endpoints are the faces of a supporting slab of maximal width, and the diameter segment is the unique contact chord of the slab. Conversely, each supporting slab of maximal width has a unique contact chord that is a diameter segment perpendicular to the faces of the slab. (E)

We already have the notion of a regular point of a surface, namely, one at which a tangent plane exists, and the surface is said to be *differentiable* at such points. But in addition to regular points and corner points, the following property of surfaces is useful.

DEFINITION (Strictly convex surfaces). A closed convex surface is a *strictly convex closed surface* if it does not contain a segment.

The curves in $\mathscr{E}^2$ that one thinks of as "smooth ovals" are closed, convex curves that are differentiable, strictly convex closed curves. Although it is not true that a surface of constant width is necessarily differentiable, the next theorem is a simple consequence of Theorem 7.

THEOREM 8. A closed convex surface of constant width is a strictly convex, closed surface. (E)

We conclude this section with an important concept suggested by supporting half-spaces. Consider a plane convex body $\mathscr{K}$. Each supporting half-

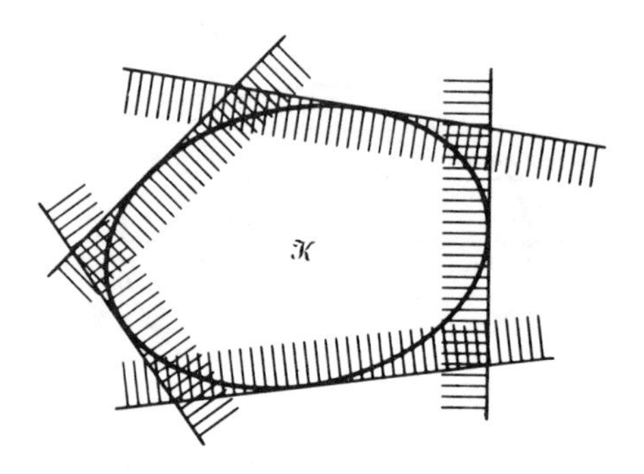

plane contains $\mathscr{K}$, and as we intersect more and more of these half-planes we obtain a better and better approximation of $\mathscr{K}$. It is not difficult to establish the following generalization.

THEOREM 9. If $\mathscr{R}$ is a non-empty, proper subset of $\mathscr{E}^n$ that is closed and convex, then $\mathscr{R}$ is the intersection of all of its closed, supporting half-spaces.

PROOF. Let $\mathscr{R}^*$ denote the intersection of all the closed half-spaces that support $\mathscr{R}$. Because $\mathscr{R}$ is contained in each such half-space, then clearly $\mathscr{R} \subset \mathscr{R}^*$.

Next, assume that $\mathscr{R}^* \not\subset \mathscr{R}$. Then there exists $P \in \mathscr{R}^*$ and $P \notin \mathscr{R}$. Because $\mathscr{R}$ is closed and convex, and $P \notin \mathscr{R}$, then, by Theorem 3, Sec. 4.2, P has a distinct foot F in $\mathscr{R}$ and the hyperplane $\mathscr{H}$ that is perpendicular to Sg$[PF]$ at F supports $\mathscr{R}$, as does the closed non-P side of $\mathscr{H}$. Since P is not in that closed half-space, then $P \notin \mathscr{R}^*$. The contradiction that P is and is not in $\mathscr{R}^*$ shows that $\mathscr{R}^* \not\subset \mathscr{R}$ is false. Hence $\mathscr{R}^* \subset \mathscr{R}$, and this, with $\mathscr{R} \subset \mathscr{R}^*$, implies $\mathscr{R} = \mathscr{R}^*$. $\square$

There is a general notion related to Theorem 9 which can be defined as follows.

DEFINITION (Minimal and maximal sets). In the sense of set theory, a set $\mathscr{S}$ is *the minimal set with a property* P if $\mathscr{S}$ has the property P and is contained in every other set with property P. Similarly, $\mathscr{S}$ is *the maximal set with a property* P if $\mathscr{S}$ has property P and contains every other set with property P.

It is clear that if a property P is preserved under all intersections, then the set theoretic minimal set with the property P is simply the intersection of all the sets with the property. Similarly, if the property is preserved under arbitrary unions, then the maximal set with the property is the union of all sets with the property.

Now consider an arbitrary set $\mathscr{R}$, and let a property P for a set $\mathscr{S}$ be defined by: $\mathscr{S}$ is convex and contains $\mathscr{R}$. If any two sets contain $\mathscr{R}$, then so does their intersection. If the two sets are convex, then so is their intersection. Each of the conditions, containing $\mathscr{R}$ and being convex, is preserved by arbitrary intersections, so the minimal convex set containing $\mathscr{R}$ is simply the intersection of all the convex sets that contain $\mathscr{R}$. This intersection has a special name.

DEFINITION (Convex hull). The *convex hull*, or *convex cover*, of a set $\mathscr{R}$ is the intersection of all the convex sets that contain $\mathscr{R}$ and is denoted by $\mathrm{CH}(\mathscr{R})$.

One immediate consequence of this new concept is the following characterization of convexity.

THEOREM 10. A set $\mathscr{R}$ is a convex set if and only if $\mathscr{R} = \mathrm{CH}(\mathscr{R})$. (E)

With each nonconvex set $\mathscr{R}$ there is associated a convex set $\mathrm{CH}(\mathscr{R})$ which is the minimal convex cover of $\mathscr{R}$. Through this association, it is sometimes possible to use the ideas of convexity theory to obtain interesting properties of nonconvex sets.

Exercises – Section 4.3

1. Find the point or segment that is the orthogonal projection of $\mathrm{Sg}[AB]$ into the line $\mathrm{Ln}(OC)$, where $A = (1, 1, 4)$, $B = (-1, 3, 2)$, and $C = (3, 0, -4)$.
2. If $\mathscr{C}$ is a differentiable curve in $\mathscr{E}^2$, then corresponding to a fixed point P there is a foot of P in each line tangent to $\mathscr{C}$ and the set of all these feet is called the "*pedal curve*" of $\mathscr{C}$ with respect to the *pedal point* P. This

notion, due to C. MacLaurin, can be modified by using a closed, convex curve $\mathscr{K}^0$ and taking the pedal curve to consist of the feet of P in the supporting lines to $\mathscr{K}^0$. When the pedal point P is the origin, then the pedal curve of $\mathscr{K}^0$ is just the set $H(u)u$, where H is the supporting function of $\mathscr{K}^0$.

If a pedal point P is taken interior to $\mathscr{K}$, then the length of each chord through P, of the pedal curve $\mathscr{K}^0_P$, is the width of $\mathscr{K}^0$ in the directions of the chord. Explain informally a construction for the curve $\mathscr{K}^0_P$ when $\mathscr{K}^0 = \triangle ABC$ and when $P \in \text{In}(\triangle^+ ABC)$, and note the chords through P that correspond to the minimum and maximum widths of $\mathscr{K}^0$.

3. In $\mathscr{E}^2$, a ball B(Q, a) is called a "disc" and the sphere S(Q, a) is a "circle." If the origin is interior to the disc B(Q, a) and $d(O, Q) = b > 0$, show that the pedal curve of the circle S(Q, a) with respect to the pedal point O has a polar coordinate equation of the form $r = a + b \cos \alpha$. Since the circle has constant width $2a$, every chord of the pedal curve that passes through the origin must have length $2a$. Check that this is so.
4. Prove Theorem 7.
5. Prove Theorem 8.
6. The simplest of all plane convex bodies of constant width, other than a disc, is a *reuleaux triangle*, which can be obtained as the intersection of three discs of radius a centered, respectively, at the vertices of an equilateral triangle whose sides have length a. The boundary $\mathscr{K}^0$ is thus the union of three 60° congruent, circular arcs. Explain informally why $\mathscr{K}^0$

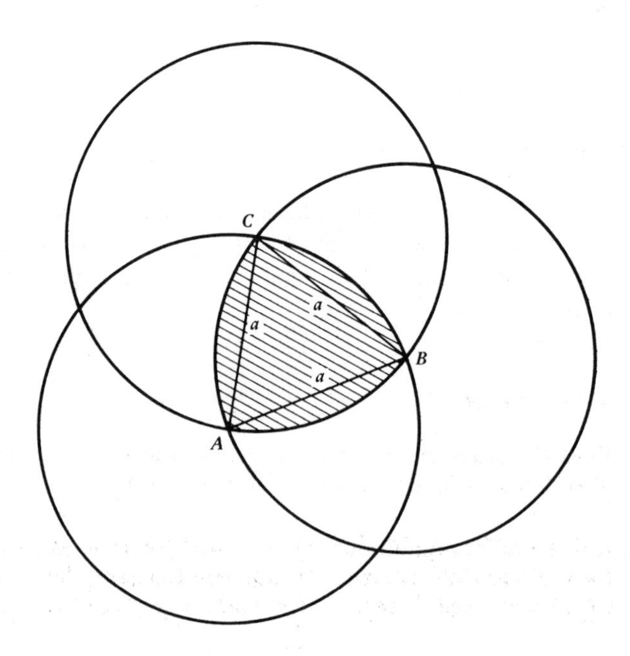

has constant width a. If a circle has width a, then its radius is $a/2$, so its circumference is $2\pi(a/2) = \pi a = (\pi) \times$ (width). Show that $(\pi) \times$ (width) is also the circumference of the realeaux triangle.

7. The reuleaux triangle is a constant width curve with three corner points, called its vertices. If the vertices of a regular pentagon are taken as the centers of five discs, each with radius equal to the length of a pentagon

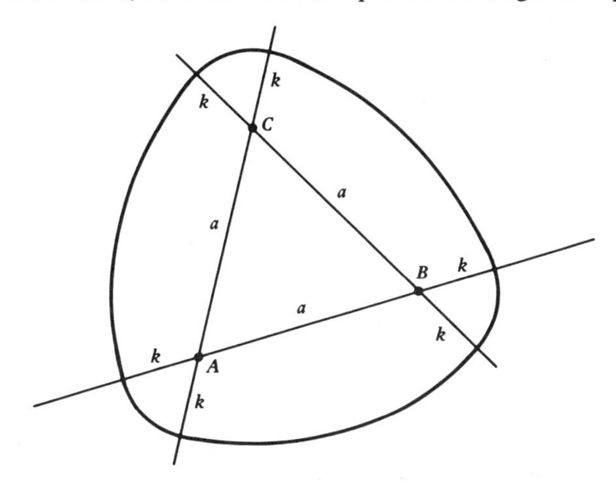

diagonal, will the intersection of these discs be a convex body of constant width whose boundary has five corner points?

8. The construction in Exercise 6 can be modified to give a differentiable curve of constant width. Corresponding to an equilateral triangle $\triangle ABC$ with side length a, let k be any positive number and let $r = a + k$. The $\measuredangle CAB$ intercepts a 60° arc on the circle $S(A, r)$, and the angle opposite to $\measuredangle CAB$ intercepts a 60° arc on the circle $S(A, k)$. The $\measuredangle ABC$ intercepts a 60° arc on $S(B, r)$, and its opposite angle intercepts a 60° arc on $S(B, k)$. Similarly, $\measuredangle BCA$ and its opposite angle intercept 60° arcs, respectively, on the circle $S(C, r)$ and $S(C, k)$. The union of the six arcs is the desired curve $\mathscr{K}^0$. Explain why it has constant width $a + 2k$ and why it has no corner points. Show that the circumference is again $(\pi) \times$ (width).

9. Give an example of a closed, convex curve that is differentiable but not strictly convex.

10. Prove Theorem 10.

11. The *closed convex hull* of an arbitrary set $\mathscr{R}$ is defined to be the intersection of all the *closed* convex sets that contain $\mathscr{R}$. Is the closed convex hull of $\mathscr{R}$ the same set as $\text{Cl}[\text{CH}(\mathscr{R})]$?

12. If a set $\mathscr{R}$ is non-empty, compact, and convex, why is its convex hull the intersection of all of its closed supporting slabs?

13. Is the interior of a convex body $\mathscr{K}$ the intersection of all the open supporting half-spaces of $\mathscr{K}$?

SECTION 4.4. CONVEX CONES; SUPPORT AND NORMAL CONES

In this section we show how an important class of sets called "convex cones" appear in a natural way in connection with convex sets and convex surfaces, and establish a few basic properties of such cones.

To begin with an example, suppose that $\mathscr{S}$ is a convex set and that point P is not in $\mathscr{S}$. Then P and each point X in $\mathscr{S}$ determine a ray $\text{Ry}[P, X)$, and these rays form a natural cone $\mathscr{C}$, with vertex P, associated with $\mathscr{S}$ and P. It

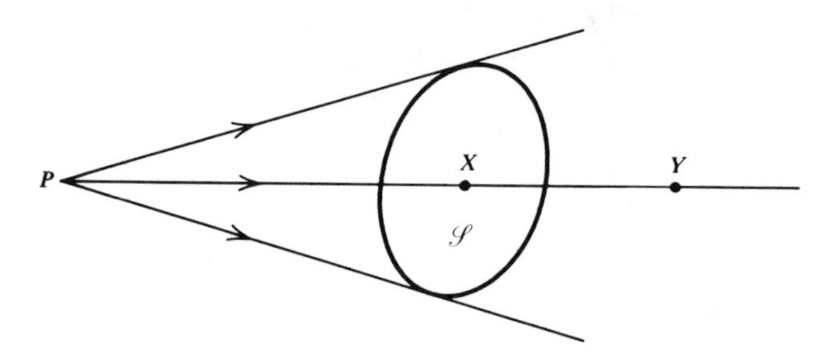

is not difficult to show that the convexity of $\mathscr{S}$ implies that of $\mathscr{C}$.* It is also apparent that $\mathscr{C}$ has the "vertex-ray" property that if $Y \in \mathscr{C}$ and $Y \neq P$, then $\text{Ry}[P, Y) \subset \mathscr{C}$. We can use convexity and the vertex-ray property to define a very general class of cones.

DEFINITION (Convex cone, cone vertex). A set $\mathscr{C}$ in $\mathscr{E}^n$ is a *convex cone*, or simply a cone, if $\mathscr{C}$ is a *proper, convex subset* of $\mathscr{E}^n$ that is *not a singleton set* and there exists a point V in $\mathscr{C}$ such that $\text{Ry}[V, X) \subset \mathscr{C}$ for all $X \in \mathscr{C}$, $X \neq V$. Each point V in $\mathscr{C}$ with this ray property is *a vertex* of the cone.

It is clear that the definition classifies a number of sets as cones that we do not ordinarily regard as cones. For example, a line is a cone and a closed ray is a cone. Also, by our definition, a closed half-space is a cone and its vertex set is the hyperplane that is its face.

Since any nonsingular affine transformation maps rays onto rays and preserves convexity, it is clear that the affine image of a cone in such a mapping is again a cone. The following particular case of this fact enables one to simplify many proofs.

THEOREM 1. A set $\mathscr{C}$ in $\mathscr{E}^n$ is a cone with vertex V if and only if the $-v$ translate of $\mathscr{C}$ is a cone with the origin as a vertex. (E)

* Although one can define nonconvex cones, we do not do so in this text.

THEOREM 2. If set $\mathscr{C}$ is a non-empty, nonsingleton proper subset of $\mathscr{E}^n$, then $\mathscr{C}$ is a cone with the origin as a vertex if and only if (i) $\lambda x \in \mathscr{C}$ for all $X \in \mathscr{C}$ and all $\lambda \geq 0$; and (ii) $x + y \in \mathscr{C}$ for all $X, Y \in \mathscr{C}$.

PROOF. Assume that $\mathscr{C}$ is a cone with the origin as a vertex. Then $O \in \mathscr{C}$. Consider any $X \in \mathscr{C}$ and any number $\lambda \geq 0$. If $X = 0$, then $\lambda x = o$, so $\lambda x \in \mathscr{C}$. If $X \neq 0$, then $\text{Ry}[O, X) \subset \mathscr{C}$ implies that $\lambda x \in \mathscr{C}$. Thus $\lambda x \in \mathscr{C}$ for all $X \in \mathscr{C}$ and all $\lambda \geq 0$. Next, consider $X, Y \in \mathscr{C}$. Because $\mathscr{C}$ is convex, then $z = \frac{1}{2}x + \frac{1}{2}y$ is in $\mathscr{C}$. We have shown that this implies $\lambda z \in \mathscr{C}$ for all $\lambda \geq 0$, so $2z \in \mathscr{C}$, hence $x + y \in \mathscr{C}$.

For the converse, assume that $\mathscr{C}$ has properties (i) and (ii). Because $\mathscr{C}$ is not empty, there exists $P \in \mathscr{C}$. By property (i), $Op = o$ is in $\mathscr{C}$, so $\mathscr{C}$ contains the origin. Next, if $X \in \mathscr{C}$, $X \neq 0$, then from property (i), $\lambda x \in \mathscr{C}$ for all $\lambda \geq 0$, so $\text{Ry}[O, X) \subset \mathscr{C}$. Finally, consider two points X, Y in $\mathscr{C}$ and a point Z between them. Then Z has a representation $z = \eta_1 x + \eta_2 y, \eta_1 + \eta_2 = 1, \eta_1, \eta_2 > 0$. By property (i), $u = \eta_1 x$ and $v = \eta_2 y$ are in $\mathscr{C}$, and by property (ii) $u + v$ is in $\mathscr{C}$, hence $Z \in \mathscr{C}$. Thus $\mathscr{C}$ is convex. Hence $\mathscr{C}$ is a cone with the origin as a vertex. □

COROLLARY 2. If set $\mathscr{C}$ is a non-empty, nonsingleton, proper subset of $\mathscr{E}^n$, then $\mathscr{C}$ is a cone with vertex V if and only if (i) $v + \lambda \mathbf{VX} \in \mathscr{C}$ for all $X \in \mathscr{C}$ and all $\lambda \geq 0$; and (ii) $v + \mathbf{VX} + \mathbf{VY} \in \mathscr{C}$ for all $X, Y \in \mathscr{C}$.

THEOREM 3. A set $\mathscr{C}$ in $\mathscr{E}^n$ is a cone with vertex V if and only if $\mathscr{C}$ is a non-empty, nonsingleton, proper subset of $\mathscr{E}^n$, and $v + \alpha\mathbf{VX} + \beta\mathbf{VY} \in \mathscr{C}$ for all $X, Y \in \mathscr{C}$ and all $\alpha, \beta \geq 0$.

PROOF. Suppose first that $\mathscr{C}$ is a cone with vertex V. Then $\mathscr{C}$ is a proper subset of $\mathscr{E}^n$, $V \in \mathscr{C}$ and $\mathscr{C} \neq \{V\}$. Let X, Y denote any two points of $\mathscr{C}$ and let α, β be any two nonnegative numbers. By Corollary 2,

$$z = v + \alpha\mathbf{VX} \in \mathscr{C}, \qquad \text{and} \qquad w = v + \beta\mathbf{VY} \in \mathscr{C}. \tag{1}$$

Now, since Z and W are in $\mathscr{C}$, Corollary 2 also implies that

$$v + \mathbf{VZ} + \mathbf{VW} \in \mathscr{C}, \qquad \text{so} \qquad v + \alpha\mathbf{VX} + \beta\mathbf{VY} \in \mathscr{C}. \tag{2}$$

For the converse, assume that $\mathscr{C}$ is a proper subset of $\mathscr{E}^n$, that $V \in \mathscr{C}$, that $\mathscr{C} \neq \{V\}$, and that for all $X, Y \in \mathscr{C}$ and all $\alpha, \beta \geq 0$,

$$v + \alpha\mathbf{VX} + \beta\mathbf{VY} \in \mathscr{C}. \tag{3}$$

Setting $\beta = 0$ in (3), it follows that if $X \in \mathscr{C}$ and $\alpha \geq 0$, then

$$v + \alpha\mathbf{VX} \in \mathscr{C}. \tag{4}$$

Moreover, from $\alpha = \beta = 1$ in (3), it follows that if X, $Y \in \mathscr{C}$, then

$$v + \mathbf{VX} + \mathbf{VY} \in \mathscr{C}. \tag{5}$$

Thus, from (4), (5), and Corollary 2, it follows that $\mathscr{C}$ is a cone with vertex V. □

For $v = o$, Theorem 3 has the following form.

COROLLARY 3. A set $\mathscr{C}$ in $\mathscr{E}^n$ is a cone with the origin as a vertex if and only if $\mathscr{C}$ is a non-empty, nonsingleton proper subset of $\mathscr{E}^n$, and $\alpha x + \beta y \in \mathscr{C}$ for all X, $Y \in \mathscr{C}$ and all $\alpha, \beta \geq 0$.

Our next theorem gives some ray properties associated with interior and boundary points of a cone.

THEOREM 4. If V is a vertex of a cone $\mathscr{C}$ in $\mathscr{E}^n$, then

(i) $V \in \mathrm{Bd}(\mathscr{C})$
(ii) $X \in \mathrm{In}(\mathscr{C})$ implies $\mathrm{Ry}(V, X) \subset \mathrm{In}(\mathscr{C})$
(iii) $X \in \mathrm{Bd}(\mathscr{C})$ and $X \neq V$ imply $\mathrm{Ry}[V, X) \subset \mathrm{Bd}(\mathscr{C})$

PROOF. Assume that $V \in \mathrm{In}(\mathscr{C})$. Then there exists $\mathrm{N}(V, \delta) \subset \mathrm{In}(\mathscr{C})$. If Y is any point of $\mathscr{E}^n$ distinct from V, the open ray $\mathrm{Ry}(V, Y)$ intersects $\mathrm{N}(V, \delta)$ and so intersects $\mathscr{C}$. By the vertex-ray property, it follows that $\mathrm{Ry}[V, Y) \subset \mathscr{C}$, so $Y \in \mathscr{C}$. This, with $V \in \mathscr{C}$, implies that every point is in $\mathscr{C}$, so $\mathscr{C} = \mathscr{E}^n$, contradicting the fact that $\mathscr{C}$ is a cone. The contradiction shows that $V \notin \mathrm{In}(\mathscr{C})$, and, since $V \in \mathscr{C}$, it follows that $V \in \mathrm{Bd}(\mathscr{C})$, which establishes (i).

Next, consider $X \in \mathrm{In}(\mathscr{C})$ and let Y be any point of the open ray $\mathrm{Ry}(V, X)$. By property (i), $X \neq V$. If $d(V, Y) < d(V, X)$, then Y is between an interior point of $\mathscr{C}$ and point V of $\mathscr{C}$, so, by Theorem 5, Sec. 3.1, $Y \in \mathrm{In}(\mathscr{C})$. If $d(V, Y) = d(V, X)$, then $Y = X$ implies $Y \in \mathrm{In}(\mathscr{C})$. Finally, if $d(V, Y) > d(V, X)$, let Z denote any point on $\mathrm{Ry}(V, X)$ such that $d(V, Z) > d(V, Y)$. Then Y is between $X \in \mathrm{In}(\mathscr{C})$ and $Z \in \mathscr{C}$, so, again by Theorem 5, Sec. 3.1, $Y \in \mathrm{In}(\mathscr{C})$. In all cases, $Y \in \mathrm{In}(\mathscr{C})$, hence $\mathrm{Ry}(V, X) \subset \mathrm{In}(\mathscr{C})$, so (ii) is established.

Now let X be a boundary point of $\mathscr{C}$, $X \neq V$, and let Y be any point of $\mathrm{Ry}[V, X)$ distinct from X and V. To show that $Y \in \mathrm{Bd}(\mathscr{C})$, we consider the cases $X \in \mathscr{C}$ and $X \notin \mathscr{C}$ separately.

If $X \in \mathscr{C}$, then $\mathrm{Ry}(V, X) \subset \mathscr{C}$ implies that $Y \in \mathscr{C}$. If Y belonged to $\mathrm{In}(\mathscr{C})$, then, by property (ii), $\mathrm{Ry}(V, Y)$ would belong to $\mathrm{In}(\mathscr{C})$, hence X would belong to $\mathrm{In}(\mathscr{C})$, which is false. Thus $Y \notin \mathrm{In}(\mathscr{C})$, so $Y \in \mathrm{Bd}(\mathscr{C})$.

If $X \notin \mathscr{C}$, then, from the ray-vertex property of cones, it follows that $\mathrm{Ry}(V, X) \cap \mathscr{C} = \varnothing$, so $Y \notin \mathscr{C}$. Now consider an arbitrary deleted neighborhood of Y, $\mathrm{N}_o(Y, \delta)$. Because $Y \in \mathrm{Ry}(V, X)$, it has a representation

$$y = (1 - \eta)v + \eta x, \qquad \eta > 0. \tag{1}$$

Since $X \in \mathrm{Bd}(\mathscr{C})$ and $X \notin \mathscr{C}$, X is a limit point to $\mathscr{C}$, so there exists $Z \in N_o(X, (1/\eta)\delta) \cap \mathscr{C}$. Because $Z \in \mathscr{C}$, then, by Corollary 2, $v + \lambda \mathbf{VZ} \in \mathscr{C}$ for all $\lambda \geq 0$, hence W defined by

$$w = (1 - \eta)v + \eta z \tag{2}$$

is in $\mathscr{C}$. From (1) and (2),

$$d(Y, W) = |y - w| = \eta|x - z| < \eta\left(\frac{1}{\eta}\delta\right) = \delta, \tag{3}$$

hence $W \in \mathrm{N}(Y, \delta)$. Because $W \in \mathscr{C}$, $W \neq Y$, hence $W \in \mathrm{N}_o(Y, \delta) \cap \mathscr{C}$. Since an arbitrary deleted neighborhood of Y intersects $\mathscr{C}$, it follows that $Y \in \mathrm{Lp}(\mathscr{C})$. This, with $Y \notin \mathscr{C}$, implies $Y \in \mathrm{Bd}(\mathscr{C})$.

In all cases, then, $Y \in \mathrm{Bd}(\mathscr{C})$, so $\mathrm{Ry}[V, X) \subset \mathrm{Bd}(\mathscr{C})$, which establishes (iii). □

COROLLARY 4. The closure of a cone with vertex V is a cone with vertex V. (E)

Because of the invariance of convexity and closedness under intersections, cones have the following important property.

THEOREM 5. The intersection of any number of cones with vertex V is either $\{V\}$ or else is a cone with vertex V. The intersection of any number of closed cones with vertex V is either $\{V\}$ or is a closed cone with vertex V. (E)

If one thinks of a cone in $\mathscr{E}^3$ as a collection of rays $\mathrm{Ry}[V, X)$, but not the lines $\mathrm{Ln}(V, X)$, it is clear that all the supporting planes pass through V. This property holds for the general cones we have been discussing and in fact characterizes closed cones in the following sense.

THEOREM 6. If set $\mathscr{C}$ is a closed, convex, proper subset of $\mathscr{E}^n$, and $\mathscr{C}$ is neither empty nor a singleton, then $\mathscr{C}$ is a closed cone if and only if there is at least one point common to all the hyperplanes that support $\mathscr{C}$. Moreover, if the intersection set of all such hyperplanes is non-empty, then it is the vertex set of the cone.

PROOF. Let $\mathscr{I}$ denote the intersection set of all the hyperplanes that support $\mathscr{C}$. Assume that $\mathscr{I}$ is not empty, and let V denote an arbitrary point of $\mathscr{I}$. By Theorem 9, Sec. 4.3, $\mathscr{C}$ is the intersection of its closed supporting half-spaces. Each of these half-spaces is a cone and is a cone with vertex V, since, by hypothesis, V is in the face of each such half-space. Since $\mathscr{C}$ is contained in each of the closed, supporting half-spaces, and $\mathscr{C}$ is not a singleton set, the intersection of these closed half-spaces is not $\{V\}$. Thus, by Theorem 5, the intersection set $\mathscr{C}$ is a closed cone with vertex V. Moreover, if $\mathscr{S}$ denotes the vertex set of $\mathscr{C}$, we have shown that $\mathscr{I} \subset \mathscr{S}$.

Conversely, suppose that $\mathscr{C}$ is a closed cone and that V is a vertex of $\mathscr{C}$. Assume that there is some hyperplane $\mathscr{H}$ that supports $\mathscr{C}$ but does not contain V. Let $f(x) = a \cdot x$ be a linear functional such that $f(x) = k$ is an equation for $\mathscr{H}$, with a chosen so that $f(x) \leq k$ for $X \in \mathscr{C}$. Because $\mathscr{C}$ is closed, $\mathscr{H}$ contacts $\mathscr{C}$ at some point P. Since $P \neq V$, $\text{Ry}[V, P) \subset \mathscr{C}$, so $f(x) \leq k$ for $X \in \text{Ry}[V, P)$, that is,

$$f[v + \lambda(p - v)] \leq k \quad \text{for all} \quad \lambda \geq 0. \tag{1}$$

Because $P \in \mathscr{H}$, $f(p) = k$, and since V in $\mathscr{C}$ is not in $\mathscr{H}$, $f(v) < k$. Therefore $k - f(v) = f(p) - f(v) > 0$. Hence if $\lambda_1 > 1$, then

$$f(v) + \lambda_1[f(p) - f(v)] = f(v) + \lambda_1[k - f(v)] > f(v) + [k - f(v)] = k. \tag{2}$$

Since f is linear, (2) implies that

$$f[v + \lambda_1(p - v)] > k, \tag{3}$$

which contradicts (1). Thus the assumption that V is not in $\mathscr{H}$ cannot hold. Therefore each vertex must be in all the hyperplanes that support $\mathscr{C}$. Hence $\mathscr{S} \subset \mathscr{I}$. This, with the previous conclusion that $\mathscr{I} \subset \mathscr{S}$, shows that $\mathscr{I} = \mathscr{S}$. □

COROLLARY 6. If a hyperplane $\mathscr{H}$ supports a cone $\mathscr{C}$ at $P \in \text{Bd}(\mathscr{C})$, and P is not a vertex V of $\mathscr{C}$, then $\text{Ry}[V, P) \subset \mathscr{H}$.

PROOF. By Corollary 4, $\text{Cl}(\mathscr{C})$ is a closed cone with vertex V. Since $\mathscr{H}$ supports $\mathscr{C}$, it also supports $\text{Cl}(\mathscr{C})$. Thus, by Theorem 6, $V \in \mathscr{H}$. Since $P \in \mathscr{H}$ and $V \in \mathscr{H}$, the affine span of P and V, namely, $\text{Ln}(VP)$, is contained in $\mathscr{H}$, hence $\text{Ry}[V, P) \subset \mathscr{H}$. □

At the start of the section we described a cone associated with a convex set $\mathscr{S}$ and a point P not in $\mathscr{S}$. Theorem 5 suggests a far more general way of associating a cone with an arbitrary set $\mathscr{S}$ and point P.

DEFINITION (Cone of a set and a point). Corresponding to a non-empty set $\mathscr{S}$ and a point P in $\mathscr{E}^n$, if the intersection of all the cones that have vertex P and contain $\mathscr{S}$ is a cone, it is *the cone of* $\mathscr{S}$ *and* P and will be denoted by $\mathrm{Cn}(\mathscr{S}; P)$.

We can describe very general conditions under which $\mathrm{Cn}(\mathscr{S}; P)$ exists in terms of a notion we have not needed up to now but which appears in many parts of convexity theory.

DEFINITION (Weak separation of sets). Two non-empty sets $\mathscr{R}$ and $\mathscr{S}$ are *weakly separated* if there exists a hyperplane $\mathscr{H}$ such that one closed side of $\mathscr{H}$ contains $\mathscr{R}$ and the other closed side contains $\mathscr{S}$. The sets are said to be weakly separated *by* $\mathscr{H}$. A point P is weakly separated from set $\mathscr{S}$ if $\{P\}$ and $\mathscr{S}$ are weakly separated sets.

Weak separation need not correspond to any intuitive notion of separatedness. For example, any non-empty subset $\mathscr{R}$ of a hyperplane $\mathscr{H}$ is weakly separated from itself, since both closed sides of $\mathscr{H}$ contain $\mathscr{R}$. But, as we shall see, even in nonintuitive cases the following weak separation property is useful. A hyperplane $\mathscr{H}$ is the zero set of some affine functional $f(x) = a \cdot x + k$, $a \neq o$. If $\mathscr{R}$ and $\mathscr{S}$ are weakly separated by $\mathscr{H}$, then f must be nonnegative on one of the sets and nonpositive on the other, even though it may be zero on either or both.

In terms of weak separation, we can now establish the following existence theorem.

THEOREM 7. If point V is weakly separated from the non-empty set $\mathscr{S}$, and $\mathscr{S}$ is not $\{V\}$, then the cone of $\mathscr{S}$ and V exists.

PROOF. Since V is weakly separated from $\mathscr{S}$ there exists a hyperplane $\mathscr{H}$, with opposite closed sides $\mathscr{H}'$ and $\mathscr{H}''$ such that $\mathscr{S} \subset \mathscr{H}'$ and $P \in \mathscr{H}''$. If V is in the face $\mathscr{H}$, then the closed half-space $\mathscr{H}'$ is a cone with vertex V that contains $\mathscr{S}$. If V is not in $\mathscr{H}$, then V lies in the open non-$\mathscr{H}'$ side of $\mathscr{H}$. Then there exists a hyperplane $\mathscr{H}_1$ that is parallel to $\mathscr{H}$ and contains V. Thus the closed side of $\mathscr{H}_1$ that contains $\mathscr{H}$ is a cone with vertex V that contains $\mathscr{S}$. In all cases, then, there exists a cone with vertex V containing $\mathscr{S}$. Thus the intersection of all the cones that contain $\mathscr{S}$ and have vertex V is not empty. Because $\mathscr{S} \neq \{V\}$, and $\mathscr{S}$ is in all the cones, the intersection is not $\{V\}$. Thus, by Theorem 5, the intersection set is a cone and hence is the cone $\mathrm{Cn}(\mathscr{S}; V)$ that also contains $\mathscr{S}$ and has vertex V. □

The cone $\mathrm{Cn}(\mathscr{S}; V)$ may be generated by V and sets other than $\mathscr{S}$. In particular, the following relation is always true.

THEOREM 8. The cone of $\mathscr{S}$ and V, if it exists, is also the cone of $\text{CH}(\mathscr{S})$ and V.

PROOF. Let $\mathscr{F}_1$ denote the family of all V-vertex cones that contain $\mathscr{S}$, and let $\mathscr{F}_2$ be the family of all V-vertex cones that contain $\text{CH}(\mathscr{S})$. Since $\mathscr{S} \subset \text{CH}(\mathscr{S})$, then clearly $\mathscr{F}_2 \subset \mathscr{F}_1$. If $\mathscr{C}$ is a cone in $\mathscr{F}_1$ then $\mathscr{C}$ is a convex set that contains $\mathscr{S}$. By its definition, $\text{CH}(\mathscr{S})$ is contained in every convex set that contains $\mathscr{S}$, hence $\text{CH}(\mathscr{S}) \subset \mathscr{C}$. Thus $\mathscr{C}$ is a V-vertex cone that contains $\text{CH}(\mathscr{S})$, so $\mathscr{C} \in \mathscr{F}_2$. Thus $\mathscr{F}_2 \subset \mathscr{F}_1$, so $\mathscr{F}_1 = \mathscr{F}_2$. The intersection set of the family $\mathscr{F}_1 = \mathscr{F}_2$ is therefore both $\text{Cn}(\mathscr{S}; V)$ and $\text{Cn}[\text{CH}(\mathscr{S}); V]$. □

We can now relate $\text{Cn}(\mathscr{S}; V)$ with the cone of a set and a point that we described at the beginning of the section.

THEOREM 9. If the cone $\text{Cn}(\mathscr{S}; V)$ exists, and if $\mathscr{S}$ is convex, then

$$\text{Cn}(\mathscr{S}; V) = \bigcup \{\text{Ry}[V, X): X \in \mathscr{S}, X \neq V\}.$$

PROOF. By Corollary 1, it suffices to prove the theorem with V taken to be the origin O. Let $\mathscr{C} = \bigcup \{\text{Ry}[O, X): X \in \mathscr{S}, X \neq 0\}$. We show first that $\mathscr{C}$ is a cone and then that $\mathscr{C} = \text{Cn}(\mathscr{S}; O)$.

Consider any $X \in \mathscr{C}$. Since $\text{Cn}(\mathscr{S}; O)$ exists, there exists $P \in \mathscr{S}$ $P \neq O$. By the definition of $\mathscr{C}$, $\text{Ry}[O, P) \subset \mathscr{C}$, hence $O \in \mathscr{C}$. If $X = O$, then $\lambda x = o \in \mathscr{C}$ for all λ. If $X \neq O$, then $X \in \mathscr{C}$ implies that $X \in \text{Ry}(O, Y)$ for some $Y \in \mathscr{S}$. From $\text{Ry}[O, Y) \subset \mathscr{C}$ and $\text{Ry}[O, X) = \text{Ry}[O, Y)$, $\text{Ry}[O, X) \subset \mathscr{C}$. Therefore $\lambda x \in \mathscr{C}$ for all $\lambda \geq 0$.

Next, consider $X, Y \in \mathscr{C}$. If $X = Y$, then $x + y = 2x \in \mathscr{C}$ because, by the previous argument, $\lambda x \in \mathscr{C}$ for all $\lambda \geq 0$. If $X \neq Y$, but one of the points X, Y is O, then $x + o = x \in \mathscr{C}$ and $o + y = y \in \mathscr{C}$ imply that $x + y \in \mathscr{C}$. Finally, suppose that X and Y are distinct and both different from O. Then X and Y are in $\mathscr{C}$ because there exist points X_o, Y_o in $\mathscr{S}$, $X_o \neq O$, $Y_o \neq O$, such that $X \in \text{Ry}(O, X_o)$ and $Y \in \text{Ry}(O, Y_o)$. Thus there exist numbers $\lambda_1 > 0$ and $\lambda_2 > 0$ such that

$$x = \lambda_1 x_o \qquad \text{and} \qquad y = \lambda_2 y_o. \tag{1}$$

Because $\mathscr{S}$ is convex, it contains any convex combination of X_o, Y_o, so

$$z = \frac{\lambda_1}{\lambda_1 + \lambda_2} x_o + \frac{\lambda_2}{\lambda_1 + \lambda_2} y_o \in \mathscr{S}. \tag{2}$$

If $Z = O$ then $Z \in \mathscr{C}$, and if $Z \neq O$, then $\text{Ry}[O, Z) \subset \mathscr{C}$ implies $Z \in \mathscr{C}$ Therefore $Z \in \mathscr{C}$, so $\lambda z \in \mathscr{C}$ for all $\lambda \geq 0$. In particular, then, $(\lambda_1 + \lambda_2)z \in \mathscr{C}$. Thus

$$(\lambda_1 + \lambda_2)z = \lambda_1 x_o + \lambda_2 y_o = x + y \in \mathscr{C}. \tag{3}$$

Since $X \in \mathscr{C}$ implies that $\lambda x \in \mathscr{C}$ for all $\lambda \geq 0$, and since $X, Y \in \mathscr{C}$ implies that $x + y \in \mathscr{C}$, then, by Theorem 2, $\mathscr{C}$ is a cone with the origin as a vertex.

Now, let X be any point of $\mathscr{C}$, and let $\mathscr{C}'$ denote any O-vertex cone that contains $\mathscr{S}$. If $X = O$ then clearly $O \in \mathscr{C}'$. If $X \neq O$, then there exists $X_o \in \mathscr{S}$, $X_o \neq O$ such that $X \in \text{Ry}(O, X)$. But, since $X_o \in \mathscr{S}$, then $X_o \in \mathscr{C}'$. Since $X_o \neq O$, then $\text{Ry}[O, X_o) \subset \mathscr{C}'$. Thus $X \in \mathscr{C}'$. Since $X \in \mathscr{C}$ implies $X \in \mathscr{C}'$, then $\mathscr{C} \subset \mathscr{C}'$. Because $\mathscr{C}$ is contained in every O-vertex cone that contains $\mathscr{S}$, then $\mathscr{C} \subset \text{Cn}(\mathscr{S}; O)$. But $\mathscr{C}$ itself is an O-vertex cone that contains $\mathscr{S}$, hence $\text{Cn}(\mathscr{S}; O) \subset \mathscr{C}$. From the opposite inclusions, it follows that $\mathscr{C} = \text{Cn}(\mathscr{S}; O)$. □

COROLLARY 9. If $\mathscr{S}$ is any non-empty, nonsingleton set in $\mathscr{E}^n$ and if V is weakly separated from $\mathscr{S}$ by some hyperplane, then

$$\text{Cn}(\mathscr{S}; V) = \bigcup \{\text{Ry}[V, X)\colon X \in \text{CH}(\mathscr{S}), X \neq V\}. \tag{E}$$

If set $\mathscr{S}$ is a ball, and if the hyperplane $\mathscr{H}$ is tangent to $\mathscr{S}$ at point V, then it is clear from Theorem 9 that $\text{Cn}(\mathscr{S}; V)$ is the union of $\{V\}$ and an open side of $\mathscr{H}$. Thus $\text{Cn}(\mathscr{S}; V)$ is not a closed cone, despite the fact that $\mathscr{S}$ is closed. However, the following modification does determine a closed cone.

THEOREM 10. If a set $\mathscr{S}$ is non-empty and compact, and if point V is not in $\mathscr{S}$, then $\text{Cn}(\mathscr{S}; V)$ is a closed cone.

PROOF. Since the convex hull of a compact set is compact, then $\text{CH}(\mathscr{S})$ is compact and of course is convex. Because $\text{Cn}(\mathscr{S}; V)$ is also $\text{Cn}[\text{CH}(\mathscr{S}); V]$, there is no loss of generality in supposing that $\mathscr{S}$ is convex as well as compact. Also, since closedness, compactness, and convexity are all invariant under translations, it suffices to consider the case in which V is the origin.

To show that $\text{Cn}(\mathscr{S}; O)$ is closed, we show that $\text{Cp}[\text{Cn}(\mathscr{S}; O)]$ is open. For this purpose, consider an arbitrary point P in the complement of $\text{Cn}(\mathscr{S}; O)$. From Theorem 9, $\text{Ry}(O, P) \cap \mathscr{S} = \varnothing$, and since $V \notin \mathscr{S}$, then $\text{Ry}[O, P) \cap \mathscr{S} = \varnothing$. Thus $\text{Ry}[O, P)$ and $\mathscr{S}$ are non-empty, disjoint sets that are closed and convex, and, since $\mathscr{S}$ is compact, it follows from Theorem 8, Sec. 4.2, that there exists a slab separating $\text{Ry}[O, P)$ and $\mathscr{S}$. If $w > 0$ is the width of the slab, then $d(X, Y) \geq w$ for all $X \in \text{Ry}[O, P)$ and all $Y \in \mathscr{S}$. Thus, in equivalent terms,

$$|\lambda p - y| \geq w, \qquad \text{for all } Y \in \mathscr{S} \text{ and all } \lambda \geq 0. \tag{1}$$

Because $\mathscr{S}$ is bounded, there exists $m > 0$ such that

$$|y| \leq m, \qquad \text{for all } Y \in \mathscr{S}. \tag{2}$$

We now define

$$\varepsilon = \frac{w|p|}{w + m} \quad \text{and} \quad \eta_o = \frac{|p|}{w + m} = \frac{\varepsilon}{w}. \tag{3}$$

Since $P \neq 0$, $\varepsilon > 0$ and $\eta_o > 0$. Also, by direct calculation,

$$|p| - \eta_o m = \varepsilon. \tag{4}$$

Now let Z denote an arbitrary point of $\mathrm{Cn}(\mathscr{S}; O)$. If $Z = O$, then from $\mathrm{Ry}(O, Z) \subset \mathrm{Cn}(\mathscr{S}; O)$ it follows that there exists $Y \in \mathscr{S}$ and $\eta_1 > 0$ such that $z = \eta_1 y$. If $Z = O$, then clearly $z = Oy$ for any $Y \in \mathscr{S}$. Thus, in all cases, Z has a representation

$$z = \eta y, \qquad Y \in \mathscr{S}, \qquad \eta \geq 0. \tag{5}$$

If $0 \leq \eta \leq \eta_o$, then, from $\eta|y| \leq \eta_o m$ and (4), we have

$$d(P, Z) = |p - \eta y| \geq |p| - \eta|y| \geq |p| - \eta_o m = \varepsilon. \tag{6}$$

If $\eta > \eta_o$, then $\eta > 0$, so $1/\eta > 0$, and from (1) and (3) we have

$$d(P, Z) = |p - \eta y| = \eta|(1/\eta)p - y| \geq \eta w > \eta_o w = \varepsilon. \tag{7}$$

In all cases, then, $d(P, Z) \geq \varepsilon$. Since Z was an arbitrary point of $\mathrm{Cn}(\mathscr{S}; O)$, it follows that $\mathrm{N}(P, \varepsilon) \cap \mathrm{Cn}(\mathscr{S}; O) = \varnothing$. Thus each point P in the complement of $\mathrm{Cn}(\mathscr{S}; O)$ is interior to the complement. Therefore $\mathrm{Cp}[\mathrm{Cn})\mathscr{S}; O)]$ is open, so $\mathrm{Cn}(\mathscr{S}; O)$ is closed. □

Since the closure of a V-vertex cone is a closed V-vertex cone, one would expect that the closure of $\mathrm{Cn}(\mathscr{S}; V)$ is the intersection of the closed V-vertex cones that contain $\mathscr{S}$, and this is in fact the case.

THEOREM 11. If the cone $\mathrm{Cn}(\mathscr{S}; V)$ exists, then its closure is the intersection of all the closed cones that contain $\mathscr{S}$ and have vertex V.

PROOF. Let $\mathscr{C}^*$ denote the intersection of all the closed, V-vertex cones that contain $\mathscr{S}$. Since $\mathrm{Cn}(\mathscr{S}; V)$ exists, $\mathscr{C}^* \neq \{V\}$, so, by Theorem 5, $\mathscr{C}^*$ is a closed cone with vertex V. The family of cones intersected to obtain $\mathscr{C}^*$ is a subset of the cones intersected to obtain $\mathrm{Cn}(\mathscr{S}; V)$, so $\mathrm{Cn}(\mathscr{S}; V) \subset \mathscr{C}^*$. By Corollary 4, $\mathrm{Cl}[\mathrm{Cn}(\mathscr{S}; V)]$ is a particular closed V-vertex cone that contains $\mathscr{S}$, and therefore $\mathscr{C}^* \subset \mathrm{Cl}[\mathrm{Cn}(\mathscr{S}; V)]$. Thus

$$\mathrm{Cn}(\mathscr{S}; V) \subset \mathscr{C}^* \subset \mathrm{Cl}[\mathrm{Cn}(\mathscr{S}; V)]. \tag{1}$$

Since the closure of a set is the smallest closed set containing it, (Exercise 11, Sec. 1.3), (1) implies that $\mathscr{C}^* = \text{Cl}[\text{Cn}(\mathscr{S}; V)]$. □

As Theorems 10 and 11 suggest, the following relations are valid.

THEOREM 12. If $\mathscr{S}$ is a non-empty, compact, convex set in $\mathscr{E}^n$, and $P \notin \mathscr{S}$, then $\text{Cn}(\mathscr{S}; V)$ exists and is the intersection of the closed half-spaces that support $\mathscr{S}$ and whose faces contain the point P. (E)

If the set $\mathscr{S}$ in Theorem 12 is a convex body, then the cone $\text{Cn}(\mathscr{S}; V)$ is the kind of object that we normally think of as a cone, and we introduce a name for such special cones.

DEFINITION (Regular cone). A cone in $\mathscr{E}^n$ is a *regular cone* if it is closed and n-dimensional and has a unique vertex.

THEOREM 13. If P is not a point of a convex body $\mathscr{K}$, then the cone of $\mathscr{K}$ and P is a regular cone. (E)

Regular cones can also be obtained in the following way.

THEOREM 14. If set $\mathscr{S}$ in a hyperplane $\mathscr{H}$ of $\mathscr{E}^n$ is an $(n-1)$-dimensional, compact, convex set, and if $P \notin \mathscr{H}$, then the cone $\text{Cn}(\mathscr{S}; P)$ is regular.

PROOF. By Theorem 10, $\text{Cn}(\mathscr{S}; P)$ exists and is closed. Since $\mathscr{S}$ is $(n-1)$-dimensional, it contains n independent points whose affine span is $\mathscr{H}$. These n points and P form a set of $n+1$ independent points because $P \notin \mathscr{H}$. Because the $n+1$ points are in $\text{Cn}(\mathscr{S}; P)$, the cone is n-dimensional. Now let $\mathscr{H}_1$ denote the hyperplane through P that is parallel to $\mathscr{H}$, and let $\mathscr{R}$ be the closed side of $\mathscr{H}_1$ that contains $\mathscr{H}$ and so contains $\mathscr{S}$. From Theorems 9 and 10, $\text{Cn}(\mathscr{S}; P) \subset \mathscr{R}$, and, since $P \in \mathscr{H}_1$, $\mathscr{H}_1$ supports $\text{Cn}(\mathscr{S}; P)$. If Q is a second vertex to $\text{Cn}(\mathscr{S}; P)$, then, by Theorem 6, Q is also in $\mathscr{H}_1$. Then, by Theorem 9, $\text{Ry}[P, Q)$ must intersect $\mathscr{S}$. But, since $\text{Ry}[P, Q)$ and $\mathscr{S}$ lie in parallel hyperplanes, they do not intersect. Thus P is the only vertex to $\text{Cn}(\mathscr{S}; P)$, and the cone is regular. □

We turn now to the description of a pair of related cones that appear in connection with a boundary point of a convex body $\mathscr{K}$. First, if P is such a boundary point, then each closed half-space that supports $\mathscr{K}$ at P is a closed cone with vertex P. The intersection of these cones is therefore a closed cone with vertex P, and the following is a rather natural name for this cone.

DEFINITION (Support cone). If P is a boundary point of a convex body $\mathscr{K}$, the *support cone to $\mathscr{K}$ at P* is the intersection of the closed half-spaces that support $\mathscr{K}$ at P.

Even a simple example shows how varied the support cones can be. If $\mathscr{K}$ is a solid tetrahedron in $\mathscr{E}^3$, that is, a 3-simplex, and if P in the boundary of $\mathscr{K}$ does not belong to an edge, then P is a regular point of $\mathscr{K}^0$ in the relative interior of a face of $\mathscr{K}$. In this case, the support cone at P is a closed half-space whose face contains a face of $\mathscr{K}$. If P belongs to an edge but is not a vertex, then the support cone is a solid dihedral angle. Finally, if P is a vertex of $\mathscr{K}$, then the support cone at P is a solid trihedral angle.

To describe the second cone, related to the support cone, we first introduce some new terminology.

DEFINITION (Outer normal direction, normal ray). If P is a boundary point of a convex body $\mathscr{K}$, the *outer normal directions* to $\mathscr{K}$ *at* P are the outer normal directions of the closed half-spaces that support $\mathscr{K}$ at P. A ray at P with an outer normal direction is a *ray that is normal to* $\mathscr{K}$ *at* P.

Corresponding to each closed half-space $\mathscr{H}'$, with face $\mathscr{H}$, that supports a convex body $\mathscr{K}$ at P, there is clearly a unique ray $\text{Ry}[P, X)$ normal to $\mathscr{K}$ in the outer normal direction of $\mathscr{H}'$, and this ray is perpendicular to $\mathscr{H}$. We now want to show that these normal rays form a cone.

THEOREM 15. The closed rays normal to a convex body $\mathscr{K}$ at a boundary point P form a closed cone with vertex P.

PROOF. Let $\mathscr{R}$ denote the union of all the closed rays normal to $\mathscr{K}$ at P, let $\mathscr{C}$ denote the support cone to $\mathscr{K}$ at P, and let $\mathscr{S}$ be the set defined by

$$\mathscr{S} = \{X : (x - p) \cdot (y - p) \leq 0 \qquad \text{for all } Y \in \mathscr{C}\}.$$

We propose to show that $\mathscr{R} = \mathscr{S}$ and then that $\mathscr{S}$ is a closed cone with vertex P.

Because $(p - p) \cdot (y - p) = 0$ for all Y, $P \in \mathscr{S}$. Since there is at least one closed ray normal to $\mathscr{K}$ at P, $P \in \mathscr{R}$. Now consider $Q \in \mathscr{R}$, $Q \neq P$. Since $\text{Ry}[P, Q)$ is normal to $\mathscr{K}$, $\mathbf{PQ}$ is an outer normal to the closed half-space $\mathscr{H}_1$: $\mathbf{PX} \cdot \mathbf{PQ} \leq 0$ that supports $\mathscr{K}$ at P. Because $\mathscr{C}$ is the intersection of all such half-spaces, $\mathscr{C} \subset \mathscr{H}_1$, so $\mathbf{PQ} \cdot \mathbf{PY} \leq 0$ for all $Y \in \mathscr{C}$, hence $Q \in \mathscr{S}$. Therefore $\mathscr{R} \subset \mathscr{S}$. Conversely, suppose that $Q \in \mathscr{S}$, $Q \neq P$. Since $\mathbf{PQ} \neq o$, the graph of $\mathbf{PX} \cdot \mathbf{PQ} \leq 0$ is a closed half-space $\mathscr{H}_2$ with outer normal direction $\mathbf{PQ}$. Because $Q \in \mathscr{S}$, $\mathbf{PQ} \cdot \mathbf{PY} \leq 0$ for all $Y \in \mathscr{C}$, so $\mathscr{C} \subset \mathscr{H}_2$, so $\mathscr{K} \subset \mathscr{H}_2$. Since $\mathbf{PP} \cdot \mathbf{PQ} = 0$, P is in the face of $\mathscr{H}_2$, so $\mathscr{H}_2$ is a closed half-space that supports $\mathscr{K}$ at P. Thus $\text{Ry}[P, Q)$ is normal to $\mathscr{K}$ at P, so $Q \in \mathscr{R}$. Therefore $\mathscr{S} \subset \mathscr{R}$. Since $\mathscr{R} \subset \mathscr{S}$ and $\mathscr{S} \subset \mathscr{R}$, then $\mathscr{R} = \mathscr{S}$. Also, since $\mathscr{R} \neq \{P\}$, $\mathscr{S} \neq \{P\}$.

To show that $\mathscr{S}$ is a P-vertex cone, we show that it satisfies the two conditions of Corollary 2. First, consider $X \in \mathscr{S}$ and $z = p + \lambda\mathbf{PX}$, where $\lambda \geq 0$.

If $X = P$ or if $\lambda = 0$, then clearly $Z \in \mathscr{S}$, so we suppose that $X \neq P$ and that $\lambda > 0$. Then, for any $Y \in \mathscr{C}$, $X \in \mathscr{S}$ implies that $\mathbf{PX} \cdot \mathbf{PV} \leq 0$, hence that $((1/\lambda)\mathbf{PZ}) \cdot \mathbf{PY} \leq 0$, hence that $\mathbf{PZ} \cdot \mathbf{PX} \leq 0$, hence that $Z \in \mathscr{S}$. Thus $p + \lambda\mathbf{PX}$ is in $\mathscr{S}$ for all $X \in \mathscr{S}$ and all $\lambda \geq 0$. Next, let X_1, X_2 be any two points of $\mathscr{S}$, and let $z = p + \mathbf{PX}_1 + \mathbf{PX}_2$. Then, for any $Y \in \mathscr{C}$, $\mathbf{PX}_1 \cdot \mathbf{PY} \leq 0$ and $\mathbf{PX}_2 \cdot \mathbf{PY} \leq 0$, which implies that $(\mathbf{PX}_1 + \mathbf{PX}_2) \cdot \mathbf{PY} = \mathbf{PZ} \cdot \mathbf{PY} \leq 0$, hence that $Z \in \mathscr{S}$. Thus $\mathscr{S}$ satisfies the conditions of Corollary 2, so $\mathscr{S}$ is a cone with vertex P.

Finally, to see that $\mathscr{S}$ is a closed cone, consider an arbitrary point $A \in \mathrm{Cp}(\mathscr{S})$. Since $A \notin \mathscr{S}$, there must exist some point Y_o in $\mathscr{C}$ such that $(a - p) \cdot (y_o - p) > 0$. Let $\varepsilon_o = (a - p) \cdot (y_o - p)$. The function g defined by $g(x) = (x - p) \cdot (y_o - p)$ is an affine functional and hence is continuous on $\mathscr{E}^n$. Corresponding to $\varepsilon = \frac{1}{2}\varepsilon_o$, the continuity of g at A implies the existence of $\mathrm{N}(A, \delta)$ such that $|g(x) - g(a)| < \varepsilon = \frac{1}{2}\varepsilon_o$ for $X \in \mathrm{N}(A, \delta)$. Since $g(a) = \varepsilon_o$, then $X \in \mathrm{N}(A, \delta)$ implies $|g(x) - \varepsilon_o| < \frac{1}{2}\varepsilon_o$, hence that $g(x) > \frac{1}{2}\varepsilon_o > 0$. Thus $(x - p) \cdot (y_o - p) > 0$ for all $X \in \mathrm{N}(A, \delta)$, so $\mathrm{N}(A, \delta) \subset \mathrm{Cp}(\mathscr{S})$, which shows that $A \in \mathrm{In}[\mathrm{Cp}(\mathscr{S})]$. Since all points of $\mathrm{Cp}(\mathscr{S})$ are interior points, $\mathrm{Cp}(\mathscr{S})$ is an open set, so $\mathscr{S}$ is a closed set, hence $\mathscr{S}$ is a closed cone. □

We now introduce a rather natural name for the cone in Theorem 15.

DEFINITION (Normal cone). If P is a boundary point of a convex body $\mathscr{K}$, the closed cone formed by the closed rays normal to $\mathscr{K}$ at P is the *normal cone* to $\mathscr{K}$ at P.

In the proof for Theorem 15, a set $\mathscr{S}$ was defined in terms of $\mathscr{C}$, a support cone to a convex body, and then $\mathscr{S}$ was shown to be a closed cone. With only slight modification, the same proof yields the following more general fact.

THEOREM 16. If $\mathscr{C}$ is a cone in $\mathscr{E}^n$, with vertex V, and if $\mathscr{S}$ is the set defined by

$$\mathscr{S} = \{X : (x - v) \cdot (y - v) \leq 0 \text{ for all } Y \in \mathscr{C}\},$$

then $\mathscr{S}$ is a closed cone with vertex V and $\{V\} \subset \mathscr{C} \cap \mathscr{S}$. (E)

The cone $\mathscr{S}$ in Theorem 16 has the following name in relation to $\mathscr{C}$.

DEFINITION (Supplementary cone). If $\mathscr{C}$ is a cone in $\mathscr{E}^n$, with vertex V, then the closed cone $\mathscr{S}$ defined by

$$\mathscr{S} = \{X : (x - v) \cdot (y - v) \leq 0 \text{ for all } Y \in \mathscr{C}\}$$

is the *supplementary cone to* $\mathscr{C}$.

As Theorem 15 indicates, the normal cone at a point P of a closed, convex surface $\mathscr{K}^0$ is the supplementary cone to the support cone at P. As one might suppose, a regular cone $\mathscr{C}_1$ in $\mathscr{E}^2$ is a closed angular region. The cone $\mathscr{C}_2$ that is supplementary to $\mathscr{C}_1$ is also a closed, angular region, and the boundaries of the cones form two angles that are supplements. The name "supplementary cone" derives from this property.

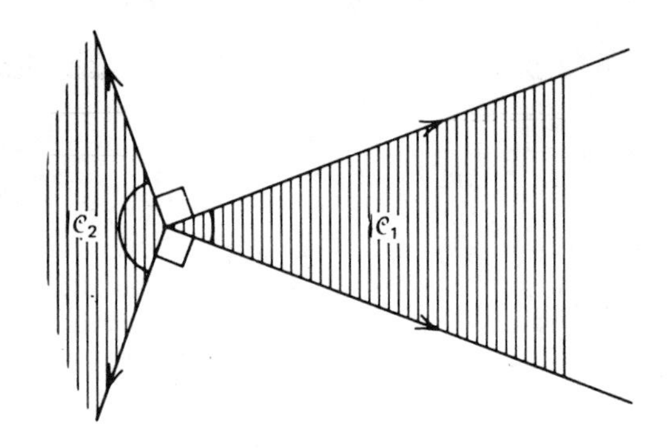

We conclude this section with a theorem about the cone structure of unbounded convex sets. In the proof we need the following fact from elementary geometry which we leave as a computational exercise.

LEMMA. If P, X, and Y are noncollinear, and if $Z \in \mathrm{Sg}(XY)$, then $C \in \mathrm{Ry}(P, X)$ and $D \in \mathrm{Ry}(P, Y)$ imply that $\mathrm{Ry}(P, Z) \cap \mathrm{Sg}(CD) \neq \varnothing$.

THEOREM 17. If $\mathscr{K}$ is an unbounded, proper subset of $\mathscr{E}^n$ that is closed and convex, then, for each point $P \in \mathscr{K}$, the closed rays that have origin P and are contained in $\mathscr{K}$ form a cone.

PROOF. Corresponding to each positive integer i, let $\mathscr{R}_i = \mathscr{K} \cap \mathrm{S}(P, i)$, and consider any set $\mathscr{R}_j$ in the sequence $\{\mathscr{R}_i\}$. Because $\mathscr{K}$ is unbounded it is not contained in $\mathrm{B}(P, j)$, so there exists $T \in \mathscr{K} \cap \mathrm{Cp}[\mathrm{B}(P, j)]$. Since $\mathscr{K}$ is convex, $\mathrm{Sg}[PT] \subset \mathscr{K}$, and since P is interior to $\mathrm{B}(P, j)$ and T is exterior, $\mathrm{Sg}[PT]$ intersects $\mathrm{S}(P, j)$. Hence $\mathscr{R}_j \neq \varnothing$. Also, since $\mathrm{S}(P, j)$ is compact, and $\mathscr{K}$ is closed, $\mathscr{R}_j$ is a closed subset of a compact set, so $\mathscr{R}_j$ is compact. Thus $\{\mathscr{R}_i\}$ is a sequence of non-empty, compact sets.

Now let $f_i(x) = (1/i)x + (1 - 1/i)p$ be the dilatation at P with ratio $1/i$, $i = 1, 2, \ldots$, and let $\mathscr{R}'_i = f_i(\mathscr{R}_i)$. Since f_i maps $\mathrm{S}(P, i)$ onto $\mathrm{S}(P, 1)$, the sequence $\mathscr{R}'_i$ is contained in $\mathrm{S}(P, 1)$. Because $\mathscr{R}_i \neq \varnothing$, $\mathscr{R}'_i \neq \varnothing$, and since f_i is a continuous mapping, the compactness of $\mathscr{R}_i$ implies the compactness of $\mathscr{R}'_i$. Moreover, the sequence $\{\mathscr{R}'_i\}$ is decreasingly nested. To see this, consider $X \in \mathscr{R}'_{i+1}$. By the definition of $\mathscr{R}'_{i+1}$, $X' = f_{i+1}^{-1}(X)$ is in $\mathscr{R}_{i+1} = \mathscr{K} \cap \mathrm{S}(P, i + 1)$. Since $\mathscr{K}$ is convex, $\mathrm{Sg}[PX'] \subset \mathscr{K}$, and since $d(P, X') = k + 1 > i$, then

$\mathrm{Sg}[PX']$ intersects $\mathrm{S}(P, i)$ at a point Y in $\mathscr{R}_i$. Then $f_i(Y)$ is in $\mathscr{R}_i$. But X and $f_i(Y)$ are both the intersection of $\mathrm{Ry}[P, X)$ with $\mathrm{S}(P, 1)$, so they are the same point, so $X \in \mathscr{R}_i$. Thus $\mathscr{R}'_i \supset \mathscr{R}'_{i+1}$.

Now let $I = \bigcap_{i=1}^{\infty} \mathscr{R}'_i$ and let $\mathscr{C}$ be the union of all the rays $\mathrm{Ry}[P, X)$, $X \in \mathscr{I}$. Because $\{\mathscr{R}'_i\}$ is a decreasingly nested sequence of non-empty, compact sets, it follows from the Cantor intersection theorem that $\mathscr{I} \neq \varnothing$. Thus there exists at least one closed ray $\mathrm{Ry}[P, X) \subset \mathscr{C}$, which also implies $P \in \mathscr{C}$. Let Z denote an arbitrary point of the ray $\mathrm{Ry}[P, X)$, $X \in \mathscr{I}$. If $Z = P$, then $Z \in \mathscr{K}$. If $Z \neq P$, let m be a positive integer such that $m > d(P, Z)$. Because $X \in \mathscr{I}$, $X \in \mathscr{R}'_m$, so $X_m = f_m^{-1}(X)$ is in $\mathscr{R}_m = \mathscr{K} \cap \mathrm{S}(P, m)$. From the convexity of $\mathscr{K}$, $\mathrm{Sg}[PX_m] \subset \mathscr{K}$. Since Z and X_m are on $\mathrm{Ry}[P, X)$, and $d(P, X_m) = m > d(P, Z)$, then $Z \in \mathrm{Sg}[PX_m]$, so $Z \in \mathscr{K}$. Thus $\mathrm{Ry}[P, X) \subset \mathscr{K}$. Hence $\mathscr{C} \subset \mathscr{K}$. Conversely, consider any closed ray $\mathrm{Ry}[P, Y)$ that is contained in $\mathscr{K}$. This ray intersects each sphere $\mathrm{S}(P, i)$ at a point Y_i in $\mathscr{R}_i$. Since $f_i(Y_i) = Y_1$, $i = 1, 2, \ldots$, and since $f_i(Y_i) \in \mathscr{R}'_i$, $i = 1, 2, \ldots$, it follows that $Y_1 \in \mathscr{I}$. Therefore $\mathrm{Ry}[P, Y) = \mathrm{Ry}[P, Y_1) \subset \mathscr{C}$. Thus $\mathscr{C}$ is exactly the set of closed rays with origin P that are contained in $\mathscr{K}$.

To show that $\mathscr{C}$ is a cone, we first observe that $\mathscr{C} \subset \mathscr{K}$ implies that $\mathscr{C}$ is a proper subset of $\mathscr{E}^n$. Also $\mathscr{C}$ has the ray-vertex property because if $X \in \mathscr{C}$, $X \neq P$, then $X \in \mathrm{Ry}[P, Q)$ for some $Q \in \mathscr{I}$. Since $\mathrm{Ry}[P, Q) \subset \mathscr{C}$, and $\mathrm{Ry}[P, Q) = \mathrm{Ry}[P, X)$, then $\mathrm{Ry}[P, X) \subset \mathscr{C}$. It remains to be shown that $\mathscr{C}$ is convex. For this purpose, let X_1 and X_2 denote two distinct points of $\mathscr{C}$, and let Z denote a point of $\mathrm{Sg}(X_1X_2)$. If one of the points X_1, X_2 is P, say $X_1 = P$, then $X_2 \neq P$, and, since $\mathrm{Ry}[P, X_2) \subset \mathscr{C}$, and $Z \in \mathrm{Sg}(PX_2)$, then $Z \in \mathscr{C}$. If neither X_1 nor X_2 is P, but the three points are collinear, then $\mathrm{Sg}[X_1X_2] \subset \mathrm{Ry}[P, X_1) \cup \mathrm{Ry}[P, X_2) \subset \mathscr{C}$ implies $Z \in \mathscr{C}$. The only remaining case is that in which P, X_1 and X_2 are noncollinear. Assume in this case that Z is not in $\mathscr{C}$. Then $A_1 = \mathrm{Ry}[P, Z) \cap \mathrm{S}(P, 1)$ is not in $\mathscr{I}$. Thus there is some set $\mathscr{R}'_m$ such that A_1 is not in $\mathscr{R}'_m$. Consequently, $A_m = f_m^{-1}(A_1)$ is not in $\mathscr{R}_m$, that is,

$$A_m = \mathrm{Ry}[P, Z) \cap \mathrm{S}(P, m) \notin \mathscr{K}. \tag{1}$$

Now, let C_i and D_i denote the respective intersections of $\mathrm{Ry}[P, X_1)$ and $\mathrm{Ry}[P, X_2)$ with $\mathrm{S}(P, i)$ and let F_i denote the midpoint of $\mathrm{Sg}[C_iD_i]$, $i = 1, 2, \ldots$. Each of the triangles $\triangle PC_iD_i$ is isosceles with base $\mathrm{Sg}[C_iD_i]$, so F_i is the foot of P in the line $\mathrm{Ln}(C_iD_i)$, $i = 1, 2, \ldots$. Let u be the unit vector in the direction of $\mathbf{PF}_1$ and let v be the unit vector $\mathbf{PC}_1$. Since $\sphericalangle(u, v)$ is acute (its measure is half that of $\sphericalangle C_1PD_1$), and since $\mathrm{Sg}[PF_i]$ is the orthogonal projection of $\mathrm{Sg}[PC_i]$ into $\mathrm{Ln}(PF_i)$, then

$$d(P, F_i) = d(P, C_i) \cos \sphericalangle(u, v)^0 = i \cos \sphericalangle(u, v)^0, \tag{2}$$

$i = 1, 2, \ldots$. From (2) it follows that there is some integer j such that

$$d(P, F_j) > m. \tag{3}$$

Because $Z \in \text{Sg}(X_1X_2)$ and $C_j \in \text{Ry}(P, X_1)$ and $D_j \in \text{Ry}(P, X_2)$, it follows from the lemma that $\text{Ry}(P, Z)$ intersects $\text{Sg}(C_jD_j)$ at some point B. Because $C_j, D_j \in \mathscr{K}$, then $B \in \mathscr{K}$. Since F_j is the foot of P in $\text{Ln}(C_jD_j)$, $d(P, B) \geq d(P, F_j)$, so, from (3), $d(P, B) > m$. Now both B and A_m are on $\text{Ry}[P, Z)$ and $d(P, A_m) = m$, whereas $d(P, B) > m$, so $A_m \in \text{Sg}[PB]$. Since $P \in \mathscr{K}$ and $B \in \mathscr{K}$, then $\text{Sg}[PB] \subset \mathscr{K}$, so $A_m \in \mathscr{K}$. But this contradicts the conclusion in (1) that $A_m \notin \mathscr{K}$. The contradiction shows that Z must be in $\mathscr{C}$. Since in all cases $Z \in \mathscr{C}$, $\mathscr{C}$ is convex, which completes the proof that $\mathscr{C}$ is a cone. □

Exercises – Section 4.4

1. Under what conditions is a line not a cone?
2. Prove Theorem 1.
3. Prove Corollary 4.
4. Prove Theorem 5.
5. Show that if P is a boundary point of a convex set $\mathscr{S}$, then P and $\mathscr{S}$ are weakly separated.
6. If P is not weakly separated from $\mathscr{S}$, must P be either an interior point of $\mathscr{S}$ or a boundary point of $\mathscr{S}$?
7. If $\mathscr{H}$ is a hyperplane, what is the nature of the cone $\text{Cn}(\mathscr{H}; P)$ if

   a. $P \in \mathscr{H}$?
   b. $P \notin \mathscr{H}$?

8. Give an example of a set $\mathscr{S}$ and a point P for which the cone $\text{Cn}(\mathscr{S}; P)$ does not exist.
9. Must the vertex set of a cone be a convex set?
10. Prove Corollary 9.
11. Prove Theorem 12.
12. Prove Theorem 13.
13. If $\mathscr{C}_1$ is a closed cone with vertex V, and if $\mathscr{C}_2$ is its supplementary cone, prove that $\mathscr{C}_1$ is the supplementary cone to $\mathscr{C}_2$.
14. Prove Theorem 16.
15. Prove the lemma preceding Theorem 17.

5
CONVEX SPANS AND INDEPENDENCE, RELATED CLASSICAL THEOREMS

INTRODUCTION

In this chapter we consider a number of important, classical results that are directly or indirectly associated with the convex spans of finite sets or with the convex dependence or independence of such sets. In the first section, we begin with a characterization of affine independence in terms of convex spans, and this result implies the Radon theorem. We then use the Radon theorem to establish the basic Helly theorem and generalize this to other Helly properties.

As one application of these results, the second section deals briefly with central sets and a particular measure of centralness. With the use of a Helly theorem, a greatest lower bound is obtained for the centralness of all convex bodies.

In the final section, we consider the convex span of a finite set in terms of its extreme points and are led by related questions to the theorem of Càratheodory and some other classical results.

SECTION 5.1. THE RADON THEOREM AND SOME HELLY THEOREMS

We begin with the simple observation that if a finite set of points $\mathscr{R}$ is separated, or partitioned, into two non-empty, disjoint sets $\mathscr{R}_1$ and $\mathscr{R}_2$, then the convex spans of $\mathscr{R}_1$ and $\mathscr{R}_2$ may or may not intersect. For the five points shown in

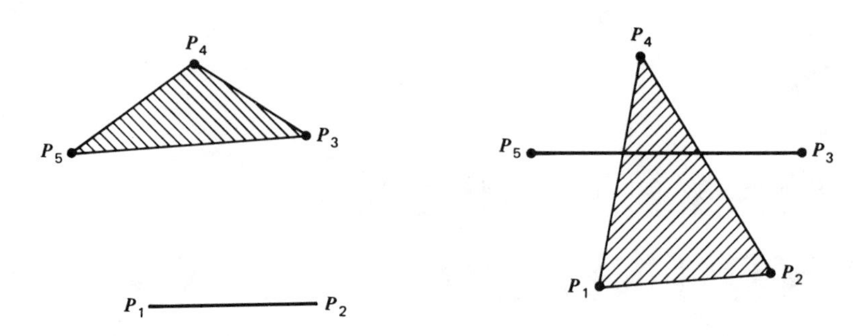

$\mathscr{E}^2$, the sets in the partition $\{P_1, P_2\}$, $\{P_3, P_4, P_5\}$ have disjoint convex spans, whereas the sets in the partition $\{P_1, P_2, P_4\}$, $\{P_3, P_5\}$ have intersecting convex spans. There are obviously different questions one can ask about this phenomenon, and one is, "when is it the case that no matter how $\mathscr{R}$ is partitioned into two non-empty, disjoint sets $\mathscr{R}_1$, $\mathscr{R}_2$, the convex spans of $\mathscr{R}_1$ and $\mathscr{R}_2$ are disjoint?"

A little experimenting with the problem just posed suggests an answer. First, $\mathscr{R}$ must have at least two points to make the problem sensible. If $\mathscr{R}$ has just two distinct points A, B, then it obviously has the disjoint-span property because only the partition $\{A\}$, $\{B\}$ is possible, and $\mathrm{CS}\{A\} \cap$

$\mathrm{CS}\{B\} = \varnothing$. If a third point C is added to $\{A, B\}$, then the new set $\mathscr{R} = \{A, B, C\}$ no longer has the property if C is on Ln(AB). For in this case one of the points, say B, is between the other two. Taking $\mathscr{R}_1 = \{B\}$, and $\mathscr{R}_2 = \{A, C\}$, it is clear that B is the intersection of $\mathrm{CS}(\mathscr{R}_1) \cap \mathrm{CS}(\mathscr{R}_2)$. On the other hand, if C is not on Ln(AB), then $\mathscr{R} = \{A, B, C\}$ has the disjoint-span property, as can be seen from the three possible partitions.

What we have found is that two distinct points, namely, two affinely independent points, have the disjoint-span property. If we add a third point, the property is preserved if the three are noncollinear, namely, if they are affinely independent, and it is lost if the three are collinear, that is, if they are

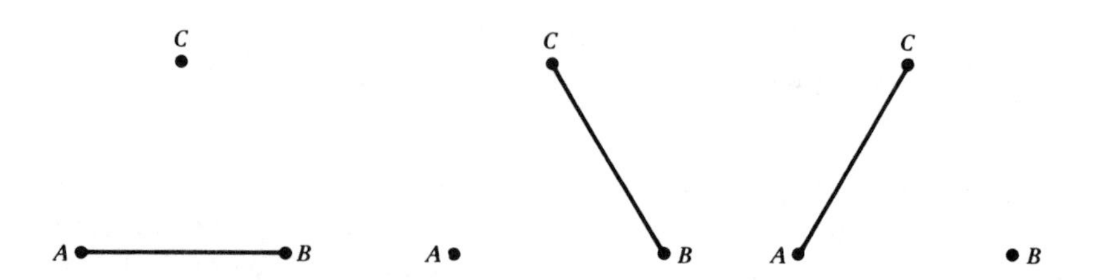

affinely dependent. The suggestion is that the disjoint-span property is associated with affinely independent sets. This suggestion is correct and gives a characterization of affinely independent sets.

THEOREM 1. A finite set $\mathscr{R}$ of m distinct points, $m > 1$, is an affinely independent set if and only if for every partition of $\mathscr{R}$ into non-empty, disjoint sets $\mathscr{R}_1, \mathscr{R}_2$, the convex spans of these sets are also disjoint.

PROOF. First consider the case in which $\mathscr{R} = \{P_1, P_2, \ldots, P_m\}$ is a set of m affinely independent points. Without loss of generality, we may suppose that an arbitrary partition of $\mathscr{R}$ into two non-empty, disjoint sets $\mathscr{R}_1$ and $\mathscr{R}_2$ is of the form

$$\mathscr{R}_1 = \{P_1, P_2, \ldots, P_k\}, \qquad \mathscr{R}_2 = \{P_{k+1}, P_{k+2}, \ldots, P_m\}. \tag{1}$$

Assume that there exists $Q \in \mathrm{CS}(\mathscr{R}_1) \cap \mathrm{CS}(\mathscr{R}_2)$. Then $Q \in \mathrm{CS}(\mathscr{R}_1)$ implies that Q has a representation

$$q = \sum_1^k \eta_i p_i, \qquad \sum_1^k \eta_i = 1, \qquad \eta_i \geq 0. \tag{2}$$

Similarly, $Q \in \mathrm{CS}(\mathscr{R}_2)$ implies that

$$q = \sum_{k+1}^m \eta_i p_i, \qquad \sum_{k+1}^m \eta_i = 1, \qquad \eta_i \geq 0. \tag{3}$$

The two representations of q in (2) and (3) imply that

$$\sum_1^k \eta_i p_i + \sum_{k+1}^m (-\eta_i) p_i = o. \tag{4}$$

Because (4) is a zero-sum linear combination of $p_1, p_2, \ldots, p_m$ representing the null vector, the affine independence of set $\mathscr{R}$ implies that all the coefficients in (4) are zero. But this is impossible, since the sum of the first k of the coefficients is 1. Thus there cannot exist a point Q in $\mathrm{CS}(\mathscr{R}_1) \cap \mathrm{CS}(\mathscr{R}_2)$, so the two convex spans are disjoint.

Next, consider the case in which $\mathscr{R} = \{P_1, P_2, \ldots, P_m\}$ has the property that for every partition of $\mathscr{R}$ into non-empty, disjoint sets $\mathscr{R}_1, \mathscr{R}_2$, the convex spans of $\mathscr{R}_1$ and $\mathscr{R}_2$ are also disjoint. Assume that $\mathscr{R}$ is an affinely dependent set. Then there exists a zero-sum linear combination of $p_1, p_2, \ldots, p_m$, representing the null vector, in which the coefficients are not all zero. That is, there exist numbers $\varphi_1, \varphi_2, \ldots, \varphi_m$, not all zero, such that

$$\sum_1^m \varphi_i p_i = o, \qquad \sum_1^m \varphi_i = o. \tag{5}$$

Since the coefficients in (5) have a zero sum, but are not all zero, then at least one is positive and at least one is negative. We may suppose the labeling chosen so that the first k are positive and the last m-k are nonpositive, where $1 \leq k < m$. Then

$$\mathscr{R}_1 = \{P_1, P_2, \ldots, P_k\}, \qquad \mathscr{R}_2 = \{P_{k+1}, P_{k+2}, \ldots, P_m\} \tag{6}$$

is a partitioning of $\mathscr{R}$ into two non-empty, disjoint sets.

The first equality in (5) can be expressed in the form

$$o = \sum_1^m \varphi_i p_i = \sum_1^k \varphi_i p + \sum_{k+1}^m \varphi_i p_i$$

and this implies that

$$\sum_1^k \varphi_i p_i = \sum_{k+1}^m (-\varphi_i) p_i. \tag{7}$$

Now, let $\varphi = \sum_1^k \varphi_i$. Since the terms in the sum are positive, $\varphi > 0$. Since $\sum_1^m \varphi_i = 0$ implies that $\sum_1^k \varphi_i = \sum_{k+1}^m (-\varphi_i)$, then

$$\varphi = \sum_1^k \varphi_i = \sum_{k+1}^m (-\varphi_i) > 0. \tag{8}$$

Dividing (7) through by φ, we obtain

$$\sum_1^k \frac{\varphi_i}{\varphi} p_i = \sum_{k+1}^m -\frac{\varphi_i}{\varphi} p_i. \tag{9}$$

The linear combinations on both sides of (9) have non-negative coefficients, and from (8) the sum of the coefficients in each case is 1. Thus both sides of (9) are convex combinations. Thus the point Q with these two representations belongs to $\text{CS}(P_1, P_2, \ldots, P_k) \cap \text{CS}(P_{k+1}, P_{k+2}, \ldots, P_m) = \text{CS}(\mathscr{R}_1) \cap \text{CS}(\mathscr{R}_2)$, contradicting the property that $\text{CS}(\mathscr{R}_1)$ and $\text{CS}(\mathscr{R}_2)$ are disjoint. Since the assumption that $\mathscr{R}$ is a dependent point set leads to a contradiction, the points of $\mathscr{R}$ must be affinely independent. □

COROLLARY 1. Any set of two or more distinct, affinely dependent points can be partitioned into two non-empty sets which are disjoint but whose convex spans intersect.

An immediate consequence of Corollary 1 is the following theorem which can be derived independently and is usually called the Radon property.

THEOREM 2 (Radon). Any finite set of $k + 2$ or more distinct points in a k-flat can be partitioned into two non-empty disjoint sets whose convex spans intersect.

We can now use Radon's theorem to establish the following Helly property.

THEOREM 3 (Basic Helly theorem). If a collection $\mathscr{F}$ of $k + 2$ convex subsets of a k-flat is such that each $k + 1$ of the sets have a non-empty intersection set, then there exists at least one point common to all the sets in $\mathscr{F}$.

PROOF. Let $\mathscr{S}_1, \mathscr{S}_2, \ldots, \mathscr{S}_{k+2}$ denote the convex sets in $\mathscr{F}$. By hypothesis, for each integer i, $i = 1, 2, \ldots, k + 2$, there exists at least one point that belongs to all the sets of $\mathscr{F}$ whose index is not i. Selecting one such point, labeled P_i, for each integer i, we obtain a set $\mathscr{R} = \{P_1, P_2, \ldots, P_{k+2}\}$. If $\mathscr{R}$ has fewer than $k + 2$ distinct points, then $P_i = P_j$ for some pair i, j, where $i \neq j$. Since $j \neq i$, $P_j \in \mathscr{S}_i$ and so $P_i = P_j$ implies that $P_i \in \mathscr{S}_i$. Since P_i also belongs to every set whose index is not i, it follows that P_i is in all the sets of $\mathscr{F}$.

Next, suppose that $\mathscr{R}$ has $k + 2$ distinct points. Then, by Radon's theorem, there exists a partition of $\mathscr{R}$ into two non-empty disjoint sets $\mathscr{R}_1$ and $\mathscr{R}_2$ whose convex spans intersect. We may suppose the labeling is such that

$$\mathscr{R}_1 = \{P_1, P_2, \ldots, P_m\}, \qquad \mathscr{R}_2 = \{P_{m+1}, P_{m+2}, \ldots, P_{k+2}\}$$

where $1 \leq m < k + 2$, and that Q represents a point in $\mathrm{CS}(\mathscr{R}_1) \cap \mathrm{CS}(\mathscr{R}_2)$. Each point P_i in $\mathscr{R}_1$ has an index less than $m + 1$ and hence unequal to any of the integers $m + 1, m + 2, \ldots, k + 2$. Thus each point of $\mathscr{R}_1$ belongs to all the sets $\mathscr{S}_{m+1}, \mathscr{S}_{m+2}, \ldots, \mathscr{S}_{k+2}$, hence

$$\mathscr{R}_1 \subset \bigcap_{m+1}^{k+2} \mathscr{S}_i. \tag{1}$$

Because the sets in $\mathscr{F}$ are convex, the intersection set in (1) is convex. Since a convex set contains the convex spans of its finite subsets (Theorem 3, Sec. 4.1), (1) implies that

$$\mathrm{CS}(\mathscr{R}_1) \subset \bigcap_{m+1}^{k+2} \mathscr{S}_i. \tag{2}$$

By the same logic, each point P_i of $\mathscr{R}_2$ has an index greater than m and so belongs to all the sets $\mathscr{S}_i$, $1 \leq i \leq m$. Thus

$$\mathscr{R}_2 \subset \bigcap_{1}^{m} \mathscr{S}_i, \tag{3}$$

which implies

$$\mathrm{CS}(\mathscr{R}_2) \subset \bigcap_{1}^{m} \mathscr{S}_i. \tag{4}$$

Now, from (2), (4) and $Q \in \mathrm{CS}(\mathscr{R}_1) \cap \mathrm{CS}(\mathscr{R}_2)$, it follows that Q belongs to all the sets in $\mathscr{F}$. □

We can extend Theorem 3 to the so-called finite Helly theorem by an inductive argument.

THEOREM 4 (Finite Helly theorem). If $\mathscr{F}$ is a finite collection of $k + 2$ or more convex subsets of a k-flat and if each $k + 1$ sets in $\mathscr{F}$ have a non-empty intersection set, then there is at least one point common to all the sets in $\mathscr{F}$.

PROOF. Let T_m denote the statement, "if m convex subsets of a k-flat are such that each $k + 1$ of them intersect, then there is at least one point common to all m sets." Then T_{k+2} is valid, since it is just Theorem 3.

We want to show that for any integer $m \geq k + 2$ the validity of T_m implies that of T_{m+1}. To do so, we suppose that $m \geq k + 2$, that T_m is valid, and we consider a collection

$$\mathscr{F} = \{\mathscr{S}_1, \mathscr{S}_2, \ldots, \mathscr{S}_{m+1}\}, \tag{1}$$

in which the sets $\mathscr{S}_i$ are convex subsets of a k-flat and are such that each $k + 1$ of them have at least one point in common.

Corresponding to $\mathscr{F}$, which has $m + 1$ sets, let $\mathscr{F}_1$ denote the collection

$$\mathscr{F}_1 = \{\mathscr{S}_1, \mathscr{S}_2, \ldots, \mathscr{S}_{m-1}, \mathscr{S}_m \cap \mathscr{S}_{m+1}\}, \tag{2}$$

which has m sets. Because the sets in $\mathscr{F}$ are convex, so are the sets in $\mathscr{F}_1$, and the k-flat that contains $\mathscr{F}$ obviously contains $\mathscr{F}_1$. Now consider any selection of $k + 1$ sets from $\mathscr{F}_1$. If these $k + 1$ sets also belong to $\mathscr{F}$, then, by hypothesis, they have a non-empty intersection set. If they do not all belong to $\mathscr{F}$, then one of them must be the set $\mathscr{S}_m \cap \mathscr{S}_{m+1}$ and the other k belong to $\mathscr{F}$. Thus the $k + 1$ sets can be represented by

$$\mathscr{S}_{i_1}, \mathscr{S}_{i_2}, \ldots, \mathscr{S}_{i_k}, \mathscr{S}_m \cap \mathscr{S}_{m+1}. \tag{3}$$

Since the $k + 2$ sets

$$\mathscr{S}_{i_1}, \mathscr{S}_{i_2}, \ldots, \mathscr{S}_{i_k}, \mathscr{S}_m, \mathscr{S}_{m+1} \tag{4}$$

all belong to $\mathscr{F}$, then, by hypothesis, each $k + 1$ of them have a non-empty intersection set. By the basic Helly theorem, namely, T_{k+2}, it follows that there is some point P that belongs to all the sets in (4). But $P \in \mathscr{S}_m$ and $P \in \mathscr{S}_{m+1}$ implies $P \in \mathscr{S}_m \cap \mathscr{S}_{m+1}$, so P belongs to all the sets in (3).

We have now shown that every selection of $k + 1$ sets from $\mathscr{F}_1$ has a non-empty intersection set. Because $\mathscr{F}_1$ has m sets, the validity of T_m implies that there is some point Q that belongs to all the sets in $\mathscr{F}_1$. But now, $Q \in \mathscr{S}_m \cap \mathscr{S}_{m+1}$ implies $Q \in \mathscr{S}_m$ and $Q \in \mathscr{S}_{m+1}$, so Q belongs to all the sets in $\mathscr{F}$.

Because T_{k+2} is valid and because T_m implies T_{m+1}, for $m \geq k + 2$, it follows that T_{k+i} is valid for every positive integer $i \geq 2$. □

The inductive proof for Theorem 4 suggests that the same conclusion might be obtained without the restriction that the collection of convex sets is finite, but a simple example shows that this is not possible. In $\mathscr{E}^2$, consider the collection of closed, upper half-planes

$$\{\mathscr{S}_i : x_2 \geq i, \qquad i = 1, 2, \ldots\}$$

Each of the sets $\mathscr{S}_i$ is convex. Moreover, if $\mathscr{S}_j$, $\mathscr{S}_k$, $\mathscr{S}_m$ are three of the sets, then $(p_1, p_2) \in \mathscr{S}_j \cap \mathscr{S}_k \cap \mathscr{S}_m$ if $p_2 \geq \max\{j, k, m\}$, hence each three of the sets have a non-empty intersection set. However, there is no point common to all the sets, because for any point $Q = (q_1, q_2)$ there exists an integer $i > q_2$, hence $Q \notin \mathscr{S}_i$.

Helly properties can be extended to infinite collections but only with extra restrictions. One of the most useful of such extensions is based on the following concept.

DEFINITION (Finite intersection property). A family of sets $\mathscr{F}$ has the *finite intersection property* if each finite subcollection of $\mathscr{F}$ has a nonnull intersection set.

If a finite collection has the finite intersection property, then, by definition, there is a point common to all the sets in the collection. Thus the next theorem, stated for infinite collections, is trivially correct for finite collections. Note that the theorem does not involve convexity and does not have any dimension restriction.

THEOREM 5. If an infinite collection of closed sets has the finite intersection property and at least one of the sets is compact, then there is a point common to all the sets in the collection.

PROOF. Let $\mathscr{F} = \{\mathscr{S}_\alpha : \alpha \in \Lambda\}$ be an infinite collection of closed sets that has the finite intersection property and that contains some set, say $\mathscr{S}^*$, that is compact. We want to show that there exists a point Q such that $Q \in \mathscr{S}_\alpha$ for every $\alpha \in \Lambda$. For an indirect proof, we assume that there is no such Q, hence that

$$\bigcap_{\alpha \in \Lambda} \mathscr{S}_\alpha = \varnothing \tag{1}$$

Now consider a point P in $\mathscr{S}^*$. From (1) it follows that there must exist at least one set in $\mathscr{F}$ to which P does not belong. Let one such set be denoted by $\mathscr{S}_P$. Thus, there exists a subcollection of $\mathscr{F}$

$$\mathscr{F}' = \{\mathscr{S}_X : X \in \mathscr{S}^*, \mathscr{S}_X \in \mathscr{F}, X \notin \mathscr{S}_X\}. \tag{2}$$

Now consider any particular set $\mathscr{S}_P$ in $\mathscr{F}'$. Because P does not belong to $\mathscr{S}_P$ and because $\mathscr{S}_P$ is a closed set, there exists a neighborhood of P that does not intersect $\mathscr{S}_P$. Let $N(P, \delta_P)$ denote one such neighborhood. Selecting one neighborhood in this way, for each $\mathscr{S}_X$ in $\mathscr{F}'$, we obtain a collection of neighborhoods

$$\mathscr{G} = \{\mathrm{N}(X, \delta_X) : X \in \mathscr{S}^*, \mathrm{N}(X, \delta_X) \cap \mathscr{S}_X = \varnothing\}. \tag{3}$$

The collection $\mathscr{G}$ is a collection of open sets that is clearly a cover of $\mathscr{S}^*$. Because $\mathscr{S}^*$ is compact, there is a finite subcollection of $\mathscr{G}$, say $\mathscr{G}_1$, that is also a cover of $\mathscr{S}^*$ and has the form

$$\mathscr{G}_1 = \{\mathrm{N}(P_1, \delta_{P_1}), \mathrm{N}(P_2, \delta_{P_2}), \ldots, \mathrm{N}(P_m, \delta_{P_m})\}. \tag{4}$$

Corresponding to each neighborhood center P_i in (4) there is a set $\mathscr{S}_{P_i}$ in $\mathscr{F}'$ and such that $\mathscr{S}_{P_i} \cap \mathrm{N}(P_i, \delta_{P_i}) = \varnothing$. These m sets $\mathscr{S}_{P_i}$, together with $\mathscr{S}^*$, form a collection

$$\mathscr{H} = \{\mathscr{S}^*, \mathscr{S}_{P_1}, \mathscr{S}_{P_2}, \ldots, \mathscr{S}_{P_m}\}. \tag{5}$$

Because $\mathscr{H}$ is a finite subcollection of $\mathscr{F}$, and $\mathscr{F}$ has the finite intersection property, it follows that there exists a point A that belongs to all the sets in $\mathscr{H}$. But now we have a self-contradicting situation. Because A is in all the sets of $\mathscr{H}$, it is in $\mathscr{S}^*$. Since the finite collection of neighborhoods in $\mathscr{G}_1$ is a cover of $\mathscr{S}^*$, it follows that A belongs to at least one of these neighborhoods, say $\mathrm{N}(P_j, \delta_{P_j})$. But $\mathrm{N}(P_j, \delta_{P_j})$ was chosen so that $\mathrm{N}(P_j, \delta_{P_j}) \cap \mathscr{S}_{P_j} = \varnothing$. Thus $A \in \mathrm{N}(P_j, \delta_{P_j})$ implies that A is not in $\mathscr{S}_{P_j}$, hence that A is not in all the sets of $\mathscr{H}$.

The contradiction just reached shows that assumption (1) must be false and hence that there does exist some point common to all the sets in $\mathscr{F}$. □

From Theorems 4 and 5 we can now obtain a Helly theorem that applies to infinite collections.

THEOREM 6 (Infinite Helly theorem). If an infinite collection $\mathscr{F}$ of closed, convex sets in a k-flat is such that each $k + 1$ of them intersect, and if at least one set in $\mathscr{F}$ is compact, then there is a point common to all the sets in $\mathscr{F}$.

PROOF. Let $\mathscr{F}_1$ denote an arbitrary finite subcollection from $\mathscr{F}$, say a subcollection of m sets. If $m \leq k + 1$, then the hypotheses of the theorem imply that $\mathscr{F}_1$ has a non-empty intersection set. If $m > k + 1$, then the finite Helly theorem (Theorem 4) implies that $\mathscr{F}_1$ has a non-empty intersection set. In all cases, then, $\mathscr{F}_1$ has a non-empty intersection set, so $\mathscr{F}$ has the finite intersection property. Therefore, by Theorem 5, there is at least one point common to all the sets in $\mathscr{F}$. □

Exercises – Section 5.1

1. Let $\mathrm{Sg}[A_iB_i]$, $i = 1, 2, \ldots, m$ be a collection of closed, collinear segments and such that each two of them intersect. Without using a Helly theorem, prove that there is a point common to all the segments.
2. Given m points in $\mathscr{E}^2$ such that each three belong to some disc of radius r, prove that there is a disc of radius r that contains all m points. What is the n-dimensional analog?
3. Prove Jung's theorem: If the diameter of a planar set of m points is 1, there exists a disc of radius $\frac{1}{3}\sqrt{3}$ that covers the set.
4. Let $\mathscr{F}$ denote a collection of closed arcs on a circle $\mathrm{S}(P, r)$, each arc smaller than a semicircle. Show that if each three of the arcs have a point in common, then there is a point common to all the arcs. Show that this result does not follow if the arcs may be as large as semicircles or if it is only assumed that each two of the arcs intersect.
5. Let $\mathscr{F}$ denote a collection of closed arcs on a circle $\mathrm{S}(P, r)$, each shorter than $\frac{1}{3}$ of the circle, that is, each less than a $120°$ arc. Prove that if each two of the arcs intersect, then there is a point common to all the arcs.
6. Let $\mathscr{F}$ denote a collection of plane convex bodies such that each two intersect. If t is an arbitrary line of the plane, show that t or a line parallel to t intersects all the sets in $\mathscr{F}$.
7. If $\mathscr{R}$ and $\mathscr{S}$ are convex sets, and $\mathrm{T}(\mathscr{R}:\mathscr{S})$ denotes the set of all points X such that the x-translate of $\mathscr{R}$ intersects $\mathscr{S}$, prove that $\mathrm{T}(\mathscr{R}:\mathscr{S})$ is a convex set.
8. If $\mathscr{R}$ and $\mathscr{S}$ are convex sets, and $\mathrm{T}(\mathscr{R}:\mathscr{S})$ denotes the set of all points X such that the x-translate of $\mathscr{R}$ is contained in $\mathscr{S}$, prove that $\mathrm{T}(\mathscr{R}:\mathscr{S})$ is a convex set.

9. Let $\mathcal{F} = \{\mathcal{S}_1, \mathcal{S}_2, \ldots, \mathcal{S}_m\}$ be a collection of convex sets in $\mathcal{E}^2$. Let $\mathcal{R}$ denote a convex set with the property that for each three sets in $\mathcal{F}$ there is some translate of $\mathcal{R}$ that intersects all three sets. Show that there must exist a translate of $\mathcal{R}$ that intersects all the sets in $\mathcal{F}$.

SECTION 5.2. DILATATIONS, CENTERS, AND CENTRALNESS

The Helly theorems have many important applications, and in this section we consider an illustrative example that concerns the centralness of convex bodies. For this purpose we need the notion of a center to a general set, and we also make use of some special similarity mappings. Thus we begin with some necessary background.

Consider a fixed point P and a positive number k. For any point $X \neq P$, the point X' such that

$$\mathbf{PX'} = k\mathbf{PX} \tag{1}$$

lies on the ray $\mathrm{Ry}[P, X)$ and its distance from P is k times the distance of X from P. For $X = P$ in (1) we have $\mathbf{PP'} = k\mathbf{PP} = k \cdot o = o$, which implies $P' = P$, so P is fixed. Regarding (1) as a mapping $X \to X'$, its effect on the points of space, namely, the *action* of the mapping, is clear. The point P is fixed, and if $k > 1$, each ray $\mathrm{Ry}[P, X)$ is stretched by a factor k with X mapping to the point on the ray k times as far from P. Thus for $k > 1$, the mapping is a uniform *expansion* at P with each sphere $\mathrm{S}(P, r)$ mapping to $\mathrm{S}(P, kr)$. For $k < 1$, the mapping is a uniform *contraction* of space at P. For $k = 1$, of course, every point is fixed and the mapping is the identity.

Next, suppose that k in (1) is taken to be a negative number. To see the action of the mapping, we can write (1) in the equivalent form

$$\mathbf{PX'} = -k(-\mathbf{PX}), \tag{2}$$

where $-k > 0$. Clearly, each vector $\mathbf{PX}$ is first reversed and then expanded or contracted by the positive factor $-k$. The sphere $\mathrm{S}(P, r)$ again maps onto $\mathrm{S}(P, kr)$, but now each ray $\mathrm{Ry}[P, X)$ maps onto its opposite ray.

To express the mapping (1) in function form, let $X' = f(X)$. Then, from $x' - p = k(x - p)$ we obtain

$$x' = f(x) = kx + (1 - k)p. \tag{3}$$

DEFINITION (Dilatation). Corresponding to a point P and a real number $k \neq 0$, the mapping of space f defined by

$$f(x) = kx + (1 - k)p$$

is a *dilatation at P with ratio k*. The dilatation is *positive* or *negative* according as k is positive or negative, and point P is the *center* of the dilatation.

DEFINITION (Central reflection). The dilatation of space at P with ratio $k = -1$, namely, the mapping

$$f(x) = -x + 2p,$$

is the *reflection of space in P*. Such a mapping is a *central reflection* with *center P*.

THEOREM 1. A dilatation of space at point P with ratio $k \neq 0$ is a one-to-one similarity of space onto itself whose ratio of similarity is $|k|$. (E)

COROLLARY 1. The reflection of space in point P is a *motion* of space, that is, an isometric mapping of space onto itself.

The reflection of space in a point P leaves P fixed and interchanges pairs of points that have P as their midpoint. Thus a set $\mathscr{S}$ maps onto itself under such a reflection if and only if it consists of pairs of points symmetric with respect to P and possibly the point P as well. We use this fact to define general centers.

DEFINITION (Center to a set). Point P is a *center to* set $\mathscr{S}$ if $\mathscr{S} \neq \varnothing$ and the reflection of space in P maps $\mathscr{S}$ onto itself. A set is a *central set* if a center to the set exists.

The familiar center P to the sphere $\mathrm{N}(P, r)$, the ball $\mathrm{B}(P, r)$, and the neighborhood $\mathrm{N}(P, r)$ is still the center in the generalized sense just defined. Now, however, an ellipse and a parallelogram are central sets. Every point of a

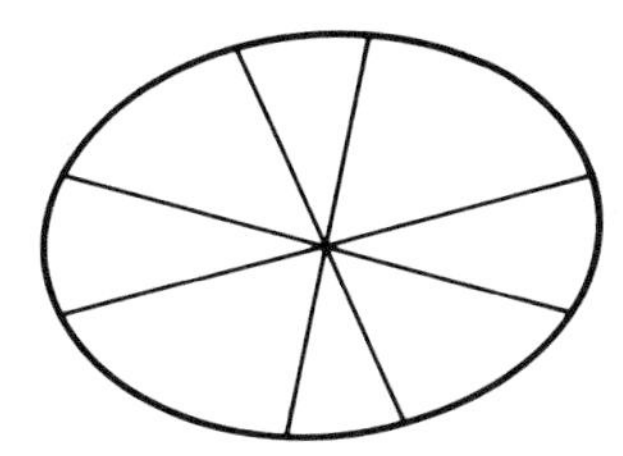

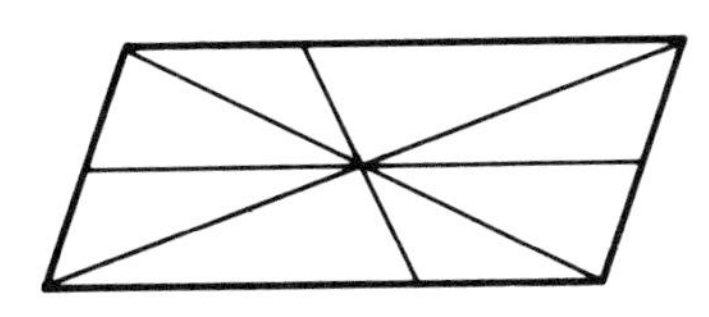

line is a center to the line. If $\mathscr{S}$ consists of two parallel hyperplanes $\mathscr{H}_1$ and $\mathscr{H}_2$, then every point in the hyperplane that is the mid-parallel to $\mathscr{H}_1$ and $\mathscr{H}_2$, is a center to $\mathscr{S}$.

As the previous examples indicate, there may be many centers to a set. However, this is not possible for bounded sets.

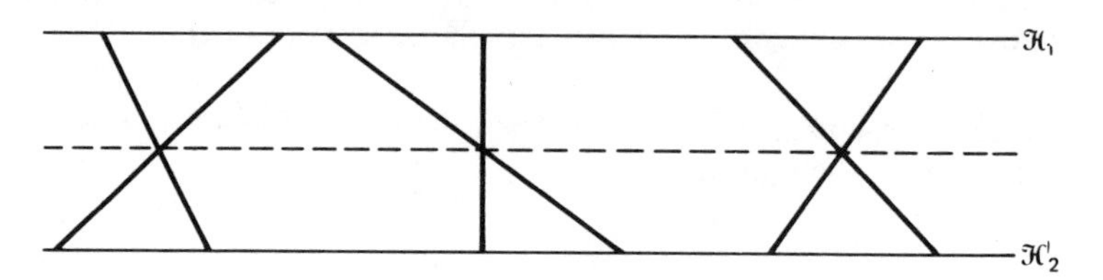

THEOREM 2. There is at most one center to a bounded set.

PROOF. Let P and Q denote two distinct centers to a set $\mathscr{S}$. Let f and g denote the reflections in P and Q, respectively, and let h denote the composition mapping $g(f)$. From

$$f(x) = -x + 2p, \quad \text{and} \quad g(x) = -x + 2q$$

it follows that

$$h(x) = g(f(x)) = g(-x + 2p) = -(-x + 2p) + 2q,$$

so

$$h(x) = x + 2(q - p) = x + 2\mathbf{PQ}.$$

Thus h is a translation by the vector $2\mathbf{PQ}$. By hypothesis, $\mathscr{S}$ is invariant under f and g, so $\mathscr{S}$ is invariant under h and hence under the iterates of h, namely $h(h)$, $h[h(h)]$, and so on. Thus if A is any point of $\mathscr{S}$, then the successive translates of $\mathscr{S}$, the points $a + 2\mathbf{PQ}$, $a + 4\mathbf{PQ}$, $a + 4\mathbf{PQ}$, $a + 6\mathbf{PQ}$, and so on are also in $\mathscr{S}$. Since the distance from A to these successive images under h increases without limit, the set $\mathscr{S}$ must be unbounded. From this it follows that if a set is bounded, then it cannot have two centers. □

COROLLARY 2. A reflection in P followed by a reflection in Q is a translation by $2\mathbf{PQ}$.

A mapping f of space that is not the identity but whose composition with itself, namely $f(f)$, is the identity is an *involution* of space. In such a mapping, each nonfixed point x is interchanged with its image because $f(f(x)) = x$. Central reflections are obviously involutions, and we can use their "interchange" property to generate central sets. If $\mathscr{S}$ is any non-empty set, and P is any point, then f, the reflection in P, maps $\mathscr{S}$ to a set $\mathscr{S}' = f(\mathscr{S})$, and P is a center to the set $\mathscr{R} = \mathscr{S} \cup \mathscr{S}'$. This is so because $f(\mathscr{R}) = f(\mathscr{S} \cup \mathscr{S}') = f(\mathscr{S}) \cup f(\mathscr{S}') = \mathscr{S}' \cup \mathscr{S} = \mathscr{R}$. By the same logic, P is a center to the set $\mathscr{S} \cap \mathscr{S}'$ if this set is not empty.

Despite the ease with which central sets can be generated, they are still very special sets. A natural question one can ask about a non-central set $\mathscr{S}$ is, "How non-central is it?" That is, if $\mathscr{S}$ does not have a center, is there a sense in which one point is more central to $\mathscr{S}$ than another, and does there exist any most central point?

For convex bodies, there is a natural answer to the previous questions in terms of what can be called the "chord centralness" of the body, defined in the following way. If P is an interior point of the convex body $\mathscr{K}$, then for each unit vector u, the line $x = p + \eta u, -\infty < \eta < \infty$, intersects the boundary of $\mathscr{K}$ in a pair of points Y, Y^*. We can think of P as central to $\mathscr{K}$ *in the direction* u if P is the midpoint of Y, Y^*. In this case, each of the segments $\mathrm{Sg}[PY]$, $\mathrm{Sg}[PY^*]$ has length $\frac{1}{2}$ that of the chord $\mathrm{Sg}[YY^*]$. If P is not the midpoint of the chord, then one of the ratios $\mathrm{d}(P, Y)/\mathrm{d}(Y, Y^*)$, $\mathrm{d}(P, Y^*)/\mathrm{d}(Y, Y^*)$ is less than $\frac{1}{2}$ and the other is greater than $\frac{1}{2}$. We arbitrarily select the smaller as a measure of the centralness of P in the direction u and define

$$r(P, u) = \min\left\{\frac{\mathrm{d}(P, Y)}{\mathrm{d}(Y, Y^*)}, \frac{\mathrm{d}(P, Y^*)}{\mathrm{d}(Y, Y^*)}\right\}. \tag{1}$$

It is clear from its definition that $\mathrm{r}(P, u) = \mathrm{r}(P, -u)$. If we consider $\mathrm{r}(P, u)$ for all possible directions u, then the minimum value, say $\mathrm{r}(P, u_o)$, corresponds to a direction in which P is least central. So P must be at least this central in every direction. The existence of a "minimum direction" depends on a continuity argument. To avoid going through this, we make use of the fact that $\mathrm{r}(P, u)$ is a bounded function, and define the *centralness of* P, with respect to $\mathscr{K}$, as

$$\mathrm{r}(P) = \mathrm{glb}\{\mathrm{r}(P, u) \colon U \in \mathrm{S}(0, 1)\}. \tag{2}$$

The relation (2) provides a sense in which we can say that one interior point A is more or less central than a second interior point B according as $\mathrm{r}(A)$ is greater or smaller than $\mathrm{r}(B)$. Clearly, then, a most central point is one at which $\mathrm{r}(P)$ is maximal. Again, to avoid going through a proof for the existence of such a point, we make use of the fact that $\mathrm{r}(P)$ is a bounded function, and define the *chord centralness* of $\mathscr{K}$ by

$$\mathrm{c}(\mathscr{K}) = \mathrm{lub}\{\mathrm{r}(P) \colon P \in \mathrm{In}(\mathscr{K})\}. \tag{3}$$

Although the definition rests on rather simple ideas, it is not an easy matter in general to determine the chord centralness of any specific convex body. It is somewhat surprising that the most central point of any simplex is easily determined and that all simplices of the same dimension have the same chord centralness. To show this, we recall a concept mentioned in the exercises of Sec. 4.1.

DEFINITION (Centroid). The *centroid* of a finite set of points $A_1, A_2, \ldots, A_m$ is the point G defined by

$$g = \frac{1}{m}\sum_{1}^{m} a_i.$$

The *centroid of a simplex* is the centroid of the set of vertices.

THEOREM 3. The chord centralness of an n-simplex is $1/(n + 1)$, and the centroid is the unique most central point.

PROOF. Let X denote an interior point of the n-simplex

$$\mathscr{K} = \mathrm{CS}(A_1, A_2, \ldots, A_{n+1}).$$

The theorem is trivially correct for $n = 1$, so we suppose that $n > 1$. Let v_i denote a unit vector in the direction $\mathbf{XA}_i$ and let $t_i = \mathrm{Ln}(XA_i)$ denote the line through X in the vertex direction v_i, $i = 1, 2, \ldots, n + 1$. The face of $\mathscr{K}$ opposite to A_i lies in a hyperplane β_i and line t_i intersects β_i at point B_i, $i = 1, 2, \ldots, n + 1$. Parallel to β_i are hyperplanes α_i and γ_i that pass through A_i and X, respectively, $n = 1, 2, \ldots, n + 1$.

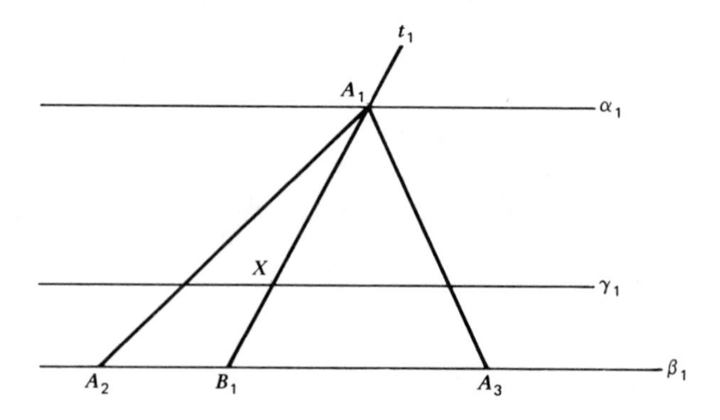

Now, consider the number $\mathrm{r}(X, v_1)$. By definition

$$\mathrm{r}(X, v_1) = \min\left\{\frac{\mathrm{d}(X, B_1)}{\mathrm{d}(A_1, B_1)}, \frac{\mathrm{d}(X, A_1)}{\mathrm{d}(A_1, B_1)}\right\}. \tag{1}$$

The parallel hyperplanes $\alpha_1, \gamma_1, \beta_1$ intercept segments on the transversal t_1 whose lengths are proportional to the distances between the hyperplanes. Hence $\mathrm{r}(X, v_1)$ can also be expressed as

$$\mathrm{r}(X, v_1) = \min\left\{\frac{\mathrm{d}(\gamma_1, \beta_1)}{\mathrm{d}(\alpha_1, \beta_1)}, \frac{\mathrm{d}(\gamma_1, \alpha_1)}{\mathrm{d}(\alpha_1, \beta_1)}\right\}. \tag{2}$$

Because X is interior to $\mathscr{K}$, it belongs to the positive convex span of the vertices, hence all $n + 1$ of its affine coordinates with respect to the basis $\{A_1, A_2, \ldots, A_{n+1}\}$ are positive. Thus X has a representation

$$x = \sum_1^{n+1} \eta_i a_i, \qquad \sum_1^{n+1} \eta_i = 1, \qquad \eta_i > 0. \tag{3}$$

From the geometric meaning of affine coordinates, Theorem 9, Sec. 4.1,

$$\eta_1 = \frac{\mathrm{d}(X, \beta_1)}{\mathrm{d}(A_1, \beta_1)} = \frac{\mathrm{d}(\gamma_1, \beta_1)}{\mathrm{d}(\alpha_1, \beta_1)}. \tag{4}$$

From (1), (2), and (4), it follows that

$$\mathrm{r}(X, v_1) = \min\{\eta_1, 1 - \eta_1\}.$$

More generally, by the same logic,

$$\mathrm{r}(X, v_i) = \min\{\eta_i, 1 - \eta_i\}, \qquad i = 1, 2, \ldots, n + 1. \tag{5}$$

Now let us consider a line t, through X, with a direction u such that neither of the points P, Q in which t intersects $\mathrm{Bd}(\mathscr{K})$ is a vertex. We may suppose that $\mathrm{d}(X, P) \leq \mathrm{d}(X, Q)$, hence that

$$\mathrm{r}(X, u) = \frac{\mathrm{d}(X, P)}{\mathrm{d}(P, Q)}. \tag{6}$$

Because P is a boundary point, it belongs to one of the hyperplanes β_i, and there is no loss of generality in supposing that $P \in \beta_1$. Let δ denote the hyperplane that passes through Q and is parallel to β_1. Because α_1 and β_1

are parallel supporting hyperplanes to $\mathscr{K}$, and because α_1 has unique contact with $\mathscr{K}$ at $A_1 \neq Q$, it follows that δ is between α_1 and β_1. Therefore $\mathrm{d}(\beta_1, \delta) < \mathrm{d}(\beta_1, \alpha_1)$. Thus we have

$$\begin{aligned} \mathrm{r}(X, u) &= \frac{\mathrm{d}(X, P)}{\mathrm{d}(P, Q)} = \frac{\mathrm{d}(\gamma_1, \beta_1)}{\mathrm{d}(\beta_1, \delta)} > \frac{\mathrm{d}(\gamma_1, \beta_1)}{\mathrm{d}(\beta_1, \alpha_1)} = \eta_1 \\ &\geq \min\{\eta_1, 1 - \eta_1\} = \mathrm{r}(X, v_1) \end{aligned} \tag{7}$$

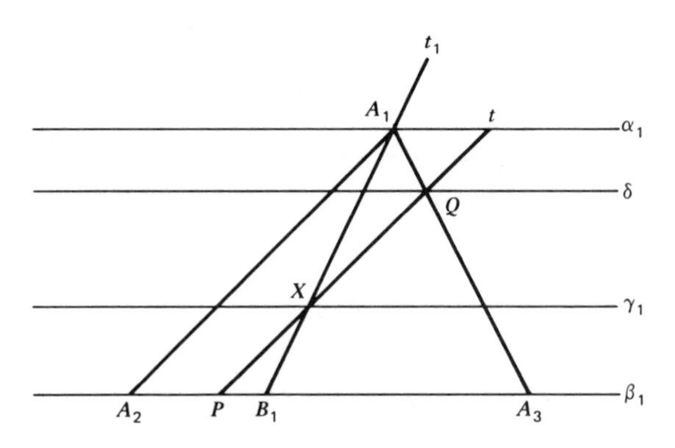

The relation (7) shows that if u is not a vertex direction, then there is always at least one vertex direction v_j such that $\mathrm{r}(X, u) > \mathrm{r}(X, v_j)$. It follows that $\mathrm{r}(X) = \mathrm{glb}\{\mathrm{r}(X, u) : U \in \mathrm{S}(0, 1)\}$ is also given by

$$\mathrm{r}(X) = \min\{\mathrm{r}(X, v_i) : i = 1, 2, \ldots, n + 1\}. \tag{8}$$

Combining (5) and (8), we obtain

$$\mathrm{r}(X) = \min\{\eta_1, \eta_2, \ldots, \eta_{n+1}, 1 - \eta_1, 1 - \eta_2, \ldots, 1 - \eta_{n+1}\}. \tag{9}$$

To further simplify (9), let $\eta_k = \min\{\eta_1, \eta_2, \ldots, \eta_{n+1}\}$, and $\eta_j = \max\{\eta_1, \eta_2, \ldots, \eta_{n+1}\}$. Then, clearly, we also have $1 - \eta_j = \min\{1 - \eta_1, 1 - \eta_2, \ldots, 1 - \eta_{n+1}\}$. If $j \neq k$, then $\eta_j + \eta_k < 1$, because these are only two of $n + 1$ positive coordinates whose sum is 1. If $j = k$, then all the coordinates are equal, so each is $1/(n + 1)$. Then $\eta_j + \eta_k = 2/(n + 1) < 1$ because $n > 1$. Thus in all cases, $\eta_j + \eta_k < 1$, hence $\eta_k < 1 - \eta_j$, which is to say that $\min\{\eta_1, \eta_2, \ldots, \eta_{n+1}\} < \min\{1 - \eta_1, 1 - \eta_2, \ldots, 1 - \eta_{n+1}\}$. Thus (9) reduces to

$$\mathrm{r}(X) = \min\{\eta_1, \eta_2, \ldots, \eta_{n+1}\}. \tag{10}$$

From (10) it follows that r(X) is maximal when the minimum affine coordinate of X is maximal. Since the sum of the $n + 1$ positive coordinates is 1, they cannot all be greater than $1/(n + 1)$, hence the minimal coordinate is always equal to or less than $1/(n + 1)$ and so is maximal when it equals $1/(n + 1)$. In this case all the coordinates have the value $1/(n + 1)$, so

$$x = \sum_{1}^{n+1} \eta_i a_i = \frac{1}{n+1} \sum_{1}^{n+1} a_i = g.$$

Thus G is the most central point of $\mathscr{K}$ and $\mathrm{r}(G) = c(\mathscr{K}) = 1/(n + 1)$. □

COROLLARY 3. The centralness of an interior point of a simplex is its minimum affine coordinate.

We now want to show that simplices are the least central of convex bodies in the sense that every n-dimensional convex body has chord centralness equal to or greater than $1/(n + 1)$. The proof is based on a Helly theorem and the properties of dilatations. We also need a consequence of the following fact.

THEOREM 4. Dilatations are affine functions. (E)

THEOREM 5. If $\mathscr{K}$ is an n-dimensional convex body, then it contains a point P for which $\mathrm{r}(P) \geq 1/(n + 1)$, hence the chord centralness of $\mathscr{K}$ is equal to or greater than $1/(n + 1)$.

PROOF. Corresponding to each point B in the boundary of $\mathscr{K}$, there is a dilatation at B with ratio $n/(n + 1)$, and we denote by $\mathscr{K}_B$ the image of $\mathscr{K}$ in this positive contraction at B. If X is any point of $\mathscr{K}$, then $\mathrm{Sg}[BX] \subset \mathscr{K}$ because $\mathscr{K}$ is convex. Because the dilatation at B is a contraction, it maps $\mathrm{Sg}[BX]$ onto a subsegment of itself, say $\mathrm{Sg}[BX']$, so $X' \in \mathrm{Sg}[BX'] \subset \mathrm{Sg}[BX] \subset \mathscr{K}$. Thus each of the sets $\mathscr{K}_B$ is a subset of $\mathscr{K}$.

All the sets $\mathscr{K}_B$ form the family

$$\mathscr{F} = \{\mathscr{K}_B : B \in \mathrm{Bd}(\mathscr{K})\}. \tag{1}$$

Since dilatations are similarities, they are continuous mappings, so the compactness of $\mathscr{K}$ implies that each of the sets $\mathscr{K}_B$ is compact. Because dilatations are affine mappings, they preserve convexity, so the convexity of $\mathscr{K}$ implies that each of the sets $\mathscr{K}_B$ is convex.

We now want to show that each $n + 1$ sets in $\mathscr{F}$ have a non-empty intersection set. For this purpose, let

$$\mathscr{F}_1 = \{\mathscr{K}_{B_1}, \mathscr{K}_{B_2}, \ldots, \mathscr{K}_{B_{n+1}}\} \tag{2}$$

be an arbitrary subcollection of $\mathscr{F}$ corresponding to $n + 1$ boundary points $B_1, B_2, \ldots, B_{n+1}$. Let G denote the centroid of the set $\{B_1, B_2, \ldots, B_{n+1}\}$, and let G_i denote the centroid of the subset of n points obtained by deleting B_i, $i = 1, 2, \ldots, n + 1$. Finally, let f_i denote the contraction at B_i such that $f_i(\mathscr{K}) = \mathscr{K}_{B_i}$, $i = 1, 2, \ldots, n + 1$.

Because a convex set contains the convex span of each of its finite subsets, and since the centroid of a finite set belongs to the convex span of the set, it follows that

$$G \in \mathscr{K}, \qquad \text{and} \qquad G_i \in \mathscr{K}, \qquad i = 1, 2, \ldots, n + 1. \tag{3}$$

Now consider $f_1(G_1)$. Since

$$g_1 = \frac{1}{n} \sum_{2}^{n+1} b_i \tag{4}$$

and

$$f_1(x) = \frac{n}{n + 1} x + \frac{1}{n + 1} b_1, \tag{5}$$

then

$$f_1(g_1) = \frac{n}{n + 1} \left[\frac{1}{n} \sum_{2}^{n+1} b_i \right] + \frac{1}{n + 1} b_1$$

$$= \frac{1}{n + 1} \sum_{1}^{n+1} b_i = g. \tag{6}$$

By the same logic,

$$f_i(G_i) = G, \qquad i = 1, 2, \ldots, n + 1. \tag{7}$$

Since $f_i(\mathscr{K}) = \mathscr{K}_{B_i}$, and $G_i \in \mathscr{K}$, then (7) implies that

$$G \in \mathscr{K}_{B_i}, \qquad i = 1, 2, \ldots, n + 1, \tag{8}$$

so the intersection set of the sets in $\mathscr{F}_1$ is not empty.

Because the sets in $\mathscr{F}$ are compact and convex, and each $n + 1$ of them intersect, it follows from the infinite form of the Helly theorem (Theorem 6, Sec. 5.1) that there exists a point P that belongs to all the sets in $\mathscr{F}$. Then P is in $\mathscr{K}$. If P were a boundary point to $\mathscr{K}$, then a line through P and an interior point would intersect the boundary at a second point Q, and P could not belong to the contracted set $\mathscr{K}_Q$. This would contradict the fact that P is in all the sets of $\mathscr{F}$. Therefore P must be an interior point of $\mathscr{K}$.

Now consider a line through P in an arbitrary direction u. This line intersects $\mathscr{K}$ in a chord Sg$[CD]$. Because $P \in \mathscr{K}_C$, $\mathrm{d}(C, P) \leq n\,\mathrm{d}(C, D)/(n + 1)$ and therefore $\mathrm{d}(P, D) \geq \mathrm{d}(C, D)/(n + 1)$. Similarly, because $P \in \mathscr{K}_D$, $\mathrm{d}(D, P) \leq n\,\mathrm{d}(D, C)/(n + 1)$, so $\mathrm{d}(P, C) \geq \mathrm{d}(D, C)/(n + 1)$. Therefore

$$\mathrm{r}(P, u) = \min\left\{\frac{\mathrm{d}(P, C)}{\mathrm{d}(C, D)}, \frac{\mathrm{d}(P, D)}{\mathrm{d}(C, D)}\right\} \geq \frac{1}{n + 1}. \tag{9}$$

Because (9) is valid for all directions u, it follows that

$$\mathrm{r}(P) = \mathrm{glb}\{\mathrm{r}(P, u) \colon U \in \mathrm{S}(0, 1)\} \geq \frac{1}{n + 1}. \tag{10}$$

Thus the chord centralness of $\mathscr{K}$ is at least $1/(n + 1)$. □

Exercises – Section 5.2

1. Prove Theorem 1.
2. Prove Theorem 4.
3. Show that if g is an affine mapping and if P is a center to set $\mathscr{S}$, then $g(P)$ is a center to $g(\mathscr{S})$.
4. Prove that if $\mathscr{S}$ is a finite set and if G is the centroid to $\mathscr{S}$, then for any affine mapping f the point $f(G)$ is the centroid to the set $f(\mathscr{S})$.
5. Prove that if P is a center to set $\mathscr{S}$ and if A, B in $\mathscr{S}$ are such that $\mathrm{d}(A, B)$ is the diameter of $\mathscr{S}$, then P must be the center of Sg$[AB]$.
6. Let $\mathscr{S}$ be a non-empty set in $\mathscr{E}^n$ and let P denote a fixed point in the space. Corresponding to each X in $\mathscr{S}$, there is a translation of space by the vector **PX** mapping $\mathscr{S}$ to a set $(\mathscr{S})_{\mathbf{PX}}$. Prove that if all the translate sets $(\mathscr{S})_{\mathbf{PX}}$, for $X \in \mathscr{S}$, have a non-empty intersection, then there exists a center to the set $\mathscr{S}$. Prove, conversely, that if a center to set $\mathscr{S}$ exists, then for any fixed point P all the translate sets $(\mathscr{S})_{\mathbf{PX}}$, $X \in \mathscr{S}$, do have a non-empty intersection set.
7. A theorem due to Blaschke asserts that if a plane convex body $\mathscr{K}$ has minimal width 1, then there exists a disc of radius $\frac{1}{3}$ contained in $\mathscr{K}$. We can prove this using Theorem 5 as follows. Let A denote an arbitrary point in the boundary of $\mathscr{K}$. At A there is at least one supporting line $\mathscr{L}_1$ and on the parallel supporting line $\mathscr{L}_2$ there is at least one contact point B. By Theorem 5, there exists a point P interior to $\mathscr{K}$ and such that $\mathrm{r}(P)$, the centralness of P with respect to $\mathscr{K}$, is equal to or greater than $\frac{1}{3}$. The line

Ln(PB) intersects the boundary of $\mathscr{K}$ again at a point C and intersects $\mathscr{L}_1$ at a point D.

a. Show that $d(P, D) \geq \frac{1}{3}d(B, D)$.

Next, consider the line through P and perpendicular to $\mathscr{L}_1$ at E and to $\mathscr{L}_2$ at F.

b. Why is $d(E, F) \geq 1$?
c. Why is $d(P, D)/d(B, D) = d(P, E)/d(E, F)$?

Use (a), (b), and (c) to show that

d. $d(P, E) \geq \frac{1}{3}$.

Why does (d) imply that

e. $d(P, A) \geq \frac{1}{3}$?

Why can we conclude from (e) that the disc with center P and radius $\frac{1}{3}$ is contained in $\mathscr{K}$?

8. Show that if the minimum width of a regular 2-simplex (an equilateral triangle) is 1, then no disc of radius greater than $\frac{1}{3}$ is contained in the simplex.
9. What analog of Blaschke's theorem for plane convex bodies would you expect might hold for n-dimensional convex bodies?
10. Prove that if a convex body $\mathscr{K}$ has constant width and a center, then $\mathscr{K}$ is a ball (cf. Exercise 5 and Theorem 7, Sec. 4.3).

SECTION 5.3. EXTREME POINTS, CONVEX HULLS, THE THEOREM OF CARATHEODORY

In this section we consider some natural questions concerning the convex spans of finite sets and see how these suggest important properties of general convex bodies and the convex hulls of general sets. In Sec. 4.1, we observed that the convex span of a finite set in $\mathscr{E}^n$ is the natural analog of a polygonal region in $\mathscr{E}^2$ or a polyhedral region in $\mathscr{E}^3$. Thus the following fact is not surprising.

THEOREM 1. The convex span of a finite set $\mathscr{R}$ is also its convex hull.

PROOF. By Theorem 1, Sec. 4.1, $CS(\mathscr{R})$ is a convex set. By definition, the convex hull of $\mathscr{R}$ is the intersection of all the convex sets that contain $\mathscr{R}$. Thus $CH(\mathscr{R})$ is a subset of every convex set that contains $\mathscr{R}$, hence $CH(\mathscr{R}) \subset CS(\mathscr{R})$. On the other hand, $\mathscr{R}$ is a finite subset of $CH(\mathscr{R})$. Since, by Theorem 3, Sec. 4.1, a convex set contains the convex spans of its finite subsets, then $CS(\mathscr{R}) \subset CH(\mathscr{R})$. The opposite inclusions imply that $CS(\mathscr{R}) = CH(\mathscr{R})$. □

If $\mathscr{R}$ is a finite set of vectors, then there exists a minimal subset of $\mathscr{R}$ whose linear span is $LS(\mathscr{R})$. Similarly, there is a minimal subset of $\mathscr{R}$ whose affine span is $AS(\mathscr{R})$. These linear and affine basis sets are the maximal subsets of $\mathscr{R}$ that are, respectively, linearly and affinely independent. From the nature of $CS(\mathscr{R})$ in $\mathscr{E}^2$ and $\mathscr{E}^3$, it is natural to expect that there is a subset of $\mathscr{R}$, namely, the "vertices" of $CS(\mathscr{R})$, that plays the role of a "convex basis" for $CS(\mathscr{R})$. To investigate this idea, we first introduce a rather natural definition.

DEFINITION (Convex independence and dependence). A finite set of two or more distinct points is *convexly independent* if no one of the points is a convex combination of the others. A singleton set is convexly independent. A finite set that is not convexly independent is *convexly dependent.*

THEOREM 2. Each non-empty subset of a convexly independent set is convexly independent. (E)

THEOREM 3. Linear independence implies affine independence, and affine independence implies convex independence. (E)

From the definition of convex independence it follows that if a finite set $\mathscr{R}$ is convexly independent, then there is no proper subset of $\mathscr{R}$ whose convex span is $CS(\mathscr{R})$. However, if $\mathscr{R}$ is convexly dependent, then a minimal generating set for $CS(\mathscr{R})$ need not be a maximal convexly independent subset of $\mathscr{R}$. This can be seen from a simple example. Let $\mathscr{R} = \{A_1, A_2, A_3, A_4, A_5\}$,

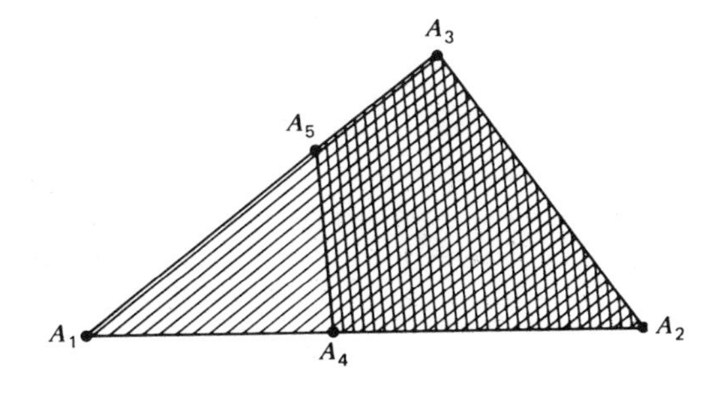

where A_1, A_2, A_3 are noncollinear, and A_4 is between A_1 and A_2, and A_5 is between A_1 and A_3. Then clearly $\{A_1, A_2, A_3\}$ is a generating set for CS($\mathscr{R}$) in the sense that CS(A_1, A_2, A_3) = CS($\mathscr{R}$). But each three of the four points A_2, A_3, A_4, A_5 are the vertices of a 2-simplex that does not contain the fourth point. Thus the four points are convexly independent. The generating set $\{A_1, A_2, A_3\}$ is therefore not a maximal, convexly independent subset of $\mathscr{R}$, although it is the vertex set that we anticipated.

In the example, there is a simple clue to what does characterize the "vertex set" or generating set for CS($\mathscr{R}$). The three points A_1, A_2, A_3 are the only points of CS($\mathscr{R}$) that do not lie between any two other points of CS($\mathscr{R}$). We introduce a name for this property.

DEFINITION (Extreme point). A point P is an extreme point of a set $\mathscr{R}$ if it belongs to $\mathscr{R}$ but does not lie between any two points of $\mathscr{R}$. The set of all extreme points of $\mathscr{R}$ is denoted by E($\mathscr{R}$).

In the example previously considered, it is the extreme point set of CS($\mathscr{R}$) that is the minimal subset of $\mathscr{R}$ which generates CS($\mathscr{R}$). Our next theorem shows that this property is general.

THEOREM 4. If $\mathscr{R}$ is a finite set of points, then the extreme points of CS($\mathscr{R}$) form the unique, convexly independent subset of $\mathscr{R}$ whose convex span is CS($\mathscr{R}$).

PROOF. Let $\mathscr{R} = \{A_1, A_2, \ldots, A_m\}$. If $\mathscr{R}$ is a convexly dependent set, then at least one of the points, say A_m, belongs to the convex span of the remaining points. Then $\mathscr{R}_1 = \{A_1, A_2, \ldots, A_{m-1}\}$ is a subset of $\mathscr{R}$ such that CS($\mathscr{R}_1$) = CS($\mathscr{R}$). If $\mathscr{R}_1$ is convexly dependent, then at least one of its points is convexly dependent on the others, and deleting this point gives a new subset $\mathscr{R}_2$ such that CS($\mathscr{R}_2$) = CS($\mathscr{R}$). Continuing in this way, deleting dependent points one at a time, we must arrive at a subset of $\mathscr{R}$, say $\mathscr{R}^* = \{A_1, A_2, \ldots, A_k\}$, that is convexly independent and such that CS($\mathscr{R}^*$) = CS($\mathscr{R}$). If $\mathscr{R}$ is itself convexly independent, then $\mathscr{R}^* = \mathscr{R}$, and $k = m$.

Now consider $P \in$ CS($\mathscr{R}^*$) and $P \notin \mathscr{R}^*$. Then, by Corollary 3, Sec. 4.1, P lies between one of the points of $\mathscr{R}^*$ and some other point of CS($\mathscr{R}^*$). Hence P is not an extreme point of CS($\mathscr{R}^*$). Therefore, E[CS($\mathscr{R}^*$)] $\subset \mathscr{R}^*$.

Next, assume that some point of $\mathscr{R}^*$ is not an extreme point of CS($\mathscr{R}^*$). We may suppose the labeling such that $A_k \notin$ E[CS($\mathscr{R}^*$)]. Then there exist two points P, Q in CS($\mathscr{R}^*$) such that $A_k \in$ Sg(PQ). Thus A_k has a representation

$$A_k = \lambda_1 P + \lambda_2 Q, \qquad \lambda_1 + \lambda_2 = 1, \qquad \lambda_1, \lambda_2 > 0. \tag{1}$$

Because P, Q belong to $\mathrm{CS}(\mathscr{R}^*)$, they have representations

$$P = \sum_1^k \eta_i A_i, \qquad \sum_1^k \eta_i = 1, \qquad \eta_i \geq 0, i = 1, 2, \ldots, k.$$

$$Q = \sum_1^k \varphi_i A_i, \qquad \sum_1^k \varphi_i = 1, \qquad \varphi_i \geq 0, i = 1, 2, \ldots, k. \tag{2}$$

Substituting from (2) in (1), we can express A_k in the form

$$A_k = \sum_1^k (\lambda_1 \eta_i + \lambda_2 \varphi_i) A_i = \sum_1^k \pi_i A_i, \tag{3}$$

where

$$\pi_i = \lambda_1 \eta_i + \lambda_2 \varphi_i, \qquad i = 1, 2, \ldots, k. \tag{4}$$

Since λ_1, λ_2 are positive, and η_i and φ_i are non-negative, it follows that $\pi_i \geq 0$, $i = 1, 2, \ldots, k$. Moreover,

$$\sum_1^k \pi_i = \lambda_1 \sum_1^k \eta_i + \lambda_2 \sum_1^k \eta_i = \lambda_1 + \lambda_2 = 1. \tag{5}$$

If $\pi_k \neq 1$, then we can solve (3) for A_k in the form

$$A_k = \sum_{i=1}^{k-1} \frac{\pi_i}{1 - \pi_k} A_i. \tag{6}$$

Since the numbers π_i are non-negative, it follows from (5) that (6) expresses A_k as a convex combination of $A_1, A_2, \ldots, A_{k-1}$, contradicting the convex independence of the set $\mathscr{R}^*$. On the other hand, if $\pi_k = 1$, then (5) and the nonnegativeness of the numbers π_i imply that $\pi_i = 0$, $i = 1, 2, \ldots, k - 1$. Then, because λ_1 and λ_2 are positive, it follows from (4) that $\eta_i = 0$ and $\varphi_i = 0$, $i = 1, 2, \ldots, k - 1$. Therefore, from (2), $\eta_k = 1$ and $\varphi_k = 1$. But then $P = A_k$ and $Q = A_k$, contradicting $P \neq Q$. In all cases, then, the assumption that some point of $\mathscr{R}^*$ is not an extreme point of $\mathrm{CS}(\mathscr{R}^*)$ leads to a contradiction. Hence it must be true that $\mathscr{R}^* \subset \mathrm{E}[\mathrm{CS}(\mathscr{R}^*)]$.

The opposite inclusion of $\mathscr{R}^*$ and $\mathrm{E}[\mathrm{CS}(\mathscr{R}^*)]$ imply that they are the same set. Since $\mathrm{CS}(\mathscr{R}^*) = \mathrm{CS}(\mathscr{R})$, it follows that

$$\mathscr{R}^* = \mathrm{E}[\mathrm{CS}(\mathscr{R})]. \tag{7}$$

The uniqueness of $\mathrm{E}[\mathrm{CS}(\mathscr{R})]$ implies the uniqueness of $\mathscr{R}^*$. $\square$

COROLLARY 4.1. The extreme points of $\text{CS}(\mathscr{R})$ belong to every finite subset of $\text{CS}(\mathscr{R})$ whose convex span is $\text{CS}(\mathscr{R})$. (E)

COROLLARY 4.2. If $\mathscr{R}$ is convexly independent, then $\mathscr{R}$ is the set of extreme points of $\text{CS}(\mathscr{R})$. (E)

By definition, the convex span of a finite set $\mathscr{R}$ is the set of all convex combinations of the points in $\mathscr{R}$. Thus it is the union of the convex spans of all subsets of $\mathscr{R}$. Since $\text{CH}(\mathscr{R}) = \text{CS}(\mathscr{R})$, it follows that the convex hull of $\mathscr{R}$ is the union of the convex spans of the subsets of $\mathscr{R}$. If $\mathscr{R}$ is convex, then it contains the convex span of each of its finite subsets and is obviously the union of all such spans. By Theorem 10, Sec. 4.3, $\text{CH}(\mathscr{R}) = \mathscr{R}$, so $\text{CH}(\mathscr{R})$ is the union of the convex spans of all finite subsets of $\mathscr{R}$. Our next theorem shows that this property of finite and convex sets holds for completely arbitrary sets.

THEOREM 5. The convex hull of any set $\mathscr{S}$ is the union of the convex spans of all the finite subsets of $\mathscr{S}$.

PROOF. Let $\mathscr{U}$ denote the union of all the convex spans of finite sets contained in $\mathscr{S}$. If $\mathscr{R}$ is any such finite set, then from $\mathscr{R} \subset \mathscr{S} \subset \text{CH}(\mathscr{S})$, and the fact that $\text{CH}(\mathscr{S})$ is convex, it follows that $\text{CS}(\mathscr{R}) \subset \text{CH}(\mathscr{S})$. Therefore $\mathscr{U} \subset \text{CH}(\mathscr{S})$.

Next, consider any two distinct points P, Q in $\mathscr{U}$. Any point X of $\text{Sg}[PQ]$ has a representation

$$X = \lambda_1 P + \lambda_2 Q, \qquad \lambda_1 + \lambda_2 = 1, \qquad 0 \le \lambda_1, \lambda_2 \le 1. \tag{1}$$

Because P and Q are in $\mathscr{U}$, each belongs to the convex span of some finite subset of $\mathscr{S}$. Thus there exist points $A_1, A_2, \ldots, A_k$ and $B_1, B_2, \ldots, B_m$ in $\mathscr{S}$ such that

$$P = \sum_1^k \eta_i A_i, \qquad \sum_1^k \eta_i = 1, \qquad Q = \sum_1^m \varphi_i B_i, \qquad \sum_1^m \varphi_i = 1, \tag{2}$$

where all the numbers η_i and φ_i are nonnegative. Substituting from (2) in (1) gives

$$X = \sum_1^k \lambda_1 \eta_i A_i + \sum_1^m \lambda_2 \varphi_i B_i. \tag{3}$$

The scalar coefficients of A_i and B_i are clearly nonnegative numbers, and

$$\sum_1^k \lambda_1 \eta_i + \sum_1^m \lambda_2 \varphi_i = \lambda_1 \sum_1^k \eta_i + \lambda_2 \sum_1^m \varphi_i = \lambda_1 + \lambda_2 = 1. \tag{4}$$

Thus (3) is a convex combination, hence $X \in \mathrm{CS}(A_1, \ldots, A_k, B_1, \ldots, B_m)$. Since X belongs to the convex span of a finite subset of $\mathscr{S}$, then $X \in \mathscr{U}$. Therefore $\mathrm{Sg}[PQ] \subset \mathscr{U}$, which shows that $\mathscr{U}$ is a convex set. Since $\mathscr{U}$ obviously contains $\mathscr{S}$, and since $\mathrm{CH}(\mathscr{S})$ is contained in every convex set that contains $\mathscr{S}$, it follows that $\mathrm{CH}(\mathscr{S}) \subset \mathscr{U}$.

The opposite inclusions of $\mathscr{U}$ and $\mathrm{CH}(\mathscr{S})$ imply that $\mathscr{U} = \mathrm{CH}(\mathscr{S})$. □

Theorem 5 shows that the convex hull of any set $\mathscr{S}$ can be built up from the inside, so to speak, by uniting the convex spans of the finite subsets of $\mathscr{S}$. With Theorem 5 and the notion of affine dependence, we can establish an important refinement of Theorem 5 that is due to Caratheodory.

THEOREM 6. If $\mathscr{S}$ is any set in $\mathscr{E}^n$, then each point X of $\mathrm{CH}(\mathscr{S})$ is a convex combination of $n + 1$ or fewer points in $\mathscr{S}$.

PROOF. Since X is in $\mathrm{CH}(\mathscr{S})$, it follows from Theorem 5 that there is at least one finite subset of $\mathscr{S}$ whose convex span contains X. Among all such finite subsets of $\mathscr{S}$, at least one has a minimal number of points. Let k denote this minimal number, and let $\mathscr{R} = \{A_1, A_2, \ldots, A_k\}$ be a subset of $\mathscr{S}$ such that $X \in \mathrm{CS}(\mathscr{R})$. Then X has a representation

$$X = \sum_1^k \eta_i A_i, \qquad \sum_1^k \eta_i = 1, \qquad \eta_i > 0, \qquad i = 1, 2, \ldots, k, \tag{1}$$

where the positiveness of η_i is due to the minimality of k.

Now assume that $k > n + 1$. Then $\mathscr{R}$ is an affinely dependent set (Theorem 5, Sec. 2.4), hence there exist numbers $\pi_1, \pi_2, \ldots, \pi_k$, not all zero such that

$$\sum_1^k \pi_i a_i = o, \qquad \text{and} \qquad \sum_1^k \pi_i = 0. \tag{2}$$

Because the numbers π_i are not all zero, but have a zero sum, it follows that at least one is positive and at least one is negative. If for each i, $1 \leq i \leq k$, such that $\pi_i < 0$, we form the ratio η_i/π_i, we obtain a finite set of negative numbers, and this set has a maximal element. That is, there exists a negative number h such that

$$h = \max\left\{\frac{\eta_i}{\pi_i} : 1 \leq i \leq k, \pi_i < 0\right\}. \tag{3}$$

We may suppose the points A_i are relabeled, if necessary, so that

$$h = \frac{\eta_k}{\pi_k} \qquad \text{or} \qquad \eta_k - h\pi_k = 0. \tag{4}$$

Because $\sum_1^k \pi_i a_i = o$, then $h \sum_1^k \pi_i a_i = o$. Thus we can express X as

$$x = \sum_1^k \eta_i a_i - h \sum_1^k \pi_i a_i = \sum_1^k (\eta_i - h\pi_i) a_i. \tag{5}$$

Now consider the scalar coefficients on the right of (5). If $\pi_i > 0$, then $-h\pi_i > 0$, and since $\eta_i > 0$, then $\eta_i - h\pi_i > 0$. If $\pi_i = 0$, then $\eta_i - h\pi_i = \eta_i > 0$. Finally, if $\pi_i < 0$, then $\eta_i/\pi_i \leq h$ implies $\eta_i \geq h\pi_i$, hence $\eta_i - h\pi_i \geq 0$. Thus in all cases,

$$\eta_i - h\pi_i \geq 0, \qquad i = 1, 2, \ldots, k. \tag{6}$$

Moreover,

$$\sum_1^k (\eta_i - h\pi_i) = \sum_1^k \eta_i - h \sum_1^k \pi_i = 1. \tag{7}$$

From (6) and (7), it follows that (5) expresses X as a convex combination of $A_1, A_2, \ldots, A_k$. But, from (4), $\eta_k - h\pi_k = 0$. Therefore (5) also expresses X as a convex combination of $A_1, A_2, \ldots, A_{k-1}$, which contradicts the fact that k is the least number of points in $\mathscr{S}$ whose convex span contains X. This contradiction implies that the assumption $k > n + 1$ is false. Hence $k \leq n + 1$. □

In Corollary 4.1, there is the suggestion for another type of generalization. If $\mathscr{R}$ is a finite set and if $\mathscr{R}_1 = \mathrm{E}[\mathrm{CS}(\mathscr{R})]$, then $\mathscr{R}_1$ is not only a minimal generating set in the sense of having the fewest number of points, it is also minimal in a set theory sense. That is, $\mathscr{R}_1$ is contained in every other finite generating subset of $\mathrm{CS}(\mathscr{R})$. If we replace convex spans by convex hulls, we can drop the finiteness condition and think of $\mathscr{S}_1 \subset \mathscr{S}$ as a "generating set" for $\mathrm{CH}(\mathscr{S})$ if $\mathrm{CH}(\mathscr{S}_1) = \mathrm{CH}(\mathscr{S})$. When $\mathscr{S}$ is finite, the hull and span definitions coincide, so from now on we use the hull definition. The question of whether or not a set $\mathscr{S}$ has a proper generating subset makes sense for any set $\mathscr{S}$. We call a generating subset "minimal" if it is contained in every other generating subset.

In the light of previous theorems, a natural candidate for a generating set of $\mathrm{CH}(\mathscr{S})$ is the extreme point set of $\mathrm{CH}(\mathscr{S})$. In the particular case when $\mathscr{S}$ is convex, $\mathrm{CH}(\mathscr{S}) = \mathscr{S}$, and the extreme points of $\mathscr{S}$ can be characterized quite simply. In this connection, it is convenient to have the following convention.

CONVENTION. The notation $\mathscr{R} - \mathscr{S}$ denotes the set of those points of $\mathscr{R}$ that are not in $\mathscr{S}$, that is, the set $\mathscr{R} \cap \mathrm{Cp}(\mathscr{S})$. In particular, $\mathscr{R} - \{P\}$ is also written as $\mathscr{R} - P$.

THEOREM 7. Point P of a convex set $\mathscr{S}$ is an extreme point of $\mathscr{S}$ if and only if the set $\mathscr{S} - P$ is convex.

PROOF. First, suppose that $P \in \mathrm{E}(\mathscr{S})$, and let A, B denote any two distinct points of $\mathscr{S} - P$. Thus $A \neq P$ and $B \neq P$, and A, B are points of $\mathscr{S}$. Because $\mathscr{S}$ is convex, $\mathrm{Sg}[AB] \subset \mathscr{S}$. Because P does not lie between any two points of $\mathscr{S}$, $P \notin \mathrm{Sg}(AB)$, hence $P \notin \mathrm{Sg}[AB]$. Therefore $\mathrm{Sg}[AB] \subset \mathscr{S} - P$. So $\mathscr{S} - P$ is a convex set.

Next, suppose that $\mathscr{S} - P$ is convex, and assume that P is not an extreme point of $\mathscr{S}$. The assumption implies the existence of two points A, B in $\mathscr{S}$ such that $P \in \mathrm{Sg}(AB)$. Since neither A nor B is P, then both A and B are in $\mathscr{S} - P$. But $\mathrm{Sg}[AB] \not\subset \mathscr{S} - P$ because P on $\mathrm{Sg}[AB]$ is not in $\mathscr{S} - P$. Therefore $\mathscr{S} - P$ is not convex. This contradicts the given convexity of $\mathscr{S} - P$ and shows that the assumption $P \notin \mathrm{E}(\mathscr{S})$ is false. Hence P must be an extreme point of $\mathscr{S}$. □

By an argument similar to that for Theorem 7, one can also obtain the following characterization.

THEOREM 8. Point P of a convex set $\mathscr{S}$ is an extreme point of $\mathscr{S}$ if and only if P does not belong to $\mathrm{CH}[\mathscr{S} - P]$. (E)

From the definition of a convex hull, it is clear that $\mathscr{R} \subset \mathscr{S}$ implies $\mathrm{CH}(\mathscr{R}) \subset \mathrm{CH}(\mathscr{S})$, and we use this fact in the next proof.

THEOREM 9. The extreme points of a convex set $\mathscr{S}$ are contained in every subset of $\mathscr{S}$ whose convex hull is $\mathscr{S}$.

PROOF. Let $\mathscr{R}$ denote a subset of $\mathscr{S}$ such that $\mathrm{CH}(\mathscr{R}) = \mathscr{S}$, and assume that there exists a point P in $\mathrm{E}(\mathscr{S})$ that is not in $\mathscr{R}$. Then $\mathscr{R} \subset \mathscr{S} - P$ implies that $\mathrm{CH}(\mathscr{R}) \subset \mathrm{CH}(\mathscr{S} - P)$. But $\mathrm{CH}(\mathscr{R}) = \mathscr{S}$ and $\mathscr{S} = \mathrm{CH}(\mathscr{S})$, so $\mathrm{CH}(\mathscr{S}) \subset \mathrm{CH}(\mathscr{S} - P)$. On the other hand, $\mathscr{S} - P \subset \mathscr{S}$ implies $\mathrm{CH}(\mathscr{S} - P) \subset \mathrm{CH}(\mathscr{S})$. The opposite inclusions give $\mathrm{CH}(\mathscr{S} - P) = \mathrm{CH}(\mathscr{S})$. But then

$P \in \mathrm{CH}(\mathscr{S})$ implies $P \in \mathrm{CH}(\mathscr{S} - P)$, which contradicts Theorem 8. The contradiction shows that there cannot be a point of $\mathrm{E}(\mathscr{S})$ that is not in $\mathscr{R}$. Hence $\mathrm{E}(\mathscr{S}) \subset \mathscr{R}$. □

It might appear from Theorem 9 that any convex set $\mathscr{S}$ is the convex hull of its extreme points and that $\mathrm{E}(\mathscr{S})$ is a minimal generating set for $\mathscr{S}$. But the facts are not that simple. An open convex set has no extreme points, nor does a closed space or closed half-space. It is also not obvious that any minimal generating set must exist, even if the set is convex. If $\mathscr{S}$, for example, is a

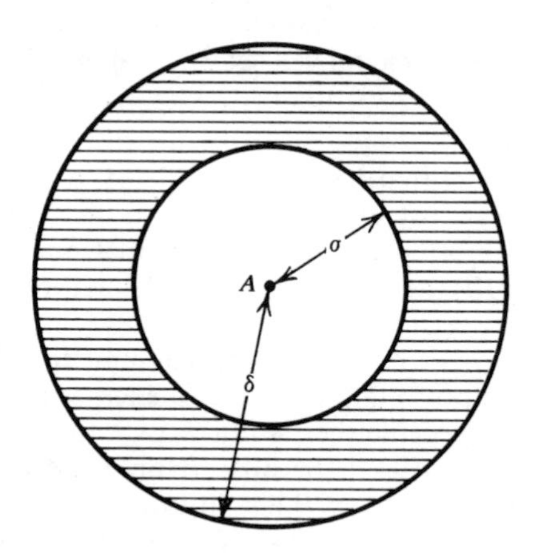

neighborhood $\mathrm{N}(A, \delta)$, consider the ring set $\mathscr{R} = \mathrm{N}(A, \delta) - \mathrm{N}(A, \sigma)$, where $\sigma < \delta$. Clearly, $\mathscr{R} \subset \mathscr{S}$ and $\mathrm{CH}(\mathscr{R}) = \mathscr{S}$, but there is no value of σ for which $\mathscr{R}$ is a minimal generating set.

The kind of generalization suggested by Theorem 9 can be established for convex sets that are also compact. Such a set is the convex hull of its extreme points, and it was shown by Krein and Milman that this is so not only in euclidean space but in more general spaces. Even for euclidean space, a careful proof requires considerable detail. We begin with a lemma that we use in the proof. This lemma in fact substitutes for the use of the kind of relative topology of k-flats that was described in Sec. 3.1.

LEMMA 1. If for some positive integer k, each k-dimensional, convex, compact set in $\mathscr{E}^k$ is the convex hull of its extreme points, then the same property holds for the k-dimensional, convex, compact sets in $\mathscr{E}^n$, where $n > k$.

PROOF. Let $\mathscr{R}$ denote a k-dimensional, convex, compact subset of $\mathscr{E}^n$, where $n > k$. By Theorem 3, Sec. 3.1, $\mathscr{R}$ contains $k + 1$ affinely independent points $A_1, A_2, \ldots, A_{k+1}$, and their affine span $\mathscr{A}$ is the unique k-flat that contains $\mathscr{R}$. By Theorem 7, Sec. 2.3, and Theorem 2, Sec. 3.1, there exists a one-to-one affine mapping f of $\mathscr{A}$ onto $\mathscr{E}^k$ that is an isometry of $\mathscr{A}$ and $\mathscr{E}^k$. Let $\mathscr{S} = f(\mathscr{R})$.

Because f maps the affine basis of $\mathscr{A}$ onto an affine basis for $\mathscr{E}^k$, $\mathscr{S}$ contains the $k + 1$ affinely independent points $f(A_i)$, $i = 1, 2, \ldots, k + 1$, hence $\mathscr{S}$ is k-dimensional. Since every isometry preserves convexity, $\mathscr{S}$ is convex, and since every isometry is a continuous mapping, $\mathscr{S}$ is compact. Thus $\mathscr{S}$ is a k-dimensional, convex, compact subset of $\mathscr{E}^k$, so, by the hypothesis of the lemma,

$$\mathrm{CH}[\mathrm{E}(\mathscr{S})] = \mathscr{S}. \tag{1}$$

Next, since the isometries f and f^{-1} preserve betweeness and nonbetweeness, the nonextreme points of $\mathscr{R}$ map onto the nonextreme points of $\mathscr{S}$, and the extreme points of $\mathscr{R}$ map onto the extreme points of $\mathscr{S}$. Hence

$$f[\mathrm{E}(\mathscr{R})] = \mathrm{E}(\mathscr{S}). \tag{2}$$

Finally, consider $\mathrm{CH}(\mathscr{U})$ for any set $\mathscr{U} \subset \mathscr{A}$. Corresponding to each set $\mathscr{U}'$ in $\mathscr{E}^n$ that is convex and that contains $\mathscr{U}$ there is a set $\mathscr{U}'' = \mathscr{U}' \cap \mathscr{A}$ that is a convex subset of $\mathscr{A}$ that contains $\mathscr{U}$. Thus $\mathrm{CH}(\mathscr{U})$ is also the intersection of all the convex sets in $\mathscr{A}$ that contain $\mathscr{U}$. Because f preserves convexity and set inclusion, the convex sets of $\mathscr{A}$ that contain $\mathscr{U}$ map onto the convex sets of $\mathscr{E}^k$ that contain $f(\mathscr{U})$. Hence $f[\mathrm{CH}(\mathscr{U})] = \mathrm{CH}[f(\mathscr{U})]$. Therefore,

$$f[\mathrm{CH}(\mathrm{E}(\mathscr{R}))] = \mathrm{CH}[f(\mathrm{E}(\mathscr{R}))]. \tag{3}$$

From (3), (2), and (1), $f[\mathrm{CH}(\mathrm{E}(\mathscr{R}))] = \mathrm{CH}[\mathrm{E}(\mathscr{S})] = \mathscr{S}$. Therefore $f^{-1}(\mathscr{S}) = \mathscr{R} = \mathrm{CH}[\mathrm{E}(\mathscr{R})]$, which we wished to prove. □

THEOREM 10. If $\mathscr{S}$ is a convex, compact set, then

$$\mathscr{S} = \mathrm{CH}[\mathrm{E}(\mathscr{S})].$$

PROOF. We first show that the theorem holds if $\mathscr{S} = \mathscr{K}$, where $\mathscr{K}$ is a convex body, and the proof is by induction on the dimension of the space.

First, if $n = 1$, then $\mathscr{K}$ is a closed interval $[a, b]$, and $\mathrm{E}(\mathscr{K}) = \{a, b\}$. Then $\mathrm{CH}[\mathrm{E}(\mathscr{K})] = [a, b] = \mathscr{K}$.

Next, assume that the theorem holds for spaces $\mathscr{E}^m$, $m = 1, 2, \ldots, n - 1$. To show that this implies the validity of the theorem for $\mathscr{E}^n$, consider a convex body $\mathscr{K}$ in $\mathscr{E}^n$. Let $\mathscr{H}$ be any supporting hyperplane to $\mathscr{K}$. Then, because $\mathscr{K}$ is closed, $\mathscr{K} \cap \mathscr{H}$ is not empty. Both $\mathscr{K}$ and $\mathscr{H}$ are closed and convex, so $\mathscr{K} \cap \mathscr{H}$ is closed and convex. Since $\mathscr{K}$ is compact, and $\mathscr{K} \cap \mathscr{H}$ is a closed subset of $\mathscr{K}$, then $\mathscr{K} \cap \mathscr{H}$ is compact. Because $\mathscr{K} \cap \mathscr{H}$ is a subset of $\mathscr{H}$, it is either a single point or else a set of dimension $h \le n - 1$. If $\mathscr{K} \cap \mathscr{H} = \{Q\}$, then $\text{CH}\{Q\} = \{Q\}$ and $\text{E}\{Q\} = \{Q\}$, so $\text{CH}[\text{E}(\mathscr{K} \cap \mathscr{H})] = \mathscr{K} \cap \mathscr{H}$. If $\mathscr{K} \cap \mathscr{H}$ is a set of dimension $h \le n - 1$, then, by the induction hypothesis and by Lemma 1 applied to $\mathscr{E}^h$, it follows that $\text{CH}[\text{E}(\mathscr{K} \cap \mathscr{H})] = \mathscr{K} \cap \mathscr{H}$. In all cases, then

$$\text{CH}[\text{E}(\mathscr{K} \cap \mathscr{H})] = \mathscr{K} \cap \mathscr{H}. \tag{1}$$

Next we want to show that the extreme points of $\mathscr{K} \cap \mathscr{H}$ are precisely the extreme points of $\mathscr{K}$ that lie in $\mathscr{H}$, that is, that $\text{E}(\mathscr{K} \cap \mathscr{H}) = \text{E}(\mathscr{K}) \cap \mathscr{H}$. To this end, first consider P in $\text{E}(\mathscr{K}) \cap \mathscr{H}$. Then because $\text{E}(\mathscr{K}) \subset \mathscr{K}$, it follows that $P \in \mathscr{K} \cap \mathscr{H}$. Since $\mathscr{K} \cap \mathscr{H} \subset \mathscr{K}$, and since P does not lie between any two points of $\mathscr{K}$, then P does not lie between any two points of $\mathscr{K} \cap \mathscr{H}$. Thus $P \in \text{E}(\mathscr{K} \cap \mathscr{H})$, and we have the inclusion

$$[\text{E}(\mathscr{K}) \cap \mathscr{H}] \subset \text{E}(\mathscr{K} \cap \mathscr{H}). \tag{2}$$

To obtain the opposite inclusion, consider $P \in \text{E}(\mathscr{K} \cap \mathscr{H})$. Assume that P is not an extreme point of $\mathscr{K}$. Then P lies between some pair of points $A, B \in \mathscr{K}$. If neither A nor B is in $\mathscr{H}$, then both lie in that side of $\mathscr{H}$ that contains points of $\mathscr{K}$. Since open half-spaces are convex, $\text{Sg}[AB]$ is contained in that same side of $\mathscr{H}$, so P on $\text{Sg}[AB]$ lies in that side of $\mathscr{H}$. But $P \in \text{E}(\mathscr{K} \cap \mathscr{H})$ implies that P is in $\mathscr{H}$ and hence is not in either side of $\mathscr{H}$. Thus at least one of the points A, B, say point A, must be in $\mathscr{H}$. Then, since $P \in \mathscr{H}$ and $A \in \mathscr{H}$, it follows that $\text{Ln}(PA) \subset \mathscr{H}$. Therefore B on $\text{Ln}(PA)$ is also in $\mathscr{H}$. Since A and B are both in $\mathscr{H}$ and both in $\mathscr{K}$, then P is between two points of $\mathscr{K} \cap \mathscr{H}$, which contradicts $P \in \text{E}(\mathscr{K} \cap \mathscr{H})$. Thus the assumption that $P \notin \text{E}(\mathscr{K})$ is false. Therefore $P \in \text{E}(\mathscr{K} \cap \mathscr{H})$ implies $P \in \text{E}(\mathscr{K})$, as well as $P \in \mathscr{H}$, hence

$$\text{E}(\mathscr{K} \cap \mathscr{H}) \subset [\text{E}(\mathscr{K}) \cap \mathscr{H}]. \tag{3}$$

From (2) and (3),

$$\text{E}(\mathscr{K} \cap \mathscr{H}) = \text{E}(\mathscr{K}) \cap \mathscr{H}. \tag{4}$$

Now consider $P \in \mathscr{K}$. If $P \in \text{Bd}(\mathscr{K})$, then there exists a hyperplane $\mathscr{H}$ that supports $\mathscr{K}$ and contains P. Then $P \in \mathscr{K} \cap \mathscr{H}$. From (1), $\mathscr{K} \cap \mathscr{H} = \text{CH}[\text{E}(\mathscr{K} \cap \mathscr{H})]$, so $P \in \text{CH}[\text{E}(\mathscr{K} \cap \mathscr{H})]$. From (4) and the properties of intersections, $\text{E}(\mathscr{K} \cap \mathscr{H}) = \text{E}(\mathscr{K}) \cap \mathscr{H} \subset \text{E}(\mathscr{K})$, so $\text{CH}[\text{E}(\mathscr{K} \cap \mathscr{H})] \subset \text{CH}[\text{E}(\mathscr{K})]$. Thus $P \in \text{CH}[\text{E}(\mathscr{K})]$, hence

$$\text{Bd}(\mathscr{K}) \subset \text{CH}[\text{E}(\mathscr{K})]. \tag{5}$$

Next, suppose that $P \in \text{In}(\mathscr{K})$. Then a line through P intersects $\text{Bd}(\mathscr{K})$ at two points C, D, and P lies between them (Corollary 9, Sec. 4.2). From (5), C and

D are in $\mathrm{CH}[\mathrm{E}(\mathscr{K})]$, which is a convex set, hence $\mathrm{Sg}[CD] \subset \mathrm{CH}[\mathrm{E}(\mathscr{K})]$, so $P \in \mathrm{CH}[\mathrm{E}(\mathscr{K})]$. Therefore

$$\mathrm{In}(\mathscr{K}) \subset \mathrm{CH}[\mathrm{E}(\mathscr{K})]. \tag{6}$$

Together, (5) and (6) show that

$$\mathscr{K} \subset \mathrm{CH}[\mathrm{E}(\mathscr{K})]. \tag{7}$$

By definition, $\mathrm{E}(\mathscr{K}) \subset \mathscr{K}$, and this implies that $\mathrm{CH}[\mathrm{E}(\mathscr{K})] \subset \mathrm{CH}(\mathscr{K})$. Since $\mathscr{K}$ is convex, $\mathrm{CH}(\mathscr{K}) = \mathscr{K}$, hence

$$\mathrm{CH}[\mathrm{E}(\mathscr{K})] \subset \mathscr{K}. \tag{8}$$

Together, (7) and (8) imply that

$$\mathscr{K} = \mathrm{CH}[\mathrm{E}(\mathscr{K})], \tag{9}$$

so the induction is complete and the theorem holds for convex bodies.

Now let $\mathscr{S}$ in $\mathscr{E}^n$ be a convex, compact set that is not a convex body. By Corollary 8.2, Sec. 4.1, $\mathscr{S}$ is a point, and $\mathscr{S} = \mathrm{CH}[\mathrm{E}(\mathscr{S})]$ is clearly true, or else $\mathscr{S}$ is a set of dimension $k < n$. By the proof just given, $\mathscr{K} = \mathrm{CH}[\mathrm{E}(\mathscr{K})]$ for any convex body in $\mathscr{E}^k$, so, by Lemma 1, it follows that $\mathscr{S} = \mathrm{CH}[\mathrm{E}(\mathscr{S})]$. □

COROLLARY 10. Every supporting hyperplane to a convex body $\mathscr{K}$ contains an extreme point of $\mathscr{K}$. (E)

As one might suppose, if $\mathscr{K}$ is a convex body in $\mathscr{E}^n$, then $\mathrm{E}(\mathscr{K})$ as well as $\mathscr{K}$ is n-dimensional and so contains an affine basis for $\mathscr{E}^n$. All points of $\mathscr{E}^n$ have unique affine coordinates with respect to this basis, so the same is true for the points in $\mathscr{K}$. However, points of $\mathscr{K}$ do not have unique representations as convex combinations of the basis unless $\mathscr{K}$ is a simplex. This property, in fact, characterizes simplices.

Comment. Two linear, affine, or convex combinations of points in a set $\mathscr{S}$ are essentially the same if any point that occurs in both does so with the same coefficient, and any point in one that is not in the other has a zero coefficient. To say that P is a unique linear (affine, convex) combination of points in $\mathscr{S}$ is to say that all linear (affine, convex) combinations of points in $\mathscr{S}$ that represent P are essentially the same.

THEOREM 11. If $\mathscr{K}$ is a convex body in $\mathscr{E}^n$, then $\mathrm{E}(\mathscr{K})$ is an n-dimensional set. Each point of $\mathscr{K}$ is a unique convex combination of points in $\mathrm{E}(\mathscr{K})$ if and only if $\mathscr{K}$ is a simplex.

PROOF. From Theorem 10, it follows that $E(\mathscr{K})$ is not empty or a single point and hence is a set of some dimension m, $1 \leq m \leq n$. Then there is a unique m-flat that contains $E(\mathscr{K})$, say the affine subspace $\mathscr{A}$. Because this subspace $\mathscr{A}$ contains all affine combinations of its points, it contains all affine combinations of points in $E(\mathscr{K})$, hence it also contains all convex combinations of points in $E(\mathscr{K})$. By Theorem 5, the union of all such convex combinations is $CH[E(\mathscr{K})]$, so $CH[E(\mathscr{K})] \subset \mathscr{A}$. But, by Theorem 10, $CH[E(\mathscr{K})]$ is $\mathscr{K}$. Therefore $\mathscr{K} \subset \mathscr{A}$. Because $\mathscr{K}$ has dimension n, $\mathscr{K} \subset \mathscr{A}$ implies $\mathscr{A} = \mathscr{E}^n$. Thus $m = n$, so $E(\mathscr{K})$ has dimension n.

Because $E(\mathscr{K})$ is n-dimensional, it contains a set $\mathscr{R} = \{A_1, A_2, \ldots, A_{n+1}\}$ that is an affine basis for $\mathscr{E}^n$. The convex hull of $\mathscr{R}$ is the simplex $\mathscr{S} = CS(A_1, A_2, \ldots, A_{n+1})$. Because the points A_i are convexly independent as well as affinely independent, it follows from Corollary 4.2 that $E(\mathscr{S}) = \mathscr{R}$.

If $E(\mathscr{K}) = \mathscr{R}$, then $\mathscr{K}$ is the simplex $\mathscr{S}$ because $\mathscr{K} = CH[E(\mathscr{K})] = CH(\mathscr{R}) = \mathscr{S}$. The $n + 1$ points of $E(\mathscr{K})$ form a basis for E^n, so each point of space has unique affine coordinates with respect to this basis. But points of the simplex $\mathscr{S} = \mathscr{K}$ are exactly those points whose affine coordinates with respect to $E(\mathscr{K})$ are nonnegative. Thus the unique affine combination representing a point of $\mathscr{K}$ is also a unique convex combination of the points in $E(\mathscr{K})$.

Next, suppose that $E(\mathscr{K}) \neq \mathscr{R}$. Then there exists a point P in $E(\mathscr{K})$ that is not in $\mathscr{R}$. If P were in $\mathscr{S}$, it would be a nonextreme point of $\mathscr{S}$ because it is not in $\mathscr{R} = E(\mathscr{S})$. This, with $\mathscr{S} \subset \mathscr{K}$, would imply $P \notin E(\mathscr{K})$. Therefore P is not in $\mathscr{S}$. Thus, if $(\eta_1, \eta_2, \ldots, \eta_{n+1})$ are the affine coordinates of P with respect to the basis $\mathscr{R}$, then at least one of the coordinates is negative. Since the sum of the coordinates is 1, at least one is positive. We may suppose the basis relabeled if necessary so that $\eta_1, \eta_2, \ldots, \eta_k$ are negative, and $\eta_{k+1}, \eta_{k+2}, \ldots, \eta_{n+1}$ are nonnegative, $1 \leq k < n + 1$. Then, from

$$p = \sum_1^k \eta_i a_i + \sum_{k+1}^{n+1} \eta_i a_i \tag{1}$$

we have

$$p + \sum_1^k (-\eta_i) a_i = \sum_{k+1}^{n+1} \eta_i a_i. \tag{2}$$

Dividing both sides of (2) by $\varphi = \sum_{k+1}^{n+1} \eta_i > 0$, we obtain

$$\frac{1}{\varphi} p + \sum_1^k \left(-\frac{\eta_i}{\varphi}\right) a_i = \sum_{k+1}^{n+1} \left(\frac{\eta_i}{\varphi}\right) a_i. \tag{3}$$

From the definition of φ, it is clear that the right side of (3) is a convex combination of the points $A_{k+1}, A_{k+2}, \ldots, A_{n+1}$ in $\mathrm{E}(\mathscr{K})$. None of the coefficients on the left of (3) is negative. Also, since

$$\sum_1^k \eta_i + \sum_{k+1}^{n+1} \eta_i = \sum_1^k \eta_i + \varphi = 1, \tag{4}$$

it follows from

$$\varphi = 1 + \sum_1^k (-\eta_i) \tag{5}$$

that

$$1 = \frac{1}{\varphi} + \sum_1^k \left(-\frac{\eta_i}{\varphi}\right). \tag{6}$$

Thus the left side of (3) is a convex combination of P and the points A_1, $A_2, \ldots, A_k$ in $\mathrm{E}(\mathscr{K})$. Thus the point Q, represented by either side of (3), is represented by two different convex combinations of points in $\mathrm{E}(\mathscr{K})$.

From the two parts of the argument, it follows that each point of $\mathscr{K}$ is a unique convex combination of points in $\mathrm{E}(\mathscr{K})$ if and only if $\mathrm{E}(\mathscr{K})$ is a set of $n + 1$ affinely independent points, hence if and only if $\mathscr{K}$ is an n-simplex. □

We showed earlier that an n-simplex in $\mathscr{E}^n$ is a convex body (Theorem 8, Sec. 4.1). However, we have not yet established the general fact that the convex span of any finite set of points is a compact set. We obtain this as a corollary of the important property that the convex hull of any compact set is compact. In the proof of this we make use of the following theorems.

THEOREM 12. The convex hull of a bounded set is bounded. (E)

THEOREM 13. If f is an affine functional, and $\mathscr{S}$ is a compact set, then f assumes the same minimal and maximal function values on the sets $\mathscr{S}$, $\mathrm{CH}(\mathscr{S})$, and $\mathrm{Cl}[\mathrm{CH}(\mathscr{S})]$.

PROOF. Since f is continuous on $\mathscr{E}^n$, it is continuous on $\mathscr{S}$. Because $\mathscr{S}$ is compact, it follows from the min–max theorem (Theorem 11, Sec. 1.7), that there is a point X_0 in $\mathscr{S}$ at which f has a minimal function value a, and there is a point X_1 in $\mathscr{S}$ at which f has a maximal function value b. Thus $f(X) \in [a, b]$ for $X \in \mathscr{S}$.

If X is a point of CH($\mathscr{S}$), then by Theorem 5 there exists a finite set $\mathscr{R} \subset \mathscr{S}$ such that X is a convex combination of the points in $\mathscr{R}$. But since f is affine, if X is a convex combination of the points P_i, $i = 1, 2, \ldots, k$, in $\mathscr{R}$, then $f(X)$ is the same convex combination of the numbers $f(P_i)$, $i = 1, 2, \ldots, k$. Since $f(P_i) \in [a, b]$, and $[a, b]$ is convex, every convex combination of the numbers $f(P_i)$ belongs to $[a, b]$. Thus $f(X) \in [a, b]$ for $X \in$ CH($\mathscr{S}$). Since X_o and X_1 are in CH($\mathscr{S}$), then $f(X_o) = a$ and $f(X_1) = b$ are the minimal and maximal values of f on CH($\mathscr{S}$).

Finally, suppose there exists $P \in$ Cl[CH($\mathscr{S}$)] such that $f(P) \notin [a, b]$. Then there exists $\varepsilon > 0$ such that N$[f(P), \varepsilon] \cap [a, b]$ is empty. Because f is continuous at P, there exists $\delta > 0$ such that $X \in$ N(P, δ) implies $f(X) \in$ N$[f(P), \varepsilon]$. Since $f(P)$ is not in $[a, b]$, P is not in CH($\mathscr{S}$). This fact, with $P \in$ Cl[CH($\mathscr{S}$)], implies $P \in$ Lp[CH($\mathscr{S}$)]. Therefore there exists $Q \in$ N$_o(P, \delta) \cap$ CH($\mathscr{S}$). But $Q \in$ N$_o(P, \delta)$ implies $f(Q) \in$ N$[f(P), \varepsilon]$, and $Q \in$ CH($\mathscr{S}$) implies $f(Q) \in [a, b]$. Therefore N$[f(P), \delta] \cap [a, b]$ is not empty. The contradiction shows that $f(P) \notin [a, b]$ is false. Hence $f(X) \in [a, b]$, for $X \in$ Cl[CH($\mathscr{S}$)]. Because X_o and X_1 are in Cl[CH($\mathscr{S}$)], f assumes a minimal value a and a maximal value b on Cl[CH($\mathscr{S}$)]. □

As in the proof of Theorem 10, we again use a lemma to replace a relative topology argument.

LEMMA 2. If for some positive integer k, the convex hull of each k-dimensional compact set in $\mathscr{E}^k$ is compact, then the same is true for each k-dimensional, compact set in $\mathscr{E}^n$, where $n > k$.

PROOF. Let $\mathscr{R}$ denote a k-dimensional compact set in $\mathscr{E}^n$, $n > k$. Then there exists a unique k-flat $\mathscr{A}$ in $\mathscr{E}^n$ that contains $\mathscr{R}$. By the argument in the proof of Lemma 1, there exists an isometric mapping of $\mathscr{A}$ onto $\mathscr{E}^k$, by an affine function f, such that $\mathscr{S} = f(\mathscr{R})$ is a k-dimensional, compact subset of $\mathscr{E}^k$, and $f[\text{CH}(\mathscr{R})] = \text{CH}(\mathscr{S})$.

Now, let P denote a limit point to CH($\mathscr{R}$). Then $N_o(P, \delta) \cap$ CH($\mathscr{R}$) is not empty, for every $\delta > 0$. Define N$'_o(P, \delta) =$ N$_o(P, \delta) \cap \mathscr{A}$. Then N$'_o(P, \delta) \subset \mathscr{A}$. Since $\mathscr{R} \subset \mathscr{A}$, and $\mathscr{A}$ is convex, then CH($\mathscr{R}$) $\subset \mathscr{A}$. Thus N$'_o(P, \delta) \cap \mathscr{A} \neq \varnothing$, for every $\delta > 0$. Because P is a limit point to CH($\mathscr{R}$), which is a subset of $\mathscr{A}$, $P \in$ Lp($\mathscr{A}$). Since $\mathscr{A}$ is closed, $P \in \mathscr{A}$. Since f is an isometry, it maps N$'_o(P, \delta)$ onto N$_o[f(P), \delta]$ in $\mathscr{E}^k$. Since $f[\text{CH}(\mathscr{R})] = \text{CH}(\mathscr{S})$, it follows that N$_o[f(P), \delta] \cap$ CH($\mathscr{S}$) $\neq \varnothing$, for every $\delta > 0$. Therefore $f(P)$ is a limit point to CH($\mathscr{S}$). By hypothesis, CH($\mathscr{S}$) is compact, so $f(P) \in$ CH($\mathscr{S}$), and therefore $P \in$ CH($\mathscr{R}$). Thus CH($\mathscr{R}$) is closed. Since $\mathscr{R}$ is compact, it is bounded, hence by Theorem 12, CH($\mathscr{R}$) is bounded. Thus CH($\mathscr{R}$) is closed and bounded and hence is compact. □

THEOREM 14. The convex hull of a compact set is compact.

PROOF. We first establish the theorem for the special case of compact sets that have the same dimension as the space that contains them. For an inductive proof, we consider the statement:

Each compact, n-dimensional set $\mathscr{S}$ in $\mathscr{E}^n$ has a compact convex hull. (*)

It is clear that the (*) statement is valid for $n = 1$. In that case, $\mathscr{S}$ is a closed, bounded set of numbers, so there exists a minimum number a in $\mathscr{S}$ and a maximum number b in $\mathscr{S}$, and $a \neq b$ because $\mathscr{S}$ is one dimensional. Thus $\mathscr{S} \subset [a, b]$ and $\mathrm{CH}(\mathscr{S})$ is $[a, b]$, which is a compact set.

Next, assume that the (*) statement is valid for $\mathscr{E}^1, \mathscr{E}^2, \ldots, \mathscr{E}^{n-1}$. Then we show, by an indirect argument, that the (*) statement must be valid for $\mathscr{E}^n$. For this purpose, assume that there exists an n-dimensional, compact set $\mathscr{S}$ in $\mathscr{E}^n$ such that $\mathrm{CH}(\mathscr{S})$ is not compact. By Theorem 12, $\mathrm{CH}(\mathscr{S})$ is bounded, so its noncompactness implies that it is not closed. Therefore there exists a point $P \in \mathrm{Cl}[\mathrm{CH}(\mathscr{S})] - \mathrm{CH}(\mathscr{S})$, that is, P is a boundary point to $\mathrm{CH}(\mathscr{S})$ that is not in $\mathrm{CH}(\mathscr{S})$. Because $\mathrm{CH}(\mathscr{S})$ is convex, then $\mathrm{Bd}[\mathrm{CH}(\mathscr{S})] = \mathrm{Bd}[\mathrm{Cl}(\mathrm{CH}(\mathscr{S}))]$ (Theorem 13, Sec. 3.1), so P is a boundary point to the set $\mathscr{K} = \mathrm{Cl}[\mathrm{CH}(\mathscr{S})]$. The set $\mathscr{K}$ is n-dimensional because it contains $\mathscr{S}$. It is obviously closed, and, since $\mathrm{CH}(\mathscr{S})$ is bounded, then its closure is bounded (cf Exercise 17, Sec. 1.7). Also, since $\mathrm{CH}(\mathscr{S})$ is convex, its closure is convex (Theorem 12, Sec. 2.6). Thus $\mathscr{K}$ is a convex body in $\mathscr{E}^n$, and $P \in \mathrm{Bd}(\mathscr{K})$.

Now, let $\mathscr{H}$ denote a supporting hyperplane to $\mathscr{K}$ that contains P. This hyperplane is the zero set of an affine functional f which can be chosen so that the closed half-space of $\mathscr{H}$ containing $\mathscr{K}$ is the graph of $f(X) \geq 0$. In particular, then,

$$f(X) \geq 0, \qquad X \in \mathscr{K}. \tag{1}$$

Since f is continuous on $\mathscr{E}^n$, it is continuous on the compact set $\mathscr{S}$ and so achieves on $\mathscr{S}$ a minimal function value a and a maximal function value b. The equality $a = b$ would imply $a = b = 0$, because $f(P) = 0$, and hence would imply that $\mathscr{K}$ was contained in $\mathscr{H} : f(X) = 0$, contradicting the n-dimensionality of $\mathscr{K}$. Thus $a \neq b$. By Theorem 13, a and b are also the minimal and maximal function values of f on $\mathrm{CH}(\mathscr{S})$ and $\mathrm{Cl}[\mathrm{CH}(\mathscr{S})]$, hence

$$f(X) \in [a, b], \qquad X \in \mathscr{K}. \tag{2}$$

Now let the set $\mathscr{S}'$ be defined by

$$\mathscr{S}' = \mathscr{S} \cap \mathscr{H}. \tag{3}$$

We want to show that $\mathscr{S}'$ is a non-empty, compact set. For the first of these properties, consider a point X_o in $\mathscr{S}$ at which f is minimal, that is, $f(X_o) = a$.

Since $X_o \in \mathscr{K}$, it follows from (1) that $a \geq 0$. Because a is the minimal function value of f on $\mathscr{K}$, and P is in $\mathscr{K}$, we also have $0 = f(P) \geq a$. Thus $a = 0$. Therefore $f(X_o) = 0$, so X_o is in $\mathscr{H}$, as well as $\mathscr{S}$, and $\mathscr{S}' \neq \varnothing$. Next, because $\mathscr{S}$ and $\mathscr{H}$ are closed, $\mathscr{S}' = \mathscr{S} \cap \mathscr{H}$ is closed. Thus $\mathscr{S}'$ is a closed subset of $\mathscr{S}$, which is compact, so $\mathscr{S}'$ is compact.

Because $\mathscr{S}' = \mathscr{S} \cap \mathscr{H}$ is a non-empty subset of $\mathscr{H}$, it is either some singleton set $\{Q\}$ or else it is a set of dimension $k \leq n - 1$. In the first case, $\text{CH}(\mathscr{S}') = \text{CH}\{Q\} = \{Q\}$, so $\text{CH}(\mathscr{S}')$ is compact. In the second case, since $k \leq n - 1$, the k-dimensional, compact sets in $\mathscr{E}^k$ have compact, convex hulls because of the induction hypothesis. Therefore, by Lemma 2, each k-dimensional, compact set in $\mathscr{E}^n$ has a compact convex hull. Hence $\text{CH}(\mathscr{S}')$ is compact.

Since $\mathscr{S}' \subset \mathscr{S}$, then $\text{CH}(\mathscr{S}') \subset \text{CH}(\mathscr{S})$. Therefore P, which is not in $\text{CH}(\mathscr{S})$, is not in $\text{CH}(\mathscr{S}')$. Thus $\{P\}$ and $\text{CH}(\mathscr{S}')$ are disjoint sets. Also, both are compact and both are convex. Thus, by the basic separation theorem (Theorem 8, Sec. 4.2), there exists a hyperplane $\mathscr{H}'$ that separates $\{P\}$ and $\text{CH}(\mathscr{S}')$. Let $\mathscr{H}_1$ denote a hyperplane that is parallel to $\mathscr{H}'$ and that contains P. Then $\text{CH}(\mathscr{S}')$ is contained in one side of $\mathscr{H}_1$. Thus an affine functional g exists, whose zero set is $\mathscr{H}_1$, such that

$$g(P) = 0, \tag{4}$$

and

$$g(X) > 0, \qquad X \in \text{CH}(\mathscr{S}'). \tag{5}$$

Now let $\mathscr{S}''$ denote the set of those points of $\mathscr{S}$ that do not lie in the $\text{CH}(\mathscr{S}')$ side of $\mathscr{H}_1$, that is,

$$\mathscr{S}'' = \mathscr{S} \cap \{X : g(X) \leq 0\}. \tag{6}$$

Because the closed half-space $g(X) \leq 0$ is a closed set, and $\mathscr{S}$ is closed, then $\mathscr{S}''$ is closed. Since $\mathscr{S}''$ is a closed subset of $\mathscr{S}$, which is compact, then $\mathscr{S}''$ is compact. Thus g, which is continuous on S'', is bounded on $\mathscr{S}''$ and in particular is bounded below on $\mathscr{S}''$. Thus there exists a negative number, $-\eta < 0$, such that

$$g(X) > -\eta, \qquad X \in \mathscr{S}''. \tag{7}$$

By definition, the points of $\mathscr{S}''$ are the points of $\mathscr{S}$ in the nonpositive side of $\mathscr{H}_1$. Since $\mathscr{S}'$, in fact $\text{CH}(\mathscr{S}')$, lies in the positive side of $\mathscr{H}_1$, then $\mathscr{S}' \cap \mathscr{S}'' = \varnothing$. Therefore $\mathscr{S}'' \cap \mathscr{H} = \varnothing$, so

$$f(X) \neq 0, \qquad X \in \mathscr{S}''. \tag{8}$$

Thus f, which is nonnegative on all of $\mathscr{K}$, is positive on $\mathscr{S}''$. Thus the minimum function value φ which f assumes on the compact set $\mathscr{S}''$ is positive, and

$$f(X) \geq \varphi > 0, \qquad X \in \mathscr{S}''. \tag{9}$$

Now let h be the function defined on $\mathscr{E}^n$ by

$$h(X) = f(X) + \frac{\varphi}{2\eta} g(X). \tag{10}$$

Since f and g are affine functionals, then h is also an affine functional, and we want to show that h must be positive on $\mathscr{S}$. First, from (1) and (9), we have

$$f(X) \geq 0, \qquad X \in \mathscr{S} - \mathscr{S}''; \qquad f(X) \geq \varphi > 0, \qquad X \in \mathscr{S}'', \tag{11}$$

and, from (6) and (7),

$$g(X) > 0, \qquad X \in \mathscr{S} - \mathscr{S}'', \qquad f(X) > -\eta, \qquad X \in \mathscr{S}''. \tag{12}$$

Therefore

$$X \in \mathscr{S} - \mathscr{S}'' \text{ implies } h(X) = f(X) + \frac{\varphi}{2\eta} g(X) > 0, \tag{13}$$

and

$$X \in \mathscr{S}'' \text{ implies } h(X) = f(X) + \frac{\varphi}{2\eta} g(X) \geq \varphi + \frac{\varphi}{2\eta}(-\eta)$$

$$= \varphi - \tfrac{1}{2}\varphi = \tfrac{1}{2}\varphi > 0. \tag{14}$$

As (13) and (14) show, h is positive on the compact set $\mathscr{S}$, so its minimum function value on $\mathscr{S}$, say π, is positive. Thus

$$h(X) \geq \pi > 0, \quad X \in \mathscr{S}. \tag{15}$$

But, since $\mathscr{S}$ is compact, the functional h has the same minimum value on $\mathscr{S}$, $\mathrm{CH}(\mathscr{S})$, and $\mathscr{K} = \mathrm{Cl}[\mathrm{CH}(\mathscr{S})]$, by Theorem 13, so

$$h(X) \geq \pi > 0, \qquad X \in \mathscr{K}. \tag{16}$$

Because $P \in \mathscr{K}$, (16) implies that $h(P) > 0$. But $f(P) = 0$ and $g(P) = 0$, so $h(P) = f(P) + (\phi/2\eta)\, g(P) = 0$.

The contradiction that $h(P)$ is both positive and zero shows that it cannot be true that the (*) statement is valid for $\mathscr{E}^1, \mathscr{E}^2, \ldots, \mathscr{E}^{n-1}$ and not valid for $\mathscr{E}^n$. Thus the induction is complete, and we have shown that in any euclidean space $\mathscr{E}^m$, the m-dimensional, compact sets have compact convex hulls.

Now consider an arbitrary compact set $\mathscr{R}$ in a space $\mathscr{E}^n$. If $\mathscr{R}$ is n-dimensional, then $\mathrm{CH}(\mathscr{R})$ is compact by the proof just given. If $\mathscr{R}$ is not n-dimensional, then it is either a singleton set, and $\mathrm{CH}(\mathscr{R})$ is compact, or else it is a set of some dimension $m < n$. Because we have proved that the m-dimensional sets in $\mathscr{E}^m$ have compact convex hulls, it follows from Lemma 2 that the m-dimensional compact sets in $\mathscr{E}^n$ have compact convex hulls. Thus in this case also $\mathscr{R}$ is compact. □

COROLLARY 14. The convex span of a finite set of points is a compact set. (E)

In concluding this section, it is worth comment that analogs of the properties in Theorems 10 and 11 have been established in much more general spaces than $\mathscr{E}^n$, and the search for other generalizations has stimulated considerable research. Thus the theory given here, in a relatively intuitive setting, is a natural background for more advanced work.

Properties of extreme points also play an important role in the optimization problems that occur in the subject of "linear programming." The next theorem, in fact, is a fundamental theorem in that subject.

THEOREM 15. If the domain $\mathscr{D}$ of an affine functional f is a compact, convex set, then f achieves its maximum value and its minimum value on the set $\mathrm{E}(\mathscr{D})$.

PROOF. Since f is continuous on a compact domain, it achieves a maximal value $f(P)$ and a minimal value $f(Q)$. Since $\mathscr{D}$ is both compact and convex, then, by Theorem 10, $\mathscr{D} = \mathrm{CH}[\mathrm{E}(\mathscr{D})]$. Thus, by Theorem 6, P is a convex combination of a finite set of points in $\mathrm{E}(\mathscr{D})$, say the set $\{A_1, A_2, \ldots, A_k\}$. Since f preserves convex combinations, it follows that in $\mathscr{E}^1$, $f(P)$ belongs to the convex span of the set of numbers $\mathscr{S} = \{f(A_1), f(A_2), \ldots f(A_k)\}$. This convex span is a closed interval $[f(A_j), f(A_h)]$, (or a point), where $f(A_j)$ and $f(A_h)$ are the minimum and maximum numbers in $\mathscr{S}$. Thus $f(P) \leq f(A_h)$. But, since $f(P)$ is a maximal value of f, $f(P) \geq f(A_h)$. Therefore $f(P) = f(A_h)$, which shows that f assumes its maximal value on an extreme point of $\mathscr{D}$. By a similar argument, it follows that $f(Q) = f(A_j)$. □

Exercises – Section 5.3

1. Prove Theorem 2.
2. Prove Theorem 3.
3. Why are any three points of $\mathscr{E}^1$ convexly dependent?

4. Give an example of a set $\mathscr{S}$ such that there is no upper bound to the number of convexly independent points in $\mathscr{S}$.
5. Describe the set $E(\mathscr{S})$ if $\mathscr{S}$ is

   a. a right circular cylinder in $\mathscr{E}^3$
   b. a right circular cone in $\mathscr{E}^3$
   c. a cube in $\mathscr{E}^3$
   d. a parabola in $\mathscr{E}^2$

6. Prove Corollary 4.1.
7. Prove Corollary 4.2.
8. Prove that $\mathscr{R} \subset \mathscr{S}$ implies $CH(\mathscr{R}) \subset CH(\mathscr{S})$.
9. Prove that $CH[CH(\mathscr{S})] = CH(\mathscr{S})$.
10. Is $E[E(\mathscr{S})] = E(\mathscr{S})$ correct always, sometimes, or never?
11. Prove Theorem 8.
12. Prove Corollary 10.
13. Prove that if P is in set $\mathscr{S}$ and P is a farthest point of $\mathscr{S}$ from some point A, then $P \in E(\mathscr{S})$.
14. A point of a closed, convex surface $\mathscr{K}^\circ$ is a regular point or a corner point according as there is exactly one or more than one supporting hyperplane at P.

   a. Must a corner point of $\mathscr{K}^\circ$ be an extreme point of $\mathscr{K}$?
   b. Must an extreme point of $\mathscr{K}$ be a corner point of $\mathscr{K}^\circ$?
   c. Must a regular point of $\mathscr{K}^\circ$ be an extreme point of $\mathscr{K}$?

15. Prove Theorem 12.
16. Prove that if f and g are affine functions, then a linear combination $\eta_1 f + \eta_2 g$ is an affine function.
17. Prove Corollary 14.
18. Prove that if $\mathscr{S}$ is convex, then $CH[Bd(\mathscr{S})] = \mathscr{S}$.

6 LINEAR COMBINATIONS OF SETS AND THE HAUSDORFF METRIC

INTRODUCTION

Among the operations that one can perform on sets, that of forming linear combinations of them is one of the most interesting and useful and has ramifications in advanced convex body theory. In this chapter, we simply want to introduce the concept and to establish some of the basic properties that are invariant under linear combinations. We illustrate the usefulness of the notion in a brief section on some separation properties. We also apply the special case of set addition to define a distance between sets, the Hausdorff distance, that is a genuine metric on the space of compact sets. In the Blaschke selection theorem we establish conditions for the existence of sets that are Hausdorff limit "points" to collections of sets. The final section illustrates the usefulness of these limit points in settling existence questions.

SECTION 6.1. LINEAR COMBINATIONS OF SETS

The multiplication of a vector by a scalar and the addition of such products suggests the following operations with sets.

DEFINITION (Linear combinations of sets). If α, β are real numbers and if $\mathscr{R}$ and $\mathscr{S}$ are sets in $\mathscr{E}^n$, then the *linear combination of* $\mathscr{R}$ *and* $\mathscr{S}$ indicated by

$\alpha\mathscr{R} + \beta\mathscr{S}$ is defined by

$$\alpha\mathscr{R} + \beta\mathscr{S} = \{\alpha x + \beta y: X \in \mathscr{R}, Y \in \mathscr{S}\}.$$

In particular, the (scalar) α-multiple of $\mathscr{R}$ is the set

$$\alpha\mathscr{R} = \{\alpha x: X \in \mathscr{R}\}.$$

If $\alpha = \beta = 1$, then $\alpha\mathscr{R} + \beta\mathscr{S} = \mathscr{R} + \mathscr{S}$ is the (vector) sum of the sets $\mathscr{R}$ and $\mathscr{S}$.

The following set properties follow in a purely formal way from familiar vector properties.

THEOREM 1. If $\mathscr{R}, \mathscr{S}, \mathscr{T}$ are sets in $\mathscr{E}^n$, and if α, β, γ are real numbers, then

(i) $\alpha\mathscr{R} + \beta\mathscr{S} = \beta\mathscr{S} + \alpha\mathscr{R}$
(ii) $(\alpha\mathscr{R} + \beta\mathscr{S}) + \gamma\mathscr{T} = \alpha\mathscr{R} + (\beta\mathscr{S} + \gamma\mathscr{T})$
(iii) $\alpha(\mathscr{R} + \mathscr{S}) = \alpha\mathscr{R} + \alpha\mathscr{S}$
(iv) $(\alpha\beta)\mathscr{R} = \alpha(\beta\mathscr{R})$ (E)

The scalar multiple of a set has a simple interpretation.

THEOREM 2. If $\alpha \neq 0$, then $\alpha\mathscr{R}$ is the image of $\mathscr{R}$ in a dilatation at the origin with ratio α. If $\alpha = 0$, and $\mathscr{R} \neq \varnothing$, then $\alpha\mathscr{R}$ is a single point which is the origin.

PROOF. If $\alpha \neq 0$, the dilatation at the origin with ratio α is the mapping of $\mathscr{E}^n$ onto $\mathscr{E}^n$ defined by $f(x) = \alpha x$. The image of $\mathscr{R}$ in this mapping is the set $f(\mathscr{R}) = \{\alpha x: X \in \mathscr{R}\}$, which is the definition of $\alpha\mathscr{R}$. If $\alpha = 0$, then $\alpha x = o$, for all X, so $\alpha x = o$ for $X \in \mathscr{R}$, hence $\alpha\mathscr{R} = \{O\}$, if $\mathscr{R} \neq \varnothing$. □

COROLLARY 2. If $\alpha = -1$, then $\alpha\mathscr{R}$ is the image of $\mathscr{R}$ in a reflection of space in the origin.

CONVENTION. $-\mathscr{R} = (-1)\mathscr{R}$.

Since $\alpha\mathscr{R} + \beta\mathscr{S}$ is just the sum of the sets $\alpha\mathscr{R}$ and $\beta\mathscr{S}$, and since we know the nature of $\alpha\mathscr{R}$ and $\beta\mathscr{S}$ from Theorem 2, the natural question now is how a sum of sets $\mathscr{R}$ and $\mathscr{S}$ is related to $\mathscr{R}$ and $\mathscr{S}$. To answer this question, we first consider the special case of adding $\mathscr{R}$ and a singleton set $\{A\}$. By definition, $Y \in \{A\}$ implies $Y = A$, so

$$\mathscr{R} + \{A\} = \{x + y: X \in \mathscr{R}, Y \in \{A\}\} = \{x + a: X \in \mathscr{R}\},$$

which is just the definition for the a-translate of $\mathscr{R}$. Thus the sum of a set and a point is the translate of the set by the vector of the point. Because of this fact, the following agreements are convenient.

CONVENTIONS. $\mathscr{R} + \{A\}$ is also denoted by $\mathscr{R} + A$ or $\mathscr{R} + a$. Also $\{A\} + \{B\}$ is denoted by $A + B$ or $a + b$.

Comment. Since the set obtained from $\mathscr{R}$ by deleting a point A, namely, $\mathscr{R} \cap \text{Cp}[\{A\}]$, has elsewhere been denoted by $\mathscr{R} - A$, we always denote the $(-a)$-translate of $\mathscr{R}$ by $\mathscr{R} - a$ or by $\mathscr{R} - \{A\}$.

We can now use the sum of a set and a point to obtain the sum of two sets in the following way. We get all the sums $X + Y$, $X \in \mathscr{R}$ and $Y \in \mathscr{S}$, if we first add all the Xs in $\mathscr{R}$ to one point in $\mathscr{S}$, then add all of them to a second point in $\mathscr{S}$, and so on. That is, $\mathscr{R} + \mathscr{S} = \bigcup \{\mathscr{R} + Y : Y \in \mathscr{S}\}$. Thus $\mathscr{R} + \mathscr{S}$ is the union of all the y-translates of $\mathscr{R}$, for $Y \in \mathscr{S}$. Put more symmetrically, the definition of $\mathscr{R} + \mathscr{S}$ implies the following facts.

THEOREM 3. The sum of two sets $\mathscr{R}$ and $\mathscr{S}$ is the union of all the translates of one by the vectors of the other, that is,

$$\mathscr{R} + \mathscr{S} = \bigcup \{\mathscr{R} + Y : Y \in \mathscr{S}\} = \bigcup \{X + \mathscr{S} : X \in \mathscr{R}\}.$$

In calculating sums of sets, the following property is extremely useful.

THEOREM 4. If $\mathscr{R}'$ is the u-translate of set $\mathscr{R}$, and $\mathscr{S}'$ is the v-translate of set $\mathscr{S}$, then $\mathscr{R}' + \mathscr{S}'$ is the $(u + v)$-translate of $\mathscr{R} + \mathscr{S}$, that is

$$(\mathscr{R} + \mathscr{S}) + (u + v) = (\mathscr{R} + u) + (\mathscr{S} + v).$$

PROOF. Consider a point Z' in $(\mathscr{R} + \mathscr{S}) + (u + v)$. Then, by definition, Z' is the $u + v$-translate of some point Z in $\mathscr{R} + \mathscr{S}$, so $z' = z + (u + v)$. Since $Z \in \mathscr{R} + \mathscr{S}$, then $z = x + y$ for some $X \in \mathscr{R}$ and some $Y \in \mathscr{S}$. Hence $z' = (x + y) + (u + v) = (x + u) + (y + v)$, which shows that $Z' \in (\mathscr{R} + u) + (\mathscr{S} + v)$. Conversely, suppose that $Z' \in (\mathscr{R} + u) + (\mathscr{S} + v)$. Then, by definition, $z' = x' + y'$ for some $X' \in (\mathscr{R} + u)$ and some $Y' \in (\mathscr{S} + v)$. Therefore $x' = x + u$ for some $X \in \mathscr{R}$ and $y' = y + v$ for some $Y \in \mathscr{S}$, so $z' = x' + y' = (x + u) + (y + v) = (x + y) + (u + v)$. Since $X + Y \in \mathscr{R} + \mathscr{S}$, it follows that $Z' \in (\mathscr{R} + \mathscr{S}) + (u + v)$. From the two parts of the argument, $(\mathscr{R} + \mathscr{S}) + (u + v) = (\mathscr{R} + u) + (\mathscr{S} + v)$. □

To illustrate the usefulness of Theorems 3 and 4 in combination, consider the problem of adding $\mathscr{R} = \text{Sg}[AB]$ and $\mathscr{S} = \text{Sg}[CD]$, where the segments are in general position. The $(-a)$-translate of $\mathscr{R}$ is a segment $\text{SG}[OB'] = \mathscr{R}'$

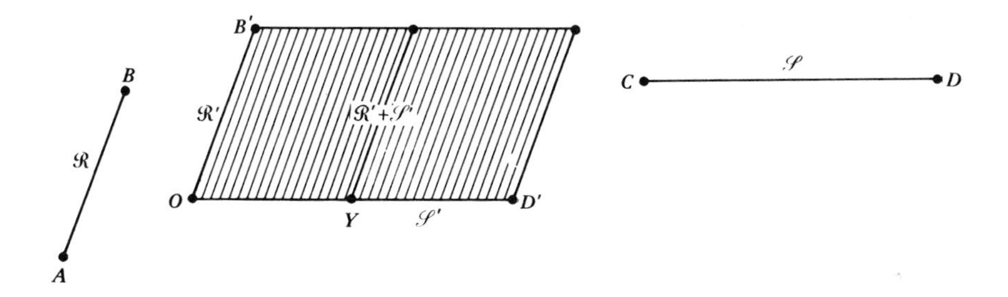

and the $(-c)$-translate of $\mathscr{S}$ is the segment $\text{Sg}[OD'] = \mathscr{S}'$. For $Y \in \text{Sg}[OD']$, $\mathscr{R}' + Y$ is just the y-translate of $\text{Sg}[OB']$. Thus $\mathscr{R}' + \mathscr{S}'$ is the full parallelogram "swept out" by $\text{Sg}[OB']$ as it moves parallel to itself with one end on $\text{Sg}[OD']$. The set $\mathscr{R} + \mathscr{S}$, by Theorem 4, is just the $(a + c)$-translate of this parallelogram.

We turn now to properties that are invariant under linear combinations and begin with one that is particularly important from the point of view of geometry.

THEOREM 5. A linear combination of two convex sets is a convex set.

PROOF. Let $\mathscr{R}$ and $\mathscr{S}$ denote convex sets, and let X, Y be two points in the linear combination set $\alpha\mathscr{R} + \beta\mathscr{S}$. A point Z between X and Y has a representation

$$z = \eta x + (1 - \eta)y, \qquad 0 < \eta < 1. \tag{1}$$

From $X, Y \in \alpha\mathscr{R} + \beta\mathscr{S}$, it follows that there exist points P, Q in $\mathscr{R}$ and points U, V in $\mathscr{S}$ such that

$$x = \alpha p + \beta u, \qquad \text{and} \qquad y = \alpha q + \beta v. \tag{2}$$

From (1) and (2),

$$\begin{aligned} z &= \eta(\alpha p + \beta u) + (1 - \eta)(\alpha q + \beta v) \\ &= \alpha[\eta p + (1 - \eta)q] + \beta[\eta u + (1 - \eta)v]. \end{aligned} \tag{3}$$

Since $\mathscr{R}$ is convex, and $P, Q \in \mathscr{R}$, then $\eta p + (1 - \eta)q \in \mathscr{R}$. Similarly, because $\mathscr{S}$ is convex and $U, V \in \mathscr{S}$, then $\eta u + (1 - \eta)v \in \mathscr{S}$. Thus (3) implies that $Z \in \alpha\mathscr{R} + \beta\mathscr{S}$, and hence that $\alpha\mathscr{R} + \beta\mathscr{S}$ is convex. □

THEOREM 6. If $\mathscr{R}$ is convex, then $\alpha\mathscr{R}$ is convex. (E)

We now turn to certain topological invariants.

THEOREM 7. If $\mathscr{S}, \mathscr{T}$ are non-empty subsets of $\mathscr{E}^n$, and if α, β are scalars, then

(i) if $\alpha \neq 0$, $\beta \neq 0$ and either $\mathscr{S}$ or $\mathscr{T}$ is open, then $\alpha\mathscr{S} + \beta\mathscr{T}$ is open.
(ii) if $\mathscr{S}$ is closed, then $\alpha\mathscr{S}$ is closed.
(iii) if $\mathscr{S}$ and $\mathscr{T}$ are bounded, then $\alpha\mathscr{S} + \beta\mathscr{T}$ is bounded.
(iv) if $\mathscr{S}$ and $\mathscr{T}$ are compact, then $\alpha\mathscr{S} + \beta\mathscr{T}$ is compact.

PROOF. Part (i). Consider $\alpha\mathscr{S} + \beta\mathscr{T}$, where $\alpha \neq 0$, $\beta \neq 0$, and one of the sets $\mathscr{S}, \mathscr{T}$, say set $\mathscr{S}$, is open. Because $\alpha \neq 0$, the dilatations $f(x) = \alpha x$ and $f^{-1}(x) = (1/\alpha)x$ exist and are one-to-one continuous mappings of $\mathscr{E}^n$ onto $\mathscr{E}^n$. By Theorem 2, $\alpha\mathscr{S} = f(\mathscr{S})$, hence $\mathscr{S} = f^{-1}(\alpha\mathscr{S})$. Since f^{-1} is continuous on the open domain $\mathscr{E}^n$ and since $\mathscr{S}$ in the range of f^{-1} is open, then, by Corollary 4, Sec. 1.5, the pre-image of $\mathscr{S}$, namely $\alpha\mathscr{S}$, is open.

Next, corresponding to a point P in $\mathscr{T}$, consider the translations $g(x) = x + \beta p$ and $g^{-1}(x) = x - \beta p$. These are also one-to-one continuous mappings of $\mathscr{E}^n$ onto $\mathscr{E}^n$, and since $\alpha\mathscr{S} + \beta p = g(\alpha\mathscr{S})$, then $\alpha\mathscr{S} = g^{-1}(\alpha\mathscr{S} + \beta p)$. As before, since g^{-1} is continuous on the open domain $\mathscr{E}^n$ and $\alpha\mathscr{S}$ is an open set in the range, then the pre-image set $\alpha\mathscr{S} + \beta p$ is open. Thus for each $P \in \mathscr{T}$, the set $\alpha\mathscr{S} + \beta p$ is an open set. But, by Theorem 3, $\alpha\mathscr{S} + \beta\mathscr{T}$ is the union of all the sets $\alpha\mathscr{S} + \beta p$, for $P \in \mathscr{T}$. Thus $\alpha\mathscr{S} + \beta\mathscr{T}$ is the union of open sets and hence is itself an open set.

Part (ii). (E)

Part (iii). (E)

Part (iv). We now consider $\alpha\mathscr{S} + \beta\mathscr{T}$, where $\mathscr{S}$ and $\mathscr{T}$ are non-empty, compact sets. Since $\mathscr{S}$ and $\mathscr{T}$ are bounded, it follows from (iii) that $\alpha\mathscr{S} + \beta\mathscr{T}$ is bounded. Also, from (ii), since $\mathscr{S}$ and $\mathscr{T}$ are closed, the sets $\mathscr{S}' = \alpha\mathscr{S}$ and $\mathscr{T}' = \beta\mathscr{T}$ are closed. Thus the compactness of $\alpha\mathscr{S} + \beta\mathscr{T}$ follows, if we can show that $\mathscr{S}' + \mathscr{T}'$ is closed. To do so, we show that each point of space has a foot in $\mathscr{S}' + \mathscr{T}'$.

Let P be an arbitrary point of space and consider the mapping $f(x) = -x + p$. This is the reflection in $\frac{1}{2}p$, hence it is a one-to-one continuous mapping of $\mathscr{E}^n$ onto $\mathscr{E}^n$. Because $\mathscr{S}'$ is non-empty and compact, its f-image $f(\mathscr{S}') = -\mathscr{S}' + p$ is also non-empty and compact. Thus both $\mathscr{T}'$ and $-\mathscr{S}' + p$ are non-empty, compact sets, so, by Corollary 7, Sec. 1.8, the nearness of the sets is achieved at some point $Y_o \in \mathscr{T}'$ and some point $Z_o \in -\mathscr{S}' + p$. That is,

$$\mathrm{n}(\mathscr{T}', -\mathscr{S}' + p) = \mathrm{glb}\{\mathrm{d}(Y, Z)\colon Y \in \mathscr{T}', Z \in -\mathscr{S}' + p\} = \mathrm{d}(Y_o, Z_o). \quad (1)$$

Since f is a one-to-one function, for each $Z \in -\mathscr{S}' + p$ there exists a unique $X \in \mathscr{S}'$ such that $z = -x + p$, and conversely. Let $X_o \in \mathscr{S}'$ denote the point such that $z_o = -x_o + p$. Then (1) implies that for all $Y \in \mathscr{T}'$ and all $X \in \mathscr{S}'$,

$$|y_o - (-x_o + p)| \leq |y - (-x + p)|, \tag{2}$$

hence that

$$|(y_o + x_o) - p| \leq |(y + x) - p|. \tag{3}$$

From (3) it follows that P has $X_o + Y_o$ as a foot in $\mathscr{S}' + \mathscr{T}'$. Thus, by Theorem 4, Sec. 1.8, $\mathscr{S}' + \mathscr{T}' = \alpha\mathscr{S} + \beta\mathscr{T}$ is a closed set. Since it is also bounded, it is compact. □

From Theorems 2, 5, 6, and 7 one can obtain the following fact.

THEOREM 8. If $\mathscr{K}_1$ and $\mathscr{K}_2$ are convex bodies and if α, β are not both zero, then $\alpha\mathscr{K}_1 + \beta\mathscr{K}_2$ is a convex body. (E)

The support and width functions of a sum figure are obtained in the following simple way.

THEOREM 9. If $\mathscr{S}_1$, $\mathscr{S}_2$, and $\mathscr{S}_3 = \mathscr{S}_1 + \mathscr{S}_2$ are non-empty, compact sets with support functions H_1, H_2, H_3 and width functions W_1, W_2, W_3, then $H_3 = H_1 + H_2$, and $W_3 = W_1 + W_2$.

PROOF. Corresponding to the unit vector u,

$$\begin{aligned} H_3(u) &= \max\{u \cdot z : Z \in \mathscr{S}_3\} = \max\{u \cdot (x_1 + x_2) : X_1 \in \mathscr{S}_1, X_2 \in \mathscr{S}_2\} \\ &= \max\{u \cdot x_1 + u \cdot x_2 : X_1 \in \mathscr{S}_1, X_2 \in \mathscr{S}_2\} \\ &= \max\{u \cdot x_1 : X_1 \in \mathscr{S}_1\} + \max\{u \cdot x_2 : X_2 \in \mathscr{S}_2\} \\ &= H_1(u) + H_2(u). \end{aligned}$$

Then,

$$\begin{aligned} W_3(u) &= H_3(u) + H_3(-u) = H_1(u) + H_2(u) + H_1(-u) + H_2(-u) \\ &= H_1(u) + H_1(-u) + H_2(u) + H_2(-u) = W_1(u) + W_2(u). \quad \square \end{aligned}$$

COROLLARY 9. If $\mathscr{K}_1$ and $\mathscr{K}_2$ are convex bodies of constant widths a_1 and a_2, then $\mathscr{K}_1 + \mathscr{K}_2$ has constant width $a_1 + a_2$.

Next we want to consider the question of how the interior and boundary of $\alpha\mathscr{R} + \beta\mathscr{S}$ are related to the interiors and boundaries of $\mathscr{R}$ and $\mathscr{S}$. If $\alpha \neq 0$, it is clear that the dilatation $f(x) = \alpha x$ maps $\text{In}(\mathscr{R})$ onto $\text{In}(\alpha\mathscr{R})$ and maps

$\text{Bd}(\mathscr{R})$ onto $\text{Bd}(\alpha\mathscr{R})$. Thus the problem really reduces to the question of how the interior and boundary of a sum $\mathscr{R} + \mathscr{S}$ relate to the interiors and boundaries of $\mathscr{R}$ and $\mathscr{S}$. One such relation is immediate. If $P \in \text{In}(\mathscr{R})$, then there exists $\text{N}(P, \delta) \subset \mathscr{R}$. For any $Y \in \mathscr{S}$, $\text{N}(P, \delta) + Y$ is just the y-translate of $\text{N}(P, \delta)$ and hence is the neighborhood $\text{N}(P + Y, \delta)$. Since $\text{N}(P + Y, \delta) \subset \mathscr{R} + \mathscr{S}$, then $P + Y$ is interior to $\mathscr{R} + \mathscr{S}$. Thus we have the following facts.

THEOREM 10. The sum of an interior point of one of the sets $\mathscr{R}$, $\mathscr{S}$ and any point of the other is an interior point of $\mathscr{R} + \mathscr{S}$.

COROLLARY 10. If $X \in \mathscr{R}$ and $Y \in \mathscr{S}$, and if $X + Y \in \text{Bd}(\mathscr{R} + \mathscr{S})$, then $X \in \text{Bd}(\mathscr{R})$ and $Y \in \text{Bd}(\mathscr{S})$.

The converse of Corollary 10 is not correct. That is, $X \in \text{Bd}(\mathscr{R})$ and $Y \in \text{Bd}(\mathscr{S})$ do not imply that $X + Y \in \text{Bd}(\mathscr{R} + \mathscr{S})$. In the example following Theorem 4, we showed that the sum of two general segments $\mathscr{R}$ and $\mathscr{S}$ is a full parallelogram. Hence, in $\mathscr{E}^2$, $\mathscr{R} + \mathscr{S}$ has a non-empty interior. But any such interior point has to be the sum of a boundary point of $\mathscr{R}$ and a boundary point of $\mathscr{S}$ because in $\mathscr{E}^2$ we have $\mathscr{R} = \text{Bd}(\mathscr{R})$ and $\mathscr{S} = \text{Bd}(\mathscr{S})$ for segments. We can, however, determine which boundary points of $\mathscr{R}$ and $\mathscr{S}$ add to produce boundary points of $\mathscr{R} + \mathscr{S}$ by making use of the following fact.

THEOREM 11. If $\mathscr{R}_1$ and $\mathscr{R}_2$ are closed half-spaces with faces $\mathscr{H}_1$ and $\mathscr{H}_2$, respectively, and with a common outer normal direction u, then $\mathscr{R}_1 + \mathscr{R}_2$ is a closed half-space with outer normal u and its face is $\mathscr{H}_1 + \mathscr{H}_2$.

PROOF. Consider the closed half-space $\mathscr{R}^*$: $u \cdot x \leq 0$ with face $\mathscr{H}^*$: $u \cdot x = 0$. For any point $P_1 \in \mathscr{H}_1$, the closed half-space $\mathscr{R}_1$ is the graph of $u \cdot \mathbf{P}_1\mathbf{X} \leq 0$. Thus if $\mathscr{R}_1'$ denotes the $(-p_1)$-translate of $\mathscr{R}_1$, then $\mathscr{R}_1' = \mathscr{R}^*$, and $\mathscr{H}_1' = \mathscr{H}^*$. Similarly, if P_2 is any point of $\mathscr{H}_2$, then a translation by $-p_2$ maps $\mathscr{R}_2$ onto $\mathscr{R}_2' = \mathscr{R}^*$ and maps $\mathscr{H}_2$ onto $\mathscr{H}_2' = \mathscr{H}^*$. Thus $\mathscr{R}_1' + \mathscr{R}_2' = \mathscr{R}^* + \mathscr{R}^*$. If $X \in \mathscr{R}^*$ and $Y \in \mathscr{R}^*$, then from $u \cdot x \leq 0$ and $u \cdot y \leq 0$ it follows that $u \cdot (x + y) \leq 0$, $X + Y \in \mathscr{R}^*$. Thus $\mathscr{R}^* + \mathscr{R}^* \subset \mathscr{R}^*$. Obviously, $\mathscr{R}^* = \mathscr{R}^* + \{0\} \subset \mathscr{R}^* + \mathscr{R}^*$, so $\mathscr{R}^* + \mathscr{R}^* = \mathscr{R}^*$. Similarly, $\mathscr{H}_1' + \mathscr{H}_2' = \mathscr{H}^* + \mathscr{H}^* = \mathscr{H}^*$. Since, by Theorem 4, $\mathscr{R}_1 + \mathscr{R}_2$ is the $(p_1 + p_2)$-translate of $\mathscr{R}_1' + \mathscr{R}_2' = \mathscr{R}^*$, then $\mathscr{R}_1 + \mathscr{R}_2$ is the closed half-space, with outer normal u, whose face contains $P_1 + P_2$. Since this face is the translate of $\mathscr{H}^* = \mathscr{H}_1' + \mathscr{H}_2'$, it is also $\mathscr{H}_1 + \mathscr{H}_2$. □

Because of Theorem 11, we introduce the following notion.

DEFINITION (Corresponding boundary points). If P_1 and P_2 are boundary points, respectively, of the compact, convex sets $\mathscr{S}_1$ and $\mathscr{S}_2$, they are *corresponding boundary points*, or simply *corresponding points*, if there exist closed half-spaces with a common outer normal direction such that one half-space supports $\mathscr{S}_1$ at P_1 and the other supports $\mathscr{S}_2$ at P_2.

THEOREM 12. If $\mathscr{S}_1$, $\mathscr{S}_2$ and $\mathscr{S}_3 = \mathscr{S}_1 + \mathscr{S}_2$ are non-empty, compact, convex sets, then a point is a boundary point of $\mathscr{S}_3$ if and only if it is the sum of corresponding boundary points of $\mathscr{S}_1$ and $\mathscr{S}_2$.

PROOF. First suppose that P_1 in the boundary of $\mathscr{S}_1$ and P_2 in the boundary of $\mathscr{S}_2$ are corresponding points. Then there exists a closed half-space $\mathscr{R}_1$ with face $\mathscr{H}_1$, and there exists a closed half-space $\mathscr{R}_2$ with face $\mathscr{H}_2$ such that $\mathscr{R}_1, \mathscr{R}_2$ have a common outer normal direction u, and $\mathscr{R}_1$ supports $\mathscr{S}_1$ at P_1 and $\mathscr{R}_2$ supports $\mathscr{S}_2$ at P_2. By Theorem 11, $\mathscr{R}_3 = \mathscr{R}_1 + \mathscr{R}_2$ is a closed half-space with outer normal u and face $\mathscr{H}_3 = \mathscr{H}_1 + \mathscr{H}_2$. Since $P_1 \in \mathscr{H}_1$ and $P_2 \in \mathscr{H}_2$, then $P_1 + P_2 \in \mathscr{H}_3$. Since $\mathscr{S}_1 \subset \mathscr{R}_1$ and $\mathscr{S}_2 \subset \mathscr{R}_2$, then $\mathscr{S}_1 + \mathscr{S}_2 \subset \mathscr{R}_1 + \mathscr{R}_2 = \mathscr{R}_3$. Thus $\mathscr{R}_3$ supports $\mathscr{S}_1 + \mathscr{S}_2$ at $P_1 + P_2$, hence $P_1 + P_2 \in \mathrm{Bd}(\mathscr{S}_1 + \mathscr{S}_2)$.

Conversely, suppose that $P \in \mathrm{Bd}(\mathscr{S}_1 + \mathscr{S}_2)$. By the basic support theorem (Theorem 6, Sec. 4.2), there exists some closed half-space $\mathscr{R}$ that supports $\mathscr{S}_1 + \mathscr{S}_2$ at P. Let u denote the outer normal direction of $\mathscr{R}$. Because $\mathscr{S}_1 + \mathscr{S}_2$ is closed, P belongs to $\mathscr{S}_1 + \mathscr{S}_2$, hence $P = P_1 + P_2$ for some $P_1 \in \mathscr{S}_1$ and $P_2 \in \mathscr{S}_2$. By Corollary 10, $P_1 \in \mathrm{Bd}(\mathscr{S}_1)$ and $P_2 \in \mathrm{Bd}(\mathscr{S}_2)$. Now let $\mathscr{R}_1$: $u \cdot (x - p_1) \leq 0$ and $\mathscr{R}_2$: $u \cdot (x - p_2) \leq 0$ be the closed half-spaces, with outer normal u, such that P_1 is in the face of $\mathscr{R}_1$ and P_2 is in the face of $\mathscr{R}_2$. Assume that $\mathscr{R}_1$ does not support $\mathscr{S}_1$. Then, because P_1 is in the face of $\mathscr{R}_1$, $\mathscr{R}_1$ must fail to support $\mathscr{S}_1$ because some point Q_1 of $\mathscr{S}_1$ is not in $\mathscr{R}_1$, that is, $u \cdot (q_1 - p_1) > 0$. But, since $\mathscr{S}_1 + \mathscr{S}_2 \subset \mathscr{R}$, then $Q_1 + P_2 \in \mathscr{R}$. Because $\mathscr{R}$ is the graph of $u \cdot (x - p) \leq 0$, then $u \cdot (q_1 + p_2 - p) \leq 0$, hence $u \cdot (q_1 + p_2 - p_1 - p_2) \leq 0$, hence $u \cdot (q_1 - p_1) \leq 0$, contradicting $u \cdot (q_1 - p_1) > 0$. Thus the assumption is false, and $\mathscr{R}_1$ does support $\mathscr{S}_1$ at P_1. By an entirely similar argument, $\mathscr{R}_2$ supports $\mathscr{S}_2$ at P_2, so P_1 and P_2 are corresponding points. □

COROLLARY 12.1. If $\mathscr{K}_1$, $\mathscr{K}_2$, and $\mathscr{K}_3 = \mathscr{K}_1 + \mathscr{K}_2$ are convex bodies, then the convex surface $\mathscr{K}_3^0$ is the sum of the corresponding points in the surfaces $\mathscr{K}_1^0$ and $\mathscr{K}_2^0$.

In Sec. 4.4 we introduced the notion of an outer normal cone at a boundary point of a convex body. We can apply this notion in describing corresponding points if we first modify it in the following way.

DEFINITION (Cone of outer normal directions). If P is a boundary point of a convex body $\mathscr{K}$, then the *cone of outer normal directions* at P is the $(-p)$-translate of the outer normal cone at P.

COROLLARY 12.2. If P_1 and P_2 are boundary points, respectively, of the convex bodies $\mathscr{K}_1$ and $\mathscr{K}_2$, then P_1 and P_2 are corresponding points of $\mathscr{K}_1$ and $\mathscr{K}_2$ if and only if the cone of outer normal directions at P_1 and the cone of outer normal directions at P_2 intersect in a cone. If the intersection is a cone, it is the cone of outer normal directions at $P_1 + P_2$. (E)

COROLLARY 12.3. If P_1 and P_2 are corresponding boundary points of convex bodies $\mathscr{K}_1$ and $\mathscr{K}_2$, respectively, and if either P_1 or P_2 is a regular point, then $P_1 + P_2$ is a regular point of $\mathscr{K}_1 + \mathscr{K}_2$.

PROOF. If P_1, for example, is a regular point of $\mathscr{K}_1$, then there is a unique closed half-space supporting $\mathscr{K}_1$ at P_1. If u is the outer normal direction of this half-space, then the cone of outer normal directions at P_1 is the ray Ry$[O, U)$. Because P_1 and P_2 are corresponding, u is also an outer normal direction at P_2, so the intersection of the two outer normal direction cones is not just the origin, hence is a cone. Since the intersection cone is contained in Ry$[O, U)$, it must be Ry$[O, U)$. Then the outer normal cone to $\mathscr{K}_1 + \mathscr{K}_2$ at $P_1 + P_2$ is just the $(p_1 + p_2)$-translate of Ry$[O, U)$, hence $P_1 + P_2$ is a regular point. □

Although the boundary of a sum figure $\mathscr{S}_1 + \mathscr{S}_2$ is formed from the sum of corresponding boundary points in $\mathscr{S}_1$ and $\mathscr{S}_2$, different pairs of corresponding points may produce the same point in $\mathscr{S}_1 + \mathscr{S}_2$. However, this cannot occur for an extreme point of the sum. In fact, the next theorem requires no restriction on $\mathscr{S}_1$ or $\mathscr{S}_2$.

THEOREM 13. If P is an extreme point of $\mathscr{S}_1 + \mathscr{S}_2$, then there is just one point P_1 in $\mathscr{S}_1$ and one point P_2 in $\mathscr{S}_2$ such that $P = P_1 + P_2$, and P_i is an extreme point of $\mathscr{S}_i$, $i = 1, 2$.

PROOF. By the definition of an extreme point, P belongs to $\mathscr{S}_1 + \mathscr{S}_2$ but is not between any two points of $\mathscr{S}_1 + \mathscr{S}_2$. Thus P has a representation $P = P_1 + P_2$, where $P_1 \in \mathscr{S}_1$ and $P_2 \in \mathscr{S}_2$. Suppose that P also has the representation $P = Q_1 + Q_2$, $Q_i \in \mathscr{S}_i$, $i = 1, 2$. Then $P_1 + Q_2 \in \mathscr{S}_1 + \mathscr{S}_2$ and $Q_1 + P_2 \in \mathscr{S}_1 + \mathscr{S}_2$, so

$$p = \tfrac{1}{2}(p_1 + p_2) + \tfrac{1}{2}(q_1 + q_2) = \tfrac{1}{2}(p_1 + q_2) + \tfrac{1}{2}(q_1 + p_2)$$

contradicts the extremeness of P unless $P_1 + Q_2$ and $Q_1 + P_2$ are the same point. But in that case, $p_1 + q_2 = q_1 + p_2$ together with $p_1 + p_2 = q_1 + q_2$ implies $p_1 = q_1$ and $p_2 = q_2$. Thus the representation $P = P_1 + P_2$ is unique.

Next, suppose that P_1 is not an extreme point of $\mathscr{S}_1$. Then P_1 is between two distinct points A_1, B_1 of $\mathscr{S}_1$. A translation by p_2 maps A_1, P_1 and B_1 onto $A_1 + P_2$, $P_1 + P_2 = P$, and $B_1 + P_2$, respectively, in $\mathscr{S}_1 + \mathscr{S}_2$.

Since a translation preserves distinctness and betweeness, P is between $A_1 + P_2$ and $B_1 + P_2$, contradicting the extremeness of P. Thus P_1 must be an extreme point of $\mathscr{S}_1$ and, by the same logic, P_2 is an extreme point of $\mathscr{S}_2$. □

We can now describe in some detail how the boundary of a sum figure is formed. In doing so, we also want to discuss a special kind of boundary point.

DEFINITION (Exposed point). Point P in a set $\mathscr{S}$ is an *exposed point* of $\mathscr{S}$ if there exists some hyperplane $\mathscr{H}$ that supports $\mathscr{S}$ at P and intersects $\mathscr{S}$ only at P. The set of all exposed points of a set $\mathscr{S}$ is denoted by $\mathrm{Ep}(\mathscr{S})$.

Now consider two non-empty, compact, convex sets $\mathscr{S}_1$, $\mathscr{S}_2$ and their sum $\mathscr{S}_3 = \mathscr{S}_1 + \mathscr{S}_2$. Corresponding to a nonnull direction u, there exist unique closed half-spaces $\mathscr{R}_1, \mathscr{R}_2$, with outer normal u, such that $\mathscr{R}_1$ supports $\mathscr{S}_1$ and $\mathscr{R}_2$ supports $\mathscr{S}_2$. If $\mathscr{H}_i$ is the face of $\mathscr{R}_i, i = 1, 2$, then $\mathscr{H}_3 = \mathscr{H}_1 + \mathscr{H}_2$ is the face of the unique closed half-space $\mathscr{R}_3$, with outer normal u, that supports $\mathscr{S}_3$, and $\mathscr{R}_3 = \mathscr{R}_1 + \mathscr{R}_2$. The contact sets $\mathscr{C}_1 = \mathscr{S}_1 \cap \mathscr{H}_1$ and $\mathscr{C}_2 = \mathscr{S}_2 \cap \mathscr{H}_2$ are sets of corresponding points of $\mathscr{S}_1$ and $\mathscr{S}_2$, and $\mathscr{C}_1 + \mathscr{C}_2$ is the contact set $\mathscr{S}_3 \cap \mathscr{H}_3$. Because $\mathscr{H}_1$ supports $\mathscr{S}_1$, a point of $\mathscr{C}_1$ that is not in $\mathrm{E}(\mathscr{S}_1)$, the extreme point set of $\mathscr{S}_1$, must be between two points of $\mathscr{S}_1 \cap \mathscr{H}_1 = \mathscr{C}_1$, which implies that an extreme point of $\mathscr{C}_1$ is also an extreme point of $\mathscr{S}_1$. Thus $\mathrm{E}(\mathscr{C}_1) = \mathrm{E}(\mathscr{S}_1) \cap \mathscr{H}_1$. Similarly, $\mathrm{E}(\mathscr{C}_2) = \mathrm{E}(\mathscr{S}_2) \cap \mathscr{H}_2$. By Corollary 10.1, Sec. 5.2, neither $\mathrm{E}(\mathscr{C}_1)$ nor $\mathrm{E}(\mathscr{C}_2)$ is empty. From Theorem 13, it follows that $\mathrm{E}(\mathscr{C}_1) + \mathrm{E}(\mathscr{C}_2) = \mathrm{E}(\mathscr{C}_1 + \mathscr{C}_2)$, and each distinct sum $X_1 + X_2$, $X_i \in \mathrm{E}(\mathscr{C}_i)$, $i = 1, 2$ produces a distinct extreme point of $\mathscr{S}_3$.

If the contact set $\mathscr{C}_1$ is a single point P_1, then P_1 is an exposed point of $\mathscr{S}_1$. It is obviously an extreme point of $\mathscr{C}_1$, so, as we saw above, it is an extreme point of $\mathscr{S}_1$. If $\mathscr{C}_2$ is also a singleton set, say $\{P_2\}$, then P_1 and P_2 are corresponding exposed points. Since $\mathscr{H}_3 \cap \mathscr{S}_3 = \mathscr{C}_1 + \mathscr{C}_2 = \{P_1 + P_2\}$, $P_1 + P_2$ is an exposed point of $\mathscr{S}_3$. The converse also holds. That is, if $\mathscr{H}_3 \cap \mathscr{S}_3$ is a singleton $\{P\}$, namely, an exposed point P, then P is an extreme point of $\mathscr{H}_3$ and so has a unique representation $P = P_1 + P_2$, $P_i \in \mathscr{S}_i$, $i = 1, 2$, where P_1, P_2 are corresponding extreme points. If P_1 and P_2 were not the only contact points in $\mathscr{H}_1$ and $\mathscr{H}_2$, respectively, then, as we saw, P would not be the only contact point in $\mathscr{H}_3$. Thus P_1 and P_2 must be corresponding exposed points.

For reference we state some of the facts established in the preceding discussion.

THEOREM 14. Every exposed point of a set is an extreme point of the set.

THEOREM 15. The sum of two corresponding exposed points of compact, convex sets $\mathscr{S}_1$, $\mathscr{S}_2$ is an exposed point of $\mathscr{S}_1 + \mathscr{S}_2$, and each exposed point of $\mathscr{S}_1 + \mathscr{S}_2$ is the unique sum of two such corresponding exposed points.

If P is a boundary point of a convex body $\mathscr{K}$, then among the hyperplanes supporting $\mathscr{K}$ at P some may intersect $\mathscr{K}$ only at P and others may intersect $\mathscr{K}$ in nonsingleton sets. We can distinguish these cases by introducing the following notion.

DEFINITION (Exposed outer normal direction). If P is a boundary point of a convex body $\mathscr{K}$, then a nonnull vector u is an *exposed outer normal direction* to $\mathscr{K}$ at P if the closed half-space with outer normal u that supports $\mathscr{K}$ is such that its face intersects $\mathscr{K}$ only at P.

Although it is clear that every nonnull direction u is an outer normal direction to a convex body $\mathscr{K}$ at at least one boundary point, there may be no boundary point for which u is an exposed outer normal direction. However, the following facts can be established.

THEOREM 16. If P is an exposed point of a convex body $\mathscr{K}$, then the union of the null vector and all vectors that are exposed outer normal directions to $\mathscr{K}$ at P is a cone with the origin as vertex. (E)

DEFINITION (Cone of exposed, outer normal directions). If P is an exposed point of a convex body $\mathscr{K}$, the cone of Theorem 16 is the *cone of exposed outer normal directions at P.*

If it exists, the cone of exposed outer normal directions to $\mathscr{K}$ at a boundary point P is clearly a subcone of the cone of outer normal directions at P. Theorems 15 and 16 also imply the following analog of Corollary 12.2.

THEOREM 17. If P_1 and P_2 are corresponding exposed points of the convex bodies $\mathscr{K}_1$ and $\mathscr{K}_2$, respectively, then the intersection of the cones of exposed outer normal directions for P_1 and P_2 is a cone and is the cone of exposed outer normal directions of $\mathscr{K}_1 + \mathscr{K}_2$ at $P_1 + P_2$. (E)

Exercises – Section 6.1

1. Prove Theorem 1.
2. Prove Theorem 6.
3. Prove parts (ii) and (iii) of Theorem 7.
4. Prove Theorem 8.
5. Prove Corollary 12.2.

6. In Exercise 8, Sec. 4.3, a construction was determined for a differentiable curve of constant width. Show that this curve can also be obtained as the boundary of the sum of a disc and a Reuleaux triangle.
7. Give an example of an extreme point that is not an exposed point.
8. Let $\mathscr{T}_1 = \text{CS}\{(0, 0), (1, 0), (0, 1)\}$, $\mathscr{T}_2 = \text{CS}\{(-1, 0), (0, 0), (0, 1)\}$, $\mathscr{T}_3 = \text{CS}\{(0, 0), (-1, 0), (0, -1)\}$, $\mathscr{T}_4 = \text{CS}\{(0, 0), (-1, 0), (1, 0)\}$. Compute $\mathscr{T}_1 + \mathscr{T}_1, \mathscr{T}_1 + \mathscr{T}_2, \mathscr{T}_1 + \mathscr{T}_3, \mathscr{T}_1 + \mathscr{T}_4$. In each case, keep track of all corresponding boundary points, making special note of extreme points.
9. If convex bodies $\mathscr{R}$ and $\mathscr{S}$ have centers P and Q, respectively, show that $\mathscr{R} + \mathscr{S}$ has center $P + Q$.
10. Explain why $\mathscr{S} + (-1)\,\mathscr{S}$ always has a center.
11. If $\mathscr{R}$ and $\mathscr{S}$ are convex bodies, show that $\text{Dm}(\mathscr{R} + \mathscr{S}) \leq \text{Dm}(\mathscr{R}) + \text{Dm}(\mathscr{S})$. Give an example in which the strict inequality holds.
12. Prove Theorem 16.
13. Prove Theorem 17.

SECTION 6.2. SOME SEPARATION PROPERTIES

This very brief section deals with some useful separation theorems and illustrates how the notion of linear combinations of sets can be used as a tool in certain proofs.

If $\mathscr{H}$ is the hyperplane $a \cdot x + h = 0$, $a \neq o$, and f is the affine functional $f(x) = a \cdot x + h$, then f maps the closed sides of $\mathscr{H}$ onto the intervals $(-\infty, 0]$ and $[0, \infty)$, respectively. If $\mathscr{S}$ is any connected set, in particular if $\mathscr{S}$ is a convex set, then $f(\mathscr{S})$ is a connected set in $\mathscr{E}^1$ and hence is a general interval. The relation of this interval $f(\mathscr{S})$ to the intervals $(-\infty, 0]$ and $[0, \infty)$ gives information about the relations of $\mathscr{S}$ and $\mathscr{H}$, and we use this idea in the next theorem.

THEOREM 1. There exists a hyperplane $\mathscr{H}$ that weakly separates two non-empty, convex sets $\mathscr{S}$ and $\mathscr{T}$ if and only if there exists a nonconstant affine functional f such that neither of the general intervals $f(\mathscr{S}), f(\mathscr{T})$ intersects the interior of the other. The hyperplane $\mathscr{H}$ is a constant set of f.

PROOF. Assume that $\mathscr{H}$: $a \cdot x + h = 0$, $a \neq o$, is a hyperplane that weakly separates $\mathscr{S}$ and $\mathscr{T}$. Let f be the affine functional $f(x) = a \cdot x + h$. The closed sides of $\mathscr{H}$ are the sets $\mathscr{H}' : f(x) \leq 0$ and $\mathscr{H}'' : f(x) \geq 0$, and, by hypothesis, one contains $\mathscr{S}$ and the other contains $\mathscr{T}$. We may suppose that $\mathscr{S} \subset \mathscr{H}'$ and $\mathscr{T} \subset \mathscr{H}''$. Then $f(\mathscr{S}) \subset f(\mathscr{H}') = (-\infty, 0]$ and $f(\mathscr{T}) \subset f(\mathscr{H}'') = [0, \infty)$. By Theorem 4, Sec. 2.5, an affine functional maps open sets onto open sets, so $\text{In}[f(\mathscr{S})] \subset \text{In}[f(\mathscr{H}')] = (-\infty, 0)$. Similarly, $\text{In}[f(\mathscr{T})] \subset (0, \infty)$. Thus neither $f(\mathscr{S})$ nor $f(\mathscr{T})$ intersects the interior of the other.

For the converse, suppose that $f(x) = a \cdot x + h$, $a \neq o$, is some affine functional such that $f(\mathscr{S}) \cap \text{In}[f(\mathscr{T})] = \varnothing$ and $f(\mathscr{T}) \cap \text{In}[f(\mathscr{S})] = \varnothing$. Then neither $f(\mathscr{S})$ nor $f(\mathscr{T})$ is $(-\infty, \infty)$. Suppose that one of the intervals $f(\mathscr{S})$, $f(\mathscr{T})$ is a singleton interval, say $f(\mathscr{S}) = \{k\}$. Then $\mathscr{S}$ is contained in the hyperplane $\mathscr{H} : f(x) = k$. If $f(\mathscr{T})$ intersected both $(-\infty, k)$ and (k, ∞), the convexity of $f(\mathscr{T})$ would imply $k \in \text{In}[f(\mathscr{T})]$, contradicting $f(\mathscr{S}) \cap \text{In}[f(\mathscr{T})] = \varnothing$. Therefore $f(\mathscr{T})$ must be contained in $(-\infty, k]$ or else in $[k, \infty)$. In either case, $\mathscr{T}$ is contained in a closed side of $\mathscr{H}$, so $\mathscr{H}$ weakly separates $\mathscr{S}$ and $\mathscr{T}$.

Next, suppose that neither $f(\mathscr{S})$ nor $f(\mathscr{T})$ is a singleton interval. For each of the intervals $f(\mathscr{S})$, $f(\mathscr{T})$ there is an endpoint set that has at least one number and at most two. Let k_1 denote that endpoint to $f(\mathscr{S})$ and k_2 that endpoint to $f(\mathscr{T})$ such that $\text{d}(k_1, k_2)$ is the nearness of the two endpoint sets. We may suppose that $k_1 \leq k_2$, and we first consider the case $k_1 < k_2$. If k_1 is a left endpoint to $f(\mathscr{S})$, then the definition of k_1 and k_2 implies that $f(\mathscr{S})$ has no right endpoint on $(k_1, k_2]$. Thus $f(\mathscr{S})$ either has a right endpoint greater than k_2 or else is unbounded above. In either case there exists $\text{N}(k_2, \delta) \subset \text{In}[f(\mathscr{S})]$. Because k_2 is a limit point to $f(\mathscr{T})$, there must be a point of $f(\mathscr{T})$ in $\text{N}(k_2, \delta)$, which contradicts $f(\mathscr{T}) \cap \text{In}[f(\mathscr{S})] = \varnothing$. Thus k_1 cannot be a left endpoint to $f(\mathscr{S})$, so it must be the right endpoint, hence $f(\mathscr{S}) \subset (-\infty, k_1]$. By exactly the same logic, k_2 cannot be a right endpoint to $f(\mathscr{T})$, so it is the left endpoint, hence $f(\mathscr{T}) \subset [k_2, \infty)$. Thus if k is any number such that $k_1 < k < k_2$, then the hyperplane $f(x) = k$ contains $\mathscr{S}$ in one of its sides and contains $\mathscr{T}$ in the other. Since it separates $\mathscr{S}$ and $\mathscr{T}$, it also separates them weakly.

Finally, suppose that $k_1 = k_2 = k$. If k were a left endpoint to both $f(\mathscr{S})$ and $f(\mathscr{T})$ there would exist $(k, x_1) \subset f(\mathscr{S})$ and also exist $(k, x_2) \subset f(\mathscr{T})$, and any point of $(k, x_1) \cap (k, x_2)$ would be interior to both $f(\mathscr{S})$ and $f(\mathscr{T})$, which is contradictory. Similarly, k cannot be a right endpoint to both $f(\mathscr{S})$ and $f(\mathscr{T})$. Thus k must be a left endpoint to one of the intervals $f(\mathscr{S})$, $f(\mathscr{T})$ and be a right endpoint to the other. Thus one closed side of the hyperplane $\mathscr{H}$: $f(x) = k$ contains $\mathscr{S}$ and the other contains $\mathscr{T}$, so $\mathscr{S}$ and $\mathscr{T}$ are weakly separated by $\mathscr{H}$. □

In Sec. 4.2 we proved that if $\mathscr{S}$ and $\mathscr{T}$ are non-empty, disjoint sets that are closed and convex, and if one of them is bounded, then there exist parallel hyperplanes $\mathscr{H}_1$ and $\mathscr{H}_2$ such that $\mathscr{H}_1$ supports $\mathscr{S}$, $\mathscr{H}_2$ supports $\mathscr{T}$, and every hyperplane parallel to $\mathscr{H}_1$, $\mathscr{H}_2$ and between them separates $\mathscr{S}$ and $\mathscr{T}$. It is natural to think of $\mathscr{S}$ and $\mathscr{T}$ as being separated by a slab, and we introduce such a name.

DEFINITION (Slab separation of sets). Two non-empty sets $\mathscr{S}$ and $\mathscr{T}$ are *separated by a slab* with faces $\mathscr{H}_1$, $\mathscr{H}_2$ if the closed, non-$\mathscr{H}_1$ side of $\mathscr{H}_2$ contains one of the sets and the closed, non-$\mathscr{H}_2$ side of $\mathscr{H}_1$ contains the other set.

The arguments for Theorem 1, with only slight modification, can be used to establish the following fact.

THEOREM 2. There exists a slab separating two non-empty, convex sets $\mathscr{S}$ and $\mathscr{T}$ if and only if there exists a nonconstant affine functional f such that some open interval lies between the intervals $f(\mathscr{S})$ and $f(\mathscr{T})$. (E).

THEOREM 3. If $\mathscr{S}$ and $\mathscr{T}$ are non-empty, convex sets, and if $\text{In}(\mathscr{T}) \neq \varnothing$, then there exists a hyperplane that weakly separates $\mathscr{S}$ and $\mathscr{T}$ if and only if $\mathscr{S} \cap \text{In}(\mathscr{T}) = \varnothing$. Moreover, all common boundary points to $\mathscr{S}$ and $\mathscr{T}$ lie in this hyperplane.

PROOF. Assume first that there is a hyperplane $\mathscr{H}$: $a \cdot x + h = 0$, $a \neq o$, that weakly separates $\mathscr{S}$ and $\mathscr{T}$, and let f be defined by $f(x) = a \cdot x = h$. Since one closed side of $\mathscr{H}$ contains $\mathscr{S}$ and the other contains $\mathscr{T}$, we may suppose that $f(x) \leq 0$ for $X \in \mathscr{S}$ and that $f(x) \geq 0$ for $X \in \mathscr{T}$, hence that $f(\mathscr{S}) \subset (-\infty, 0]$ and $f(\mathscr{T}) \subset [0, \infty)$. If $P \in \text{In}(\mathscr{T})$, there exists $\text{N}(P, \delta) \subset \text{In}(\mathscr{T})$ and, by Theorem 4, Sec. 2.5, $f[\text{N}(P, \delta)]$ is some neighborhood of $f(P)$, say $\text{N}[f(P), \eta]$. Because $\text{N}[f(P), \eta] \subset [0, \infty)$, $f(P) > 0$. Thus $f(P) \notin (-\infty, 0]$, hence $P \notin \mathscr{S}$, so $\mathscr{S} \cap \text{In}(\mathscr{T}) = \varnothing$, as we wished to show. A continuous, real valued function that has a greatest lower bound or least upper bound on a set has the same greatest lower bound and least upper bound on the closure of the set. Thus 0, which is an upper bound to $f(\mathscr{S})$, is also an upper bound to $f[\text{Cl}(\mathscr{S})]$. Hence $\text{Cl}(\mathscr{S})$ maps into $(-\infty, 0]$. By the same logic, $\text{Cl}(\mathscr{T})$ maps into $[0, \infty)$. Thus any boundary point common to $\mathscr{S}$ and $\mathscr{T}$ must map onto 0, so the point must be in $\mathscr{H}$.

For the converse part of the theorem, assume that $\mathscr{S} \cap \text{In}(\mathscr{T}) = \varnothing$. Let $\mathscr{R}$ be defined by the linear combination $\mathscr{R} = \mathscr{S} - \text{In}(\mathscr{T})$. Because $\mathscr{S}$ and $\text{In}(\mathscr{T})$ are convex, then $\mathscr{R}$ is convex (Theorem 5, Sec. 6.1), and, since $\text{In}(\mathscr{T})$ is open, then $\mathscr{R}$ is open (Theorem 7, Sec. 6.1). If the origin O belonged to $\mathscr{R}$ there would exist $X \in \mathscr{S}$ and $Y \in \text{In}(\mathscr{T})$ such that $o = x - y$, implying $X = Y$ and contradicting $\mathscr{S} \cap \text{In}(\mathscr{T}) = \varnothing$. Thus $O \notin \mathscr{R}$, so $O \notin \text{In}(\mathscr{R})$, hence $O \in \text{Bd}(\mathscr{R}) \cup \text{Ex}(\mathscr{R})$. If $O \in \text{Bd}(\mathscr{R})$, then the hyperplane supporting $\mathscr{R}$ at O weakly separates $\{O\}$ and $\mathscr{R}$. If $O \in \text{Ex}(\mathscr{R})$, then O has a foot F in $\text{Cl}(\mathscr{R})$. The hyperplane supporting $\mathscr{R}$ at F separates $\{O\}$ and $\mathscr{R}$, so the parallel hyperplane through O weakly separates $\{O\}$ and $\mathscr{R}$. In all cases, then, there exists a hyperplane $\mathscr{H}$: $a \cdot x = 0$ that weakly separates $\{O\}$ and $\mathscr{R}$, and a may be chosen so that the linear functional $f(x) = a \cdot x$ is nonnegative on $\mathscr{R}$. Then

$$f(P) \geq f(O) = 0 \text{ for all } P \in \mathscr{R}. \tag{1}$$

If $X \in \mathscr{S}$ and $Y \in \text{In}(\mathscr{T})$, then $P = X - Y$ is in $\mathscr{R}$, so, from (1), $f(P) = f(X - Y) = f(X) - f(Y) \geq 0$. Therefore

$$f(X) \geq f(Y) \text{ for } X \in \mathscr{S} \text{ and } Y \in \text{In}(\mathscr{T}). \tag{2}$$

From (2) it follows that each number in $f(\mathscr{S})$ is an upper bound to $f[\text{In}(\mathscr{T})]$. Thus there exists a least upper bound k to the open interval $f[\text{In}(\mathscr{T})]$. Because k is also a least upper bound to $f(\mathscr{T})$, it follows that $f[\text{In}(\mathscr{T})] \subset (-\infty, k)$ and $f(\mathscr{T}) \subset (-\infty, k]$. Now assume that there exists some point X_o in $\mathscr{S}$ such that $f(X_o) = k_o < k$. Then, because k is a limit point to $f[\text{In}(\mathscr{T})]$, there exists a number k_1 that belongs to $(k_o, k) \cap f[\text{In}(\mathscr{T})]$. Thus there exists $Y_1 \in \text{In}(\mathscr{T})$ such that $f(Y_1) = k_1 > k_o = f(X_o)$, which contradicts (2). Thus no such point X_o exists, and $f(X) \geq k$ for all $X \in \mathscr{S}$. This relation, with $f(Y) \leq k$ for all $Y \in \mathscr{T}$, implies that the hyperplane $\mathscr{H}' : f(x) = k$ weakly separates $\mathscr{S}$ and $\mathscr{T}$. Because f maps the closures of $\mathscr{S}$ and $\mathscr{T}$ into $[k, \infty)$ and $(-\infty, k]$, respectively, any boundary point common to $\mathscr{S}$ and $\mathscr{T}$ must map onto k, so the point must be in $\mathscr{H}'$. □

THEOREM 4. There exists a slab separating two non-empty, convex sets $\mathscr{S}$ and $\mathscr{T}$ if and only if the nearness of the sets is positive.

PROOF. First, suppose that $\mathscr{S}$ and $\mathscr{T}$ are separated by a slab, and let w denote the width of the slab. By the definition of slab separation, $\text{d}(X, Y) \geq w > 0$ for all $X \in \mathscr{S}$ and $Y \in \mathscr{T}$. Thus $\text{n}(\mathscr{S}, \mathscr{T}) = \text{glb}\{\text{d}(X, Y) : X \in \mathscr{S}, Y \in \mathscr{T}\} \geq w > 0$, so the nearness of $\mathscr{S}$ and $\mathscr{T}$ is positive.

For the converse, suppose that $\text{n}(\mathscr{S}, \mathscr{T}) = \eta > 0$. Consider the set $\mathscr{R} = \mathscr{S} - \mathscr{T}$. If X is any point of $\mathscr{R}$, it has a representation as $X = P - Q$, where $P \in \mathscr{S}$ and $Q \in \mathscr{T}$. Therefore $\text{d}(O, X) = |x| = |p - q| = \text{d}(P, Q) \geq \eta > 0$, hence $\text{d}(O, \mathscr{R}) \geq \eta > 0$. Thus, by Corollary 2, Sec. 1.8, the origin O is not in $\text{Cl}(\mathscr{R})$. Also, because $\mathscr{S}$ and $\mathscr{T}$ are convex, then $\mathscr{R}$ is convex, so $\text{Cl}(\mathscr{R})$ is convex. Thus $\{O\}$ and $\text{Cl}(\mathscr{R})$ are closed, disjoint, convex sets. Because $\{O\}$ is also a bounded set, it follows from Theorem 8, Sec. 4.2, that there exists a slab separating $\{O\}$ and $\text{Cl}(\mathscr{R})$ such that one face $\mathscr{H}$ contains $\{O\}$ and the parallel face $\mathscr{H}'$ supports $\text{Cl}(\mathscr{R})$. The faces $\mathscr{H}$ and $\mathscr{H}'$ have equations $a \cdot x = 0$ and $a \cdot x = k$, respectively, and we may suppose that $k > 0$. Then $f(x) = a \cdot x \geq k$ for $X \in \text{Cl}(\mathscr{R})$, so $f(\mathscr{R}) \subset [k, \infty)$.

For any $P \in \mathscr{S}$ and $Q \in \mathscr{T}$, the point $X = P - Q$ is in $\mathscr{R}$, so $f(X) \in [k, \infty)$ implies that $f(X) = f(P - Q) \geq k$. Because f is linear, it follows that

$$f(P) - f(Q) \geq k, \qquad P \in \mathscr{S}, \qquad Q \in \mathscr{T}. \tag{1}$$

For any particular $Q_o \in \mathscr{T}$, (1) implies that

$$f(P) \geq f(Q_o) + k, \qquad P \in \mathscr{S}, \tag{2}$$

hence $f(\mathscr{S})$ is bounded below. Thus $k_1 = \text{glb}\{f(\mathscr{S})\}$ exists. Similarly, for any particular $P_o \in \mathscr{S}$, (1) implies that

$$f(Q) \leq f(P_o) - k, \qquad Q \in \mathscr{T}, \tag{3}$$

so $f(\mathcal{T})$ is bounded above, and $k_2 = \text{lub}\{f(\mathcal{T})\}$ exists. If ε is any positive number, then, from the definitions of k_1 and k_2, it follows that there exists $P_1 \in \mathcal{S}$ and $Q_1 \in \mathcal{T}$ such that

$$f(P_1) < k_1 + \tfrac{1}{2}\varepsilon, \qquad \text{and} \qquad f(Q_1) > k_2 - \tfrac{1}{2}\varepsilon. \tag{4}$$

From (4),

$$f(P_1) - f(Q_1) < k_1 - k_2 + \varepsilon. \tag{5}$$

Also, from (1),

$$f(Q_1) - f(P_1) \leq -k. \tag{6}$$

Adding (5) and (6), we obtain

$$0 < k_1 - k_2 + \varepsilon - k, \qquad \text{or} \qquad k_1 - k_2 > k - \varepsilon. \tag{7}$$

Because k_1, k_2, and k are constants, and (7) is valid for any $\varepsilon > 0$, it follows that

$$k_1 - k_2 \geq k. \tag{8}$$

We now have $f(\mathcal{T}) \subset (-\infty, k_2)$ and $f(\mathcal{S}) \subset [k_1, \infty)$, and also $k_1 - k_2 \geq k > 0$. Thus (k_2, k_1) is an open interval that lies between the intervals $f(\mathcal{T})$ and $f(\mathcal{S})$. Therefore, by Theorem 2, the slab with parallel faces $\mathcal{H}_2 : f(x) = k_2$ and $\mathcal{H}_1 : f(x) = k_1$ separates the sets $\mathcal{S}$ and $\mathcal{T}$. □

Exercises – Section 6.2

1. Prove Theorem 2.
2. Let $\mathcal{S}$ and $\mathcal{T}$ be any two non-empty, convex sets. Prove that $\mathcal{S}$ and $\mathcal{T}$ can be weakly separated if and only if the origin O does not belong to $\text{In}(\mathcal{S} - \mathcal{T})$.

SECTION 6.3. HAUSDORFF DISTANCE, BLASCHKE SELECTION THEOREM

In Sec. 1.8, when we discussed the nearness of sets $\mathcal{S}$ and $\mathcal{T}$, we observed that the sense in which $\text{n}(\mathcal{S}, \mathcal{T})$ is like a distance between $\mathcal{S}$ and $\mathcal{T}$ is not a true metric. Using linear combinations of sets, we can now introduce a measure for pairs of compact sets that is a genuine distance function called the Hausdorff metric.

A suggestion for Hausdorff distance comes from the effect of adding a neighborhood $N(O, \varepsilon)$ and a set $\mathscr{S}$. As we saw in Sec. 6.1, the sum $N(O, \varepsilon) + \mathscr{S}$ is just the union of all the neighborhoods $N(X, \varepsilon)$ for $X \in \mathscr{S}$. Thus $N(O, \varepsilon) + \mathscr{S}$ is the set $\mathscr{S}$ "swelled out" everywhere by an amount ε. If $\mathscr{T}$ is some second set that is bounded, then $\mathscr{S}$ can be sufficiently swelled out until it contains

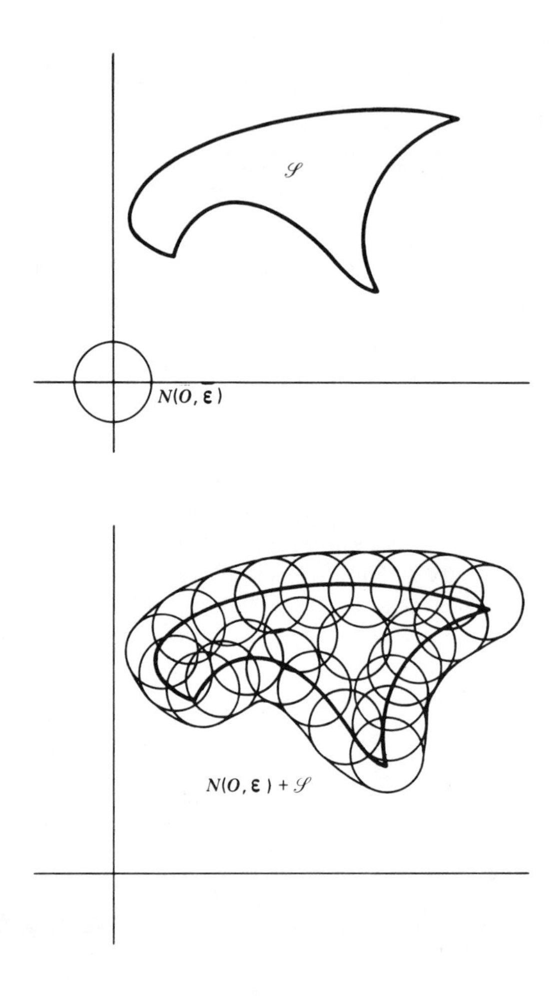

$\mathscr{T}$. That is, there is some ε_1 such that $N(O, \varepsilon_1) + \mathscr{S} \supset \mathscr{T}$. The number ε_1 gives a measure of the distance of $\mathscr{T}$ from $\mathscr{S}$ in the sense that any point Y of $\mathscr{T}$ is closer than ε_1 to some point of $\mathscr{S}$, because $Y \in N(X, \varepsilon_1)$ for some X in $\mathscr{S}$. In the same way, if $\mathscr{S}$ is bounded, then there is some ε_2 such that $N(O, \varepsilon_2) + \mathscr{T} \supset \mathscr{S}$, and ε_2 is a measure similarly of the distance of $\mathscr{S}$ from

$\mathscr{T}$. It is not true, in general, that if $\mathscr{S}$ swelled out by ε contains $\mathscr{T}$, then $\mathscr{T}$ swelled out by ε must contain $\mathscr{S}$. But clearly the larger of ε_1 and ε_2, say ε_3, is such that $\mathscr{S} + \mathrm{N}(O, \varepsilon_3) \supset \mathscr{T}$ and $\mathscr{T} + \mathrm{N}(O, \varepsilon_3) \supset \mathscr{S}$. What is of interest, of course, is how small ε_3 can be. Since we cannot expect that there will always be a smallest value of ε that satisfies both inclusions, we define Hausdorff distance in the following way.

DEFINITION (Hausdorff distance). If $\mathscr{S}$ and $\mathscr{T}$ are non-empty sets in $\mathscr{E}^n$, the *Hausdorff distance* between $\mathscr{S}$ and $\mathscr{T}$, if it exists, is the number

$$\mathrm{D}(\mathscr{S}, \mathscr{T}) = \mathrm{glb}\{\varepsilon > 0 : \mathscr{S} + \mathrm{N}(O, \varepsilon) \supset \mathscr{T} \text{ and } \mathscr{T} + \mathrm{N}(O, \varepsilon) \supset \mathscr{S}\}.$$

As a particular example, consider the ball $\mathrm{B}(A, r)$ and the concentric sphere $\mathrm{S}(A, 2r)$. Clearly, $\mathrm{B}(A, r) + \mathrm{N}(O, \varepsilon)$ contains $\mathrm{S}(A, 2r)$ if $\varepsilon > r$ and does not contain $\mathrm{S}(A, 2r)$ if $\varepsilon \leq r$. Thus $r = \mathrm{glb}\{\varepsilon > 0 : \mathrm{B}(A, r) + \mathrm{N}(O, \varepsilon) \supset \mathrm{S}(A, r)\}$.

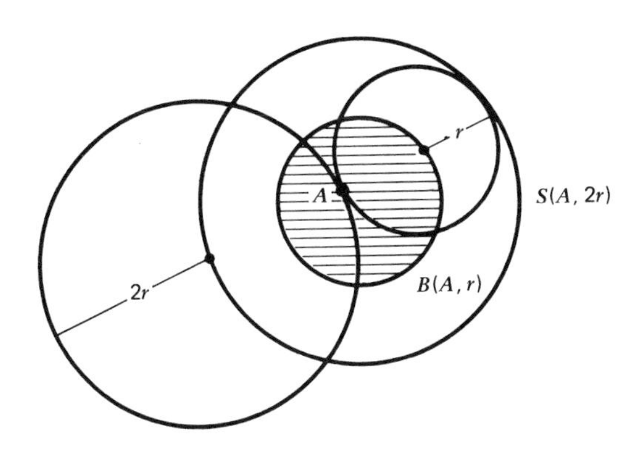

On the other hand, $\mathrm{S}(A, 2r) + \mathrm{N}(O, \varepsilon)$ contains $\mathrm{B}(A, r)$ if $\varepsilon > 2r$ and does not contain $\mathrm{B}(A, r)$ if $\varepsilon \leq 2r$. Thus $2r = \mathrm{glb}\{\varepsilon > 0 : \mathrm{S}(A, 2r) + \mathrm{N}(O, \varepsilon) \supset \mathrm{B}(A, r)\}$. Thus the greatest lower bound to the ε's for which both inclusions hold is $2r$, so $\mathrm{D}[\mathrm{B}(A, r), \mathrm{S}(A, 2r)] = 2r$.

We can think of each non-empty, compact set in $\mathscr{E}^n$ as a "point" in a new "space" and $\mathrm{D}(\mathscr{S}, \mathscr{T})$ as the distance between two such "points." We want to show that D is a metric for this space, and in the proof we need the following fact.

THEOREM 1. $\mathrm{N}(O, \varepsilon_1) + \mathrm{N}(O, \varepsilon_2) = \mathrm{N}(O, \varepsilon_1 + \varepsilon_2)$. (E)

THEOREM 2. The Hausdorff distance exists for pairs of non-empty, compact sets in $\mathscr{E}^n$ and is a metric for such pairs of sets.

PROOF. First, we observe that if $\mathscr{S}$ and $\mathscr{T}$ are non-empty, compact sets, then there exists a positive number ε such that $\mathscr{S} + \mathrm{N}(O, \varepsilon) \supset \mathscr{T}$ and $\mathscr{T} + \mathrm{N}(O, \varepsilon) \supset \mathscr{S}$. Because $\mathscr{S} \cup \mathscr{T}$ is also compact, its diameter exists, and for $\varepsilon = \mathrm{Dm}(\mathscr{S} \cup \mathscr{T})$ the two inclusions are obviously correct. Thus $\mathrm{D}(\mathscr{S}, \mathscr{T})$ is the greatest lower bound of a non-empty set of positive numbers.

Because the definition of $\mathrm{D}(\mathscr{S}, \mathscr{T})$ is symmetric with respect to $\mathscr{S}$ and $\mathscr{T}$, then $\mathrm{D}(\mathscr{S}, \mathscr{T}) = \mathrm{D}(\mathscr{T}, \mathscr{S})$.

Next, suppose that $\mathrm{D}(\mathscr{S}, \mathscr{T}) = 0$. Let ε be an arbitrary positive number and let X be any point of $\mathscr{S}$. Since $\mathrm{D}(\mathscr{S}, \mathscr{T}) = 0$, $\mathscr{S} \subset \mathscr{T} + \mathrm{N}(O, \varepsilon)$ and $X \in \mathscr{T} + \mathrm{N}(O, \varepsilon)$. Thus $X \in \mathrm{N}(Y, \varepsilon)$ for some $Y \in \mathscr{T}$, hence $\mathrm{d}(X, Y) < \varepsilon$, so $\mathrm{d}(X, \mathscr{T}) < \varepsilon$. Because ε was arbitrary, it follows that $\mathrm{d}(X, \mathscr{T}) = 0$, so, by Corollary 2, Sec. 1.8, $X \in \mathscr{T}$. Therefore $\mathscr{S} \subset \mathscr{T}$. By a symmetric argument, $\mathscr{T} \subset \mathscr{S}$, and so $\mathscr{S} = \mathscr{T}$. Conversely, if $\mathscr{S} = \mathscr{T}$, then $\mathrm{D}(\mathscr{S}, \mathscr{T}) = \mathrm{D}(\mathscr{S}, \mathscr{S})$. Since $\mathscr{S} \subset \mathscr{S} + \mathrm{N}(O, \varepsilon)$ for every $\varepsilon > 0$, $\mathrm{D}(\mathscr{S}, \mathscr{S}) = 0$, so $\mathrm{D}(\mathscr{S}, \mathscr{T}) = 0$. Thus $\mathrm{D}(\mathscr{S}, \mathscr{T}) = 0$ if and only if $\mathscr{S} = \mathscr{T}$.

It is still to be shown that D satisfies the triangle inquality. To this end, consider non-empty, compact sets $\mathscr{R}$, $\mathscr{S}$, $\mathscr{T}$, and an arbitrary positive number ε. Define

$$\varepsilon_1 = \mathrm{D}(\mathscr{S}, \mathscr{T}) + \tfrac{1}{2}\varepsilon, \qquad \text{and} \qquad \varepsilon_2 = \mathrm{D}(\mathscr{T}, \mathscr{R}) + \tfrac{1}{2}\varepsilon. \tag{1}$$

Then $\varepsilon_1 > \mathrm{D}(\mathscr{S}, \mathscr{T})$ implies that

$$\mathscr{S} + \mathrm{N}(O, \varepsilon_1) \supset \mathscr{T} \qquad \text{and} \qquad \mathscr{T} + \mathrm{N}(O, \varepsilon_1) \supset \mathscr{S}, \tag{2}$$

and $\varepsilon_2 > \mathrm{D}(\mathscr{T}, \mathscr{R})$ implies that

$$\mathscr{T} + \mathrm{N}(O, \varepsilon_2) \supset \mathscr{R} \qquad \text{and} \qquad \mathscr{R} + \mathrm{N}(O, \varepsilon_2) \supset \mathscr{T}. \tag{3}$$

The relations (2) and (3), together with Theorem 1, imply that

$$\mathscr{S} + \mathrm{N}(O, \varepsilon_1 + \varepsilon_2) = \mathscr{S} + \mathrm{N}(O, \varepsilon_1) + \mathrm{N}(O, \varepsilon_2) \supset \mathscr{T} + \mathrm{N}(O, \varepsilon_2) \supset \mathscr{R}. \tag{4}$$

Similarly,

$$\mathscr{R} + \mathrm{N}(O, \varepsilon_1 + \varepsilon_2) = \mathscr{R} + \mathrm{N}(O, \varepsilon_1) + \mathrm{N}(O, \varepsilon_2) \supset \mathscr{T} + \mathrm{N}(O, \varepsilon_1) \supset \mathscr{S}. \tag{5}$$

Now, from (4), (5), and (1),

$$\mathrm{D}(\mathscr{S}, \mathscr{R}) \leq \varepsilon_1 + \varepsilon_2 = \mathrm{D}(\mathscr{S}, \mathscr{T}) + \mathrm{D}(\mathscr{T}, \mathscr{R}) + \varepsilon. \tag{6}$$

Because (6) is valid for any $\varepsilon > 0$, it follows that

$$D(\mathscr{S}, \mathscr{R}) \leq D(\mathscr{S}, \mathscr{T}) + D(\mathscr{T}, \mathscr{R}), \tag{7}$$

which is the triangle inequality. □

The collection of all the non-empty, compact subsets of $\mathscr{E}^n$, together with the Hausdorff metric D, form a metric space associated with $\mathscr{E}^n$ that can be denoted by $\mathscr{H}-\mathscr{E}^n$. Each "point" of the space $\mathscr{H}-\mathscr{E}^n$ is thus a set of euclidean points, and a set of "points" in $\mathscr{H}-\mathscr{E}^n$ is a collection of compact subsets of $\mathscr{E}^n$. All the topological concepts of Chapter 1, developed for the space $\mathscr{E}^n$ with the euclidean metric d, make sense for the space $\mathscr{H}-\mathscr{E}^n$ with the Hausdorff metric D. For example, the δ-neighborhood of "point" $\mathscr{S}$ in $\mathscr{H}-\mathscr{E}^n$ is the set of "points" $\mathscr{X}$ such that $D(\mathscr{S}, \mathscr{X}) < \delta$. Any collection $\mathscr{F}$ of compact subsets of $\mathscr{E}^n$ is automatically a "point set" of $\mathscr{H}-\mathscr{E}^n$, and it makes sense to speak of $\mathscr{F}$ as being open, or closed, or bounded, or compact *in the Hausdorff sense.*

Geometrically speaking, if two compact sets $\mathscr{R}$, $\mathscr{S}$ in $\mathscr{E}^n$ are close together, as "points" of $\mathscr{H}-\mathscr{E}^n$, that is, if $D(\mathscr{R}, \mathscr{S})$ is small, and if $\mathscr{R}$, $\mathscr{S}$ are sets that can be pictured, then they will appear to have roughly the same shape and to be located approximately in the same portion of space. The closer together the "points" are, the more alike the corresponding euclidean sets appear to be. Thus if a sequence of "points," $\mathscr{S}_1, \mathscr{S}_2, \ldots, \mathscr{S}_n, \ldots$, converges to a "point" $\mathscr{S}$, that is, if $D(\mathscr{S}, \mathscr{S}_i) \to 0$, then the euclidean sets $\mathscr{S}_i$ approximate $\mathscr{S}$ more and more closely as i increases. We now want to define a particular collection of compact subsets of $\mathscr{E}^n$ which we show can be used to approximate any arbitrary compact set in $\mathscr{E}^n$ as closely as one wishes.

For the remainder of this section, we let $\mathscr{W}$ denote a closed cube n cell, or solid cube, of side length δ, as defined in Sec. 1.7. As in Theorem 4, Sec. 1.7, we divide $\mathscr{W}$ into 2^n subcubes, each of side length $\frac{1}{2}\delta$, and we denote this collection by $\mathscr{W}_1$. Each cube in $\mathscr{W}_1$ can be divided into 2^n subcubes of side length $(1/2^2)\delta$, and the collection of these 2^{2n} cubes is denoted by $\mathscr{W}_2$. Continuing in this way, for each positive integer i, we obtain a collection $\mathscr{W}_i$ of subcubes of $\mathscr{W}$. There are 2^{in} cubes in $\mathscr{W}_i$, each of side length $(1/2^i)\delta$. For each i, the union of the cubes in $\mathscr{W}_i$ is $\mathscr{W}$.

DEFINITION (Cubical figures, cubical approximations). A union of cubes, all of which belong to one of the collections $\mathscr{W}_i$, is a *cubical figure.* If $\mathscr{K}$ is a compact subset of $\mathscr{W}$, the union of those cubes in $\mathscr{W}_i$ that intersect $\mathscr{K}$ is the *ith cubical approximation of* $\mathscr{K}$, denoted by $C_i(\mathscr{K})$.

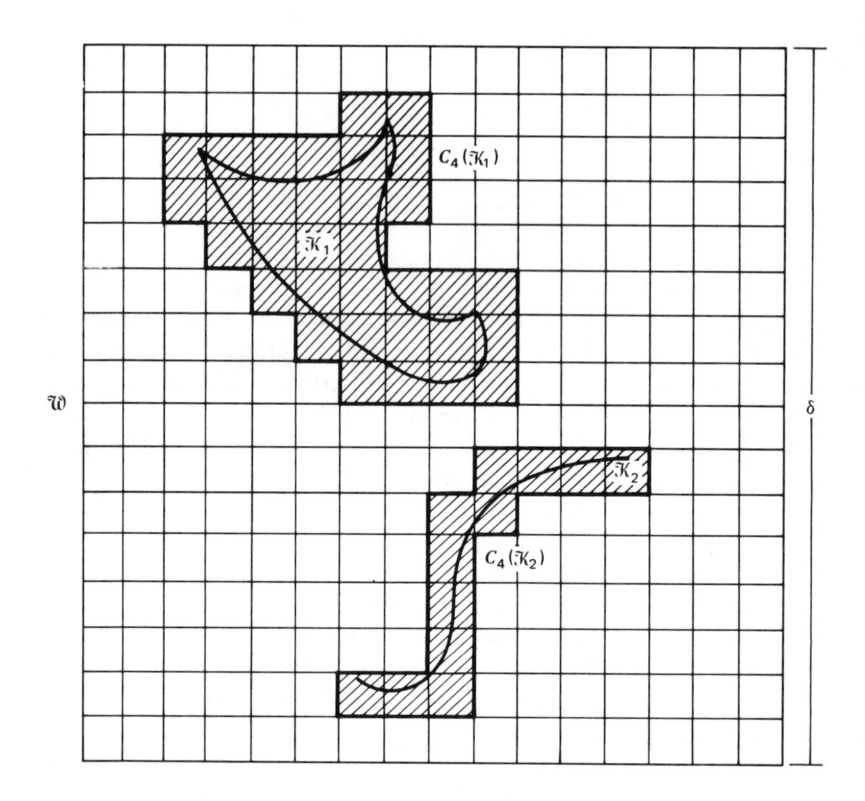

By way of illustration, we picture in $\mathscr{E}^2$ the fourth cubical approximation of two compact subsets on $\mathscr{W}$, sets $\mathscr{K}_1$ and $\mathscr{K}_2$.

THEOREM 3. Every cubical figure is compact, and if $\mathscr{K}$ is a non-empty, compact subset of $\mathscr{W}$, then

$$D(\mathscr{K}, C_i(\mathscr{K})) \leq \frac{\sqrt{n}}{2^i}\delta.$$

PROOF. By definition, a cubical figure is the union of a finite number of cubes in one of the collections $\mathscr{W}_i$. Each of these cubes is compact, and since the union of a finite number of compact sets is compact, the cubical figure is compact.

Now let P denote a point in $C_i(\mathscr{K})$. Then P belongs to some cube of $\mathscr{W}_i$ and this cube has side length $(1/2^i)\delta$. The diameter of the cube is $(\sqrt{n}/2^i)\delta$. Thus $d(P, Q) \leq (\sqrt{n}/2^i)\delta$ for some $Q \in \mathscr{K}$ because each cube in $C_i(\mathscr{K})$ intersects $\mathscr{K}$. Thus if $\varepsilon = (\sqrt{n}/2^i)\delta$, then $d(P, Q) < \varepsilon$, so $P \in N(Q, \varepsilon) = Q + N(O, \varepsilon)$, and therefore $P \in \mathscr{K} + N(O, \varepsilon)$. Since P is any point of $C_i(\mathscr{K})$, it follows that $C_i(\mathscr{K}) \subset \mathscr{K} + N(O, \varepsilon)$. On the other hand, $\mathscr{K} \subset C_i(\mathscr{K})$ implies that $\mathscr{K} \subset C_i(\mathscr{K}) + N(O, \varepsilon)$, so $D(\mathscr{K}, C_i(\mathscr{K})) \leq \varepsilon$, as the theorem asserts. □

COROLLARY 3. If $\mathscr{K}$ is any non-empty, compact subset of $\mathscr{W}$, and ε_o is any positive number, there exists a cubical approximation of $\mathscr{K}$, $C_i(\mathscr{K})$, such that $D(\mathscr{K}, C_i(\mathscr{K})) < \varepsilon_o$.

The next result is the main idea of this section and is known as the Blaschke selection theorem.

THEOREM 4. Every infinite collection of compact subsets of $\mathscr{W}$, as "points" of $\mathscr{H}$-$\mathscr{E}^n$, has a limit "point," and such a limit "point," as a euclidean set, is a compact subset of $\mathscr{W}$.

PROOF. Let $\mathscr{F}$ be an infinite collection of compact subsets of $\mathscr{W}$. We search for a limit "point" to $\mathscr{F}$ by using cubical approximations. We first observe that since $\mathscr{W}_1$ is a finite set of cubes, the set of all subsets of $\mathscr{W}_1$ is finite, so the number of cubical figures in $\mathscr{W}_1$ is finite. For each set in $\mathscr{F}$ there is a unique cubical figure in $\mathscr{W}_1$ that is the cubical approximation of the set. But, since the number of sets in $\mathscr{F}$ is infinite, and the number of cubical figures in $\mathscr{W}_1$ is finite, there must be at least one cubical figure in $\mathscr{W}_1$, call it $\mathscr{A}_1$, that is the cubical approximation of infinitely many sets in $\mathscr{F}$. These sets, with cubical approximation $\mathscr{A}_1$, form an infinite subcollection $\mathscr{F}_1$ of $\mathscr{F}$. We now repeat the argument for $\mathscr{F}_1$ and the cubes of $\mathscr{W}_2$. Since there are infinitely many sets in $\mathscr{F}_1$ and only a finite number of cubical figures in $\mathscr{W}_2$, there is some cubical figure $\mathscr{A}_2$ in $\mathscr{W}_2$ that is the cubical approximation of infinitely many sets in $\mathscr{F}_1$, and we denote this subcollection of $\mathscr{F}_1$ by $\mathscr{F}_2$. Continuing in this way, we obtain a nested sequence of infinite subcollections of $\mathscr{F}$,

$$\mathscr{F} \supset \mathscr{F}_1 \supset \mathscr{F}_2 \supset \cdots \supset \mathscr{F}_n \supset \cdots$$

and a sequence of cubical figures, $\mathscr{A}_1, \mathscr{A}_2, \ldots, \mathscr{A}_n, \ldots$ such that all the sets in $\mathscr{F}_i$ have $\mathscr{A}_i$ as their cubical approximation in $\mathscr{W}_i$, $i = 1, 2, \ldots$.

We now recall that each cube of $\mathscr{W}_i$ was divided into 2^n subcubes in producing $\mathscr{W}_{i+1}$. Thus if a compact set $\mathscr{K} \subset \mathscr{W}$ intersects a cube in $\mathscr{W}_{i+1}$, it also intersects that cube of $\mathscr{W}_i$ which was subdivided to produce the cube of $\mathscr{W}_{i+1}$ intersecting $\mathscr{K}$. Thus $C_{i+1}(\mathscr{K}) \subset C_i(\mathscr{K})$. Therefore if $\mathscr{K} \in \mathscr{F}_{i+1}$, then

$$\mathscr{A}_{i+1} = C_{i+1}(\mathscr{K}) \subset C_i(\mathscr{K}) = \mathscr{A}_i.$$

Thus the sequence of cubical figures $\{\mathscr{A}_i\}$ is also decreasingly nested. Since the figures $\mathscr{A}_i$ are non-empty and compact, it follows from the Cantor intersection theorem that

$$\mathscr{A} = \bigcap_{i=1}^{\infty} \mathscr{A}_i$$

is a non-empty, compact subset of $\mathscr{W}$ (Theorem 12, Sec. 1.4).

We now want to show that $\mathscr{A}$ and the sets in $\mathscr{F}$, regarded as "points" in the space $\mathscr{H}$–$\mathscr{E}^n$ are such that $\mathscr{A}$ is a limit "point" to $\mathscr{F}$ in the Hausdorff sense. To establish this, we show that for any $\varepsilon > 0$ there exists a "point" $\mathscr{K}$ in $\mathscr{F}$ such that $0 < \mathrm{D}(\mathscr{A}, \mathscr{K}) < \varepsilon$, and hence that every deleted Hausdorff neighborhood of $\mathscr{A}$ intersects $\mathscr{F}$.

First, corresponding to $\varepsilon > 0$ we consider the set $\mathscr{A}^* = \mathscr{A} + \mathrm{N}(O, \varepsilon)$. By Theorem 7, Sec. 6.1, $\mathscr{A}^*$ is an open set, so $\mathrm{Cp}(\mathscr{A}^*)$ is closed. If sets $\mathscr{R}_i$ are defined by $\mathscr{R}_i = \mathscr{A}_i \cap \mathrm{Cp}(\mathscr{A}^*)$, $i = 1, 2, \ldots$, then, as in the proof for Theorem 13, Sec. 1.4, it follows that $\{\mathscr{R}_i\}$ is a decreasingly nested sequence of compact sets. Hence if all the sets $\mathscr{R}_i$ were non-empty, the Cantor intersection theorem would imply that their intersection was non-empty. But

$$\bigcap_{i=1}^{\infty} \mathscr{R}_i = \bigcap_{i=1}^{\infty} [\mathscr{A}_i \cap \mathrm{Cp}(\mathscr{A}^*)] = \left[\bigcap_{i=1}^{\infty} \mathscr{A}_i\right] \cap \mathrm{Cp}(\mathscr{A}^*) = \mathscr{A} \cap \mathrm{Cp}(\mathscr{A}^*) = \varnothing.$$

Thus there must be some set $\mathscr{R}_m$ that is empty, and $\mathscr{A}_m \cap \mathrm{Cp}(\mathscr{A}^*) = \varnothing$ implies $\mathscr{A}_m \subset \mathscr{A}^*$. Because the sequence $\{\mathscr{A}_i\}$ is decreasingly nested, we then have

$$\mathscr{A}_i \subset \mathscr{A}^* = \mathscr{A} + \mathrm{N}(O, \varepsilon), \text{ for all } i \geq m. \tag{1}$$

Next, corresponding to $\varepsilon > 0$, let j be a positive integer such that

$$\frac{\sqrt{n}}{2^j}\,\delta < \varepsilon. \tag{2}$$

Now consider a positive integer $i \geq \max\{m, j\}$. The collection $\mathscr{F}_i$ is infinite, so it contains a set $\mathscr{K} \neq \mathscr{A}$. Since $\mathscr{K} \in \mathscr{F}_i$, then, by the way $\mathscr{F}_i$ was defined, $\mathrm{C}_i(\mathscr{K}) = \mathscr{A}_i$, so $\mathscr{K} \subset \mathscr{A}_i$. Now, from (1),

$$\mathscr{K} \subset \mathscr{A}_i \subset \mathscr{A} + \mathrm{N}(O, \varepsilon). \tag{3}$$

By Theorem 3, $\mathrm{D}(\mathscr{K}, \mathscr{A}_i) \leq (\sqrt{n}/2^i)\delta$, and $i \geq j$ implies $(\sqrt{n}/2^i)\delta \leq (\sqrt{n}/2^j)\delta$. These relations with (2) show that $\mathrm{D}(\mathscr{K}, \mathscr{A}_i) < \varepsilon$, hence that $\mathscr{A}_i \subset \mathscr{K} + \mathrm{N}(O, \varepsilon)$. Then, because $\mathscr{A} \subset \mathscr{A}_i$, it follows that

$$\mathscr{A} \subset \mathscr{K} + \mathrm{N}(O, \varepsilon). \tag{4}$$

From (3) and (4) and $\mathscr{K} \neq \mathscr{A}$, we have

$$0 < \mathrm{D}(\mathscr{A}, \mathscr{K}) < \varepsilon. \tag{5}$$

The "point" $\mathscr{K}$ in $\mathscr{F}_i$ is also a "point" in $\mathscr{F}$, and (5) shows that in the space $\mathscr{H}$–$\mathscr{E}^n$ every deleted neighborhood of $\mathscr{A}$ intersects $\mathscr{F}$. Thus $\mathscr{A}$ is a limit "point" to $\mathscr{F}$. □

Later we will need the following implication of the Blaschke selection theorem.

THEOREM 5. If $\mathscr{C}$ is the collection of all compact subsets of $\mathscr{W}$, then the sets in $\mathscr{C}$ as "points" of $\mathscr{H}$–$\mathscr{E}^n$ form a set that is compact in the sense of the Hausdorff metric.

PROOF. Let $\mathscr{F} = \{\mathscr{G}_\alpha : \alpha \in \Lambda\}$ be an infinite family of collections of compact sets in $\mathscr{E}^n$ such that each collection $\mathscr{G}_\alpha$, as a "point set" in $\mathscr{H}$–$\mathscr{E}^n$ is open in the Hausdorff sense, and each set in $\mathscr{C}$ belongs to at least one of the collections in $\mathscr{F}$. That is, $\mathscr{F}$ is an open cover of $\mathscr{C}$ in the sense of the Hausdorff metric. We want to show that some finite subfamily of $\mathscr{F}$ is also a cover of $\mathscr{C}$. To do so, we assume that no finite subfamily of $\mathscr{F}$ covers $\mathscr{C}$, and we show that this leads to a contradiction.

First, our assumption implies that no single collection in $\mathscr{F}$ is a cover of $\mathscr{C}$. Thus for each $\alpha \in \Lambda$ there are sets in $\mathscr{C}$ that do not belong to $\mathscr{G}_\alpha$, and we define $\mathscr{C}_\alpha$ to be the collection of these sets in $\mathscr{C}$ that are not in $\mathscr{G}_\alpha$.

Now consider any set $\mathscr{R}$ in the collection $\mathscr{C}_\alpha$. Because $\mathscr{R}$ is a subset of $\mathscr{W}$, then for each positive integer i the set $\mathscr{R}$ has a cubical approximation $\mathrm{C}_i(\mathscr{R})$ in $\mathscr{W}_i$. The ith cubical approximations of all the sets in $\mathscr{C}_\alpha$ form a new collection which we denote by $\mathscr{C}_{\alpha,i}$. Thus

$$\mathscr{C}_{\alpha,i} = \bigcup \{\mathrm{C}_i(\mathscr{R}) : \mathscr{R} \in \mathscr{C}_\alpha\}.$$

Next, we want to show that for each positive integer i there is some cubical figure in $\mathscr{W}_i$ that belongs to every collection $\mathscr{C}_{\alpha,i}$, $\alpha \in \Lambda$. Suppose that this is not so. Then if $\mathscr{K}_1, \mathscr{K}_2, \ldots, \mathscr{K}_m$ is a listing of all the cubical figures in $\mathscr{W}_i$, there must exist collections $\mathscr{C}_{\alpha_1,i}, \mathscr{C}_{\alpha_2,i}, \ldots, \mathscr{C}_{\alpha_m,i}$ such that $\mathscr{K}_1 \notin \mathscr{C}_{\alpha_1,i}$, $\mathscr{K}_2 \notin \mathscr{C}_{\alpha_2,i}, \ldots, \mathscr{K}_m \notin \mathscr{C}_{\alpha_m,i}$. Corresponding to the indices $\alpha_1, \alpha_2, \ldots, \alpha_m$ there is a subfamily of $\mathscr{F}$ consisting of the collections

$$\mathscr{G}_{\alpha_1}, \mathscr{G}_{\alpha_2}, \ldots, \mathscr{G}_{\alpha_m}.$$

This subfamily of $\mathscr{F}$ is finite, so, by our assumption, it is not a cover of $\mathscr{C}$. Thus there is some set $\mathscr{R}$ in $\mathscr{C}$ that does not belong to any of the collections $\mathscr{G}_{\alpha_1}, \mathscr{G}_{\alpha_2}, \ldots, \mathscr{G}_{\alpha_m}$. Therefore $\mathscr{R}$ is in all the collections $\mathscr{C}_{\alpha_1}, \mathscr{C}_{\alpha_2}, \ldots, \mathscr{C}_{\alpha_m}$. Therefore $\mathrm{C}_i(\mathscr{R})$ is in all of the collections $\mathscr{C}_{\alpha_1,i}, \mathscr{C}_{\alpha_2,i}, \ldots, \mathscr{C}_{\alpha_m,i}$. But $\mathrm{C}_i(\mathscr{R})$ is a cubical figure in $\mathscr{W}_i$, so $C_i(\mathscr{R})$ is one of the sets $\mathscr{K}_j$, and $\mathscr{K}_j$ is not in the collection $\mathscr{C}_{\alpha_j, i}$. The contradiction shows that at least one of the cubical figures in $\mathscr{W}_i$ must belong to $\mathscr{C}_{\alpha,i}$ for every $\alpha \in \Lambda$. Let $\mathscr{K}_i{}^*$ denote a cubical figure in $\mathscr{W}_i$ with this property. Then for each $\alpha \in \Lambda$ and each positive integer i there exists a set $\mathscr{K}_{\alpha,i} \in \mathscr{C}_{\alpha,i}$ such that $C_i(\mathscr{K}_{\alpha,i}) = \mathscr{K}_i{}^*$.

As i takes on the values of all the positive integers, the sequence $\mathscr{K}_1{}^*$, $\mathscr{K}_2{}^*$, $\ldots$, $\mathscr{K}_n{}^*$, $\ldots$ may or may not contain infinitely many different sets. If it does, then, by the Blaschke theorem (Theorem 4), there is a compact set $\mathscr{K}$ that is a limit "point" to $\{\mathscr{K}_i{}^*\}$ in the sense of the Hausdorff metric. Suppose $\mathscr{K}$ is not a subset of $\mathscr{W}$. Then there exists $P \in \mathscr{K}$ and $P \notin \mathscr{W}$, so $d(P, \mathscr{W}) > 0$. Then $\mathscr{W} + N(O, \varepsilon)$ does not contain $\mathscr{K}$ for any $\varepsilon \leq d(P, \mathscr{W})$. Since $\mathscr{K}_i{}^* \subset \mathscr{W}$, then $\mathscr{K}_i{}^* + N(O, \varepsilon)$ does not contain $\mathscr{K}$, so $D(\mathscr{K}, \mathscr{K}_i{}^*) \geq d(P, \mathscr{W})$, $i = 1, 2, \ldots$, contradicting the fact that $\mathscr{K}$ is a limit "point" to $\{\mathscr{K}_i{}^*\}$. Thus $\mathscr{K}$ must be a subset of $\mathscr{W}$, hence a set in $\mathscr{C}$.

If there are only a finite number of different sets in the sequence $\{\mathscr{K}_i{}^*\}$, then at least one set must occur with infinitely many different indices. Let $\mathscr{K}$ denote such a set. Since all the sets in $\{\mathscr{K}_i{}^*\}$ are compact subsets of $\mathscr{W}$, then $\mathscr{K}$ is again a set in $\mathscr{C}$.

For either definition of $\mathscr{K}$, corresponding to any positive number ε and any positive integer h, there exists a positive integer i such that $i > h$ and $D(\mathscr{K}, \mathscr{K}_i{}^*) < \varepsilon$.

Because the family $\mathscr{F}$ is a cover of $\mathscr{C}$, and $\mathscr{K}$ is a set in $\mathscr{C}$, there must be some collection $\mathscr{G}_\alpha$ in $\mathscr{F}$ to which the set $\mathscr{K}$ belongs. By hypothesis, $\mathscr{G}_\alpha$ is an open set in the space $\mathscr{H}$–$\mathscr{E}^n$, so there exists some ε_o-neighborhood of $\mathscr{K}$, in the Hausdorff sense, that is contained in $\mathscr{G}_\alpha$. That is, every compact set $\mathscr{S}$ in $\mathscr{E}^n$ such that $D(\mathscr{K}, \mathscr{S}) < \varepsilon_o$ belongs to the collection $\mathscr{G}_\alpha$. Corresponding to ε_o, let h be a positive integer such that

$$\frac{\sqrt{n}}{2^h}\,\delta < \tfrac{1}{2}\varepsilon_o, \tag{1}$$

and let i be a positive integer such that $i > h$ and

$$D(\mathscr{K}, \mathscr{K}_i) < \tfrac{1}{2}\varepsilon_o, \tag{2}$$

Corresponding to the indices α and i there is a set $\mathscr{K}_{\alpha,i}$ in the collection $\mathscr{C}_{\alpha,i}$ such that $C_i(\mathscr{K}_{\alpha,i}) = \mathscr{K}_i{}^*$. Since $i > h$, then, by Theorem 3 and (1), we have

$$D(\mathscr{K}_{\alpha,i}, \mathscr{K}_i{}^*) \leq \frac{\sqrt{n}}{2^i}\,\delta < \tfrac{1}{2}\varepsilon_o. \tag{3}$$

Now, by the triangle inequality and (2) and (3), it follows that

$$D(\mathscr{K}, \mathscr{K}_{\alpha,i}) \leq D(\mathscr{K}, \mathscr{K}_i{}^*) + D(\mathscr{K}_i{}^*, \mathscr{K}_{\alpha,i}) < \tfrac{1}{2}\varepsilon_o + \tfrac{1}{2}\varepsilon_o = \varepsilon_o. \tag{4}$$

Thus $\mathscr{K}_{\alpha,i}$ must belong to $\mathscr{G}_\alpha$. But $\mathscr{K}_{\alpha,i}$ is in the collection $\mathscr{C}_{\alpha,i}$ and hence is a set in $\mathscr{C}$ that is not in $\mathscr{G}_\alpha$.

We now have a contradiction stemming from our assumption that $\mathscr{C}$ could fail to be compact in the sense of the Hausdorff metric. Therefore $\mathscr{C}$ must be compact in this sense. □

The argument used for $\mathscr{E}^n$ with metric d to show that a closed subset of a compact set is compact can simply be paraphrased to establish the same result for $\mathscr{H}$–$\mathscr{E}^n$ with metric D. In turn, the latter property implies the corollary of the next theorem.

THEOREM 6. The collection $\mathscr{G}$ of all non-empty, compact, convex sets in $\mathscr{E}^n$ is a closed "point" set in $\mathscr{H}$–$\mathscr{E}^n$.

PROOF. Let $\mathscr{K}$ denote a set in $\mathscr{E}^n$ that in $\mathscr{H}$–$\mathscr{E}^n$ is a limit "point" to $\mathscr{G}$. In particular, $\mathscr{K}$ is a "point" of $\mathscr{H}$–$\mathscr{E}^n$ and hence is a non-empty, compact set in $\mathscr{E}^n$. Thus if we can show that $\mathscr{K}$ is convex, in the ordinary sense, we have established that $\mathscr{K}$ belongs to $\mathscr{G}$, hence that $\mathscr{G}$ is closed in $\mathscr{H}$–$\mathscr{E}^n$.

Consider two points A, B in $\mathscr{K}$ and a point P such that $p = \eta a + (1 - \eta)b$, $0 < \eta < 1$, and let ε be any positive number. Because $\mathscr{K}$ is a limit "point" to $\mathscr{G}$, the deleted $\frac{1}{2}\varepsilon$-neighborhood of $\mathscr{K}$ in $\mathscr{H}$–$\mathscr{E}^n$ must intersect $\mathscr{G}$. That is, there must exist a compact, convex set $\mathscr{R} \in \mathscr{G}$ such that $0 < \mathrm{D}(\mathscr{K}, \mathscr{R}) < \frac{1}{2}\varepsilon$. Then $\mathscr{K} \subset \mathscr{R} + \mathrm{N}(O, \frac{1}{2}\varepsilon)$ and $\mathscr{R} \subset \mathscr{K} + \mathrm{N}(O, \frac{1}{2}\varepsilon)$. Since $\mathscr{R}$ and $\mathrm{N}(O, \frac{1}{2}\varepsilon)$ are convex sets, then, by Theorem 5, Sec. 6.1, their sum is convex. Because A and B are in $\mathscr{K}$ and $\mathscr{K} \subset \mathscr{R} + \mathrm{N}(O, \frac{1}{2}\varepsilon)$, which is convex, then $p = \eta a + (1 - \eta)b$ is in $\mathscr{R} + \mathrm{N}(O, \frac{1}{2}\varepsilon)$. Since $\mathscr{R} \subset \mathscr{K} + \mathrm{N}(O, \frac{1}{2}\varepsilon)$, then

$$\mathscr{R} + \mathrm{N}(O, \tfrac{1}{2}\varepsilon) \subset [\mathscr{K} + \mathrm{N}(O, \tfrac{1}{2}\varepsilon)] + \mathrm{N}(O, \tfrac{1}{2}\varepsilon) = \mathscr{K} + \mathrm{N}(O, \varepsilon).$$

Therefore

$$P \in \mathscr{K} + \mathrm{N}(O, \varepsilon). \tag{1}$$

From (1) it follows that there exist $X \in \mathscr{K}$ and $Y \in \mathrm{N}(O, \varepsilon)$ such that $P = X + Y$. Thus

$$\mathrm{d}(P, X) = |p - x| = |y| < \varepsilon. \tag{2}$$

If there is some $\varepsilon > 0$ such that the X in (2) is P, then $X \in \mathscr{K}$ implies that $P \in \mathscr{K}$. If $X \neq P$ for all $\varepsilon > 0$, then (2) implies that every deleted neighborhood $\mathrm{N}_o(P, \varepsilon)$ intersects $\mathscr{K}$, hence that P is a limit point to $\mathscr{K}$. But $\mathscr{K}$ is compact, hence closed, so $P \in \mathscr{K}$. Thus in all cases, $P \in \mathscr{K}$, and $\mathscr{K}$ is a convex set. □

COROLLARY 6. If $\mathscr{C}^*$ is the collection of all non-empty, compact, convex subsets of $\mathscr{W}$, then $\mathscr{C}^*$ is a compact "point" set in $\mathscr{H}$–$\mathscr{E}^n$. (E)

Exercises – Section 6.3

1. Prove Theorem 1.
2. If $\mathscr{S}$ and $\mathscr{T}$ are compact sets in $\mathscr{E}^n$ and $\mathscr{T} \not\subset \mathscr{S}$, prove that there exists $\varepsilon_1 > 0$ such that $\mathscr{T} \subset \mathscr{S} + \mathrm{N}(O, \varepsilon_1)$ and $\mathscr{T} \not\subset \mathscr{S} + \mathrm{N}(O, \varepsilon)$ if $\varepsilon < \varepsilon_1$.
3. A singleton $\{A\}$ in $\mathscr{E}^n$ is a compact set and hence is a "point" in $\mathscr{H}$–$\mathscr{E}^n$. Explain why $\mathrm{d}(A, B) = \mathrm{D}[\{A\}, \{B\}]$.
4. Give a formula for $\mathrm{D}[\{P\}, \mathrm{S}(A, r)]$ when

   a. $P \in \mathrm{N}(A, r)$.
   b. $P \in \mathrm{Cp}[\mathrm{B}(A, r)]$.

5. If $\mathscr{R} = \mathrm{B}(A_1, r_1)$ and $\mathscr{S} = \mathrm{B}(A_2, r_2)$ and $\mathscr{R} \cap \mathscr{S} = \varnothing$, give a formula for $\mathrm{D}(\mathscr{R}, \mathscr{S})$.
6. If $\mathscr{R} = \mathrm{Sg}[AB]$ and $P \notin \mathrm{Ln}(AB)$, is it correct that

$$\mathrm{D}[\{P\}, \mathscr{R}] = \max\{\mathrm{d}(P, \mathrm{Ln}(AB), \mathrm{d}(P, A), \mathrm{d}(P, B))\}?$$

7. Let $\{\mathscr{S}_i : i = 1, 2, \ldots\}$ be a decreasingly nested sequence of non-empty, compact sets in $\mathscr{E}^n$ and let $\mathscr{S}^*$ be the intersection set of all the sets $\mathscr{S}_i$. Must $\mathscr{S}^*$ be a limit "point" in $\mathscr{H}$–$\mathscr{E}^n$ to the "point" sequence $\{\mathscr{S}_i\}$?
8. Use Corollary 12.3, Sec. 6.1, to prove that the convex bodies in $\mathscr{E}^n$ that have differentiable boundaries are "dense" in the collection of all convex bodies, that is, for any convex body $\mathscr{K}_o$ and $\varepsilon > 0$ there exists a convex body $\mathscr{K}$, with a differentiable boundary, and such that $\mathrm{D}(\mathscr{K}_o, \mathscr{K}) < \varepsilon$.
9. Prove Corollary 6.

SECTION 6.4. APPLICATIONS OF THE HAUSDORFF METRIC AND THE BLASCHKE SELECTION THEOREM

In this section we want to show how the Hausdorff metric and the space $\mathscr{H}$–$\mathscr{E}^n$ can be applied to an important aspect of problem solving. It is an interesting characteristic of this application that the problems themselves often have no apparent relation to the metric D or the space $\mathscr{H}$–$\mathscr{E}^n$. In the solutions we consider, a basic role is played by the min–max theorem, namely, that a real valued function, continuous on a compact domain, attains minimum and maximum values. This theorem, of great importance in many parts of mathematics, has been a primary tool in our study of $\mathscr{E}^n$. The discussion indicates its usefulness in the space $\mathscr{H}$–$\mathscr{G}^n$.

We begin with a rather informal statement of three notable and difficult problems in euclidean plane geometry.

(Isoperimetric problem) Among all compact, convex sets of a given perimeter (arc length of the boundary) which ones have maximum area?

(Minimun cover problem) Among all the compact, convex sets that can cover every set of a fixed diameter, which ones have minimum area?

(Maximum inclusion problem) Among the compact, convex sets that will fit inside every compact, convex set of a given minimum width, which ones have maximum area?

For each problem there are three separate questions a complete solution must answer.

1. *Existence*: Does any solution to the problem exist?
2. *Uniqueness*: How many different solutions are there?
3. *Identification*: What properties characterize and distinguish the solutions?

For example, with respect to the isoperimetric problem, let $\mathscr{F}$ denote the collection of all compact, convex sets in $\mathscr{E}^2$ with perimeter p (where p is a fixed positive number). Then (1) is the question, "Does there exist any set $\mathscr{K}_o$ in $\mathscr{F}$ such that the area of $\mathscr{K}_o$ is equal to or greater than that of $\mathscr{K}$ for all $\mathscr{K} \in \mathscr{F}$?" If we regard two sets in $\mathscr{E}^2$ as "essentially different" if they are not congruent, then (2) is the question, "How many essentially different sets are there in $\mathscr{F}$ that have the maximal area property of question (1)? Question (3) asks, "What characterizes the essentially different sets in $\mathscr{F}$ that have the maximal area property?"

It is rather natural to guess that the solution to the isoperimetric problem is a disc (a ball in $\mathscr{E}^2$) with perimeter p (hence with radius $p/2\pi$). It is now known, in fact, that this is a solution and it is the unique solution in the sense that every set in $\mathscr{F}$ that is not congruent to such a disc has smaller area than the disc. But if almost anyone can guess the solution, what made the problem famous? The answer is that the existence question was both subtle and difficult. To gain some insight about this difficulty, let $A(\mathscr{K})$ denote the area of set $\mathscr{K}$ and let

$$\mathscr{A} = \{A(\mathscr{K}): \mathscr{K} \in \mathscr{F}\}$$

be the set of area numbers for the sets in $\mathscr{F}$. It is not difficult to show that the set $\mathscr{A}$ is bounded above and hence that a least upper bound to $\mathscr{A}$, say the number a^*, exists. But does a^* belong to $\mathscr{A}$? That is just the existence question, "Is there a set $\mathscr{K}_o$ in $\mathscr{F}$ such that $A(\mathscr{K}_o) = a^*$?" In the nineteenth century, the famous geometer Jacob Steiner proved that if $\mathscr{K}$ in $\mathscr{F}$ is not a disc of perimeter p, then there exists $\mathscr{K}'$ in $\mathscr{F}$ such that $A(\mathscr{K}') > A(\mathscr{K})$. Thus he proved—actually in several different ways—that if there is a set in $\mathscr{F}$ with area a^*, it must be a disc of perimeter p. He believed that he had solved the isoperimetric problem, but he had proved only that if there is a solution, it must be a disc. He had not proved that there is a solution. The great mathematician G. F. B. Riemann made the same sort of slip in reasoning at one point in his work.

The foregoing remarks were not made to detract in any sense from the greatness of either Steiner or Riemann but simply to emphasize how subtle and difficult existence questions may be. As an indication that they may also be decisive, consider the following rather loosely described problem.

> A segment Sg$[AB]$ of length h is to be moved continuously in the plane so that in its final position it has been turned through $180°$, so B coincides with A and A coincides with B. A "turning set" $\mathscr{S}$ is defined to consist of all points that Sg$[AB]$ touches at some stage of the motion. Among all turning sets $\mathscr{S}$, find the set, or sets, of minimum area.

This problem was completely solved by A. S. Besicovitch when he proved that the existence question has a negative answer. Strange as it may seem, he proved that no matter how great h is, for any positive ε, however small, there exists a turning set $\mathscr{S}$ whose area is less than ε. In this case, zero is the greatest lower bound to the set of area numbers, but does not belong to the set, and the problem has no solution.

Having indicated the importance, and sometimes the difficulty, of existence questions, we want to show how the methods of the book can be used to settle the existence question for each of the three problems listed. We are not able to do so with the same degree of rigor, in all respects, as in the previous sections. An accurate modern treatment of area and arc length would require a good portion of the theory of Lebesgue measure and integration, which is outside the scope of the book. As a compromise, we assume certain necessary facts from that theory—ones that are quite plausible on intuitive grounds—but exercise the same care as before in the use of our methods in combination with these facts.

We now give assumptions about area and arc length that we will use.

ASSUMPTIONS. To each non-empty, compact, convex subset $\mathscr{K}$ of $\mathscr{E}^2$ there is associated a nonnegative number $A(\mathscr{K})$, the *area* of $\mathscr{K}$, and a nonnegative number $L(\mathscr{K})$, the *arc length* or *perimeter* of $Bd(\mathscr{K})$, with the following properties.

1. Congruent sets have the same area and arc length, that is, $\mathscr{K} \cong \mathscr{K}'$ implies that $A(\mathscr{K}) = A(\mathscr{K}')$ and $L(\mathscr{K}) = L(\mathscr{K}')$.
2. If $\mathscr{K}_1 \subset \mathscr{K}_2$, then $A(\mathscr{K}_1) \leq A(\mathscr{K}_2)$ and $L(\mathscr{K}_1) \leq L(\mathscr{K}_2)$.
3. $Dm(\mathscr{K}) \leq L(\mathscr{K})$.
4. The functions A and L are continuous in the Hausdorff metric on the domain $\mathscr{H}$-$\mathscr{E}^2$, that is, corresponding to $\mathscr{K}_o \in \mathscr{H}$-$\mathscr{E}^2$ and $\varepsilon > 0$ there exists a $\delta > 0$ such that $D(\mathscr{K}_o, \mathscr{K}) < \delta$ implies that $|A(\mathscr{K}_o) - A(\mathscr{K})| < \varepsilon$ and $|L(\mathscr{K}_o) - L(\mathscr{K})| < \varepsilon$.
5. If $\mathscr{K}$ is a basic figure such as a full triangle (2-simplex) or a full square (square 2-cell), the numbers $A(\mathscr{K})$ and $L(\mathscr{K})$ are given by the familiar formulas of elementary euclidean geometry.

We also need the following facts, which we need not assume.

THEOREM 1. Every isometry is a continuous, one-to-one mapping that preserves convexity and compactness. (E)

COROLLARY 1. A set that is congruent to a compact, convex set is itself a compact, convex set.

We now turn to the part that was missing from Steiner's "solution" to the isoperimetric problem, namely, the proof that the problem has a solution. The lemma simplifies the proof.

LEMMA 1. If $\mathscr{F}$ is the collection of all compact, convex sets in $\mathscr{E}^2$ with fixed, positive perimeter p, there exists a full square $\mathscr{W}$ such that for each $\mathscr{K} \in \mathscr{F}$ there exists $\mathscr{K}' \subset \mathscr{W}$ and $\mathscr{K}' \cong \mathscr{K}$.

PROOF. By Assumption 3, p is an upper bound to the diameter set of $\mathscr{F}$, that is, to the set $\{Dm(\mathscr{K}) : \mathscr{K} \in \mathscr{F}\}$. Let $\mathscr{W}$ be the full square

$$\mathscr{W} = \{(x_1, x_2) : -p \leq x_1 \leq p, -p \leq x_2 \leq p\}.$$

Clearly, the disc $B(O, p)$ is contained in $\mathscr{W}$. Now consider $\mathscr{K} \in \mathscr{F}$. Because $\mathscr{K} \neq \varnothing$, there exists $A \in \mathscr{K}$. For any point X in $\mathscr{K}$, $d(A, X) \leq Dm(\mathscr{K}) \leq p$ implies that $X \in B(A, p)$, hence $\mathscr{K} \subset B(A, p)$. The translation of $\mathscr{E}^2$ by the vector $-a$ maps $B(A, p)$ onto $B(O, p)$. It therefore maps $\mathscr{K} \subset B(A, p)$ onto an image set $\mathscr{K}' \subset B(O, p) \subset \mathscr{W}$, and $\mathscr{K}' \cong \mathscr{K}$ because translations are motions of $\mathscr{E}^2$. $\square$

THEOREM 2. The isoperimetric problem has a solution.

PROOF. Let $\mathscr{F}$ be the collection of all compact, convex sets in $\mathscr{E}^2$ with fixed, positive perimeter p. Let $\mathscr{G}$ be the collection of all non-empty, compact, convex subsets of $\mathscr{E}^2$. Thus $\mathscr{F}$ is a subset of $\mathscr{G}$, and both $\mathscr{F}$ and $\mathscr{G}$ are "point" sets in $\mathscr{H}$–$\mathscr{E}^2$. In particular, the "point" set $\mathscr{G}$, by Theorem 6, Sec. 6.3, is a closed set in $\mathscr{H}$–$\mathscr{E}^2$. The arc length function L is a mapping $\mathrm{L}: \mathscr{G} \to \mathscr{E}^1$ and, by Assumption 4, L is a continuous function. The singleton set $\{p\}$ is a closed set in $\mathscr{E}^1$. Thus since $\mathscr{F}$ is the subset of $\mathscr{G}$ defined by

$$\mathscr{F} = \{\mathscr{K} : \mathrm{L}(\mathscr{K}) = p\} = \{\mathscr{K} : \mathrm{L}(\mathscr{K}) \in \{p\}\},$$

it follows from Corollary 4.2, Sec. 1.5, that $\mathscr{F}$ is a closed "point" set in $\mathscr{H}$–$\mathscr{E}^2$.

Now let $\mathscr{W}$ be the full square of Lemma 1 and let $\mathscr{G}_1$ be the collection of all non-empty, compact, convex subsets of $\mathscr{W}$. By Corollary 6, Sec. 6.3, $\mathscr{G}_1$ as a "point" set in $\mathscr{H}$–$\mathscr{E}^2$ is compact in $\mathscr{H}$–$\mathscr{E}^2$. Thus the "point" set $\mathscr{F}_1$ in $\mathscr{H}$–$\mathscr{E}^2$ defined by $\mathscr{F}_1 = \mathscr{F} \cap \mathscr{G}_1$ is compact in $\mathscr{H}$–$\mathscr{E}^2$, since it is the intersection of the closed set $\mathscr{F}$ and the compact set $\mathscr{G}_1$ (Theorem 11, Sec. 1.4). By Assumption 4, the real valued area function A is continuous on $\mathscr{H}$–$\mathscr{E}^2$ and hence is continuous on $\mathscr{F}_1$. Since $\mathscr{F}_1$ is compact, the min–max theorem implies that A attains a maximum value at some "point" $\mathscr{K}_o$ in $\mathscr{F}_1$.

Let $\mathscr{K}$ denote an arbitrary set in $\mathscr{F}$. By Lemma 1, there exists $\mathscr{K}' \subset \mathscr{W}$ and $\mathscr{K}' \cong \mathscr{K}$. By Corollary 1, $\mathscr{K}'$ is compact and convex, so $\mathscr{K}' \in \mathscr{G}$. By Assumption 1, $\mathrm{L}(\mathscr{K}') = \mathrm{L}(\mathscr{K}) = p$, so $\mathscr{K}' \in \mathscr{F}$. Also, $\mathscr{K}' \subset \mathscr{W}$, so $\mathscr{K}' \in \mathscr{G}_1$. Thus $\mathscr{K} \in \mathscr{F} \cap \mathscr{G}_1 = \mathscr{F}_1$. Since A attains its maximum value on $\mathscr{F}_1$ at the "point" $\mathscr{K}_o$, then $\mathrm{A}(\mathscr{K}') \leq \mathrm{A}(\mathscr{K}_o)$. But, by Assumption 1, $\mathscr{K}' \cong \mathscr{K}$ implies $\mathrm{A}(\mathscr{K}') = \mathrm{A}(\mathscr{K})$. Therefore $\mathrm{A}(\mathscr{K}) \leq \mathrm{A}(\mathscr{K}_o)$. Hence $\mathscr{K}_o$ is a solution to the isoperimetric problem. □

Turning now to the minimum cover problem, the second in our list, let $\mathscr{F}$ denote the collection of all sets in $\mathscr{E}^2$ with a fixed, positive diameter h. The problem, as we stated it, is to find the compact, convex sets of least area that can cover every set in $\mathscr{F}$. But what does it mean to say that a set $\mathscr{R}$ "can cover" a set $\mathscr{S}$? In intuitive terms, it means that $\mathscr{R}$ either covers $\mathscr{S}$ ($\mathscr{R} \supset \mathscr{S}$) or else $\mathscr{R}$ can be moved to a new position where it covers $\mathscr{S}$, that is, some copy of $\mathscr{R}$ covers $\mathscr{S}$. To express this idea mathematically, we introduce some new terms. First, since the congruence of sets is reflexive, symmetric, and transitive, it is an equivalence relation. Thus if $\mathscr{S}$ is any set, it is rather natural to call the *collection of all sets congruent to* $\mathscr{R}$ the "congruence equivalence class of $\mathscr{R}$". We now define an "equivalence cover."

DEFINITION (Equivalence cover). Set $\mathscr{R}$ is an *equivalence cover* of set $\mathscr{S}$ if some set in the congruence equivalence class of $\mathscr{R}$ covers $\mathscr{S}$. Set $\mathscr{R}$ is an equivalence cover of a collection of sets $\mathscr{F}$ if it is an equivalence cover of each set in $\mathscr{F}$.

It is not difficult to see that the definition of an equivalence cover implies the following properties.

THEOREM 3. If $\mathscr{F}$ is a collection of sets and if $\mathscr{R}$ is an equivalence cover of $\mathscr{F}$, then every set that contains $\mathscr{R}$ or is congruent to $\mathscr{R}$ is also an equivalence cover of $\mathscr{F}$. (E)

THEOREM 4. If $\mathscr{F}$ is a collection of sets, set $\mathscr{R}$ is an equivalence cover of $\mathscr{F}$ if and only if for each $\mathscr{K} \in \mathscr{F}$ there is a corresponding subset of $\mathscr{R}$ that is congruent to $\mathscr{K}$. (E)

A slight but useful modification of the minimum cover problem follows from a simple observation. If a set $\mathscr{R}$ is closed, then $\mathscr{R}$ covers a set $\mathscr{S}$ if and only if it covers the closure of $\mathscr{S}$. Thus a closed set is an equivalence cover of all sets of diameter h if and only if it is an equivalence cover of all closed sets of diameter h. We can now restate the minimum cover problem in more precise form.

Minimum cover problem. Let $\mathscr{F}_h$ be the collection of all closed sets in $\mathscr{E}^2$ that have diameter h. Among all compact, convex equivalence covers of $\mathscr{F}_h$, find the covers of minimal area.

CONVENTION. By "cc-equivalence cover" we mean a compact, convex equivalence cover.

The notion of an equivalence cover, which is the heart of the minimum cover problem, depends on congruence, and this, in turn, depends on the motions of $\mathscr{E}^2$, namely, the isometries of $\mathscr{E}^2$ onto itself. We must establish some properties of these motions before we take up the problem itself. First, we recall from linear algebra that the motions of $\mathscr{E}^2$ that leave the origin fixed are mappings $x \to x'$ of the form

$$\begin{pmatrix} x_1' \\ x_2' \end{pmatrix} = \begin{pmatrix} a & b \\ c & d \end{pmatrix} \begin{pmatrix} x_1 \\ x_2 \end{pmatrix} \tag{1}$$

where

$$a^2 + c^2 = 1, \qquad b^2 + d^2 = 1, \qquad ab + cd = 0. \tag{2}$$

The composition of these motions with translations generate all the motions of $\mathscr{E}^2$. Thus each motion of $\mathscr{E}^2$ has a unique representation of the form

$$\begin{pmatrix} x_1' \\ x_2' \end{pmatrix} = \begin{pmatrix} a & b \\ c & d \end{pmatrix} \begin{pmatrix} x_1' \\ x_2' \end{pmatrix} + \begin{pmatrix} e \\ f \end{pmatrix}, \tag{3}$$

where a, b, c, d satisfy (2). Conversely, each mapping (3) that satisfies (2) is a motion of $\mathscr{E}^2$. For reasons that will soon be apparent, we state this characterization in the following form.

THEOREM 5. If $f: \mathscr{E}^2 \to \mathscr{E}^2$ is a motion of $\mathscr{E}^2$, then there exists a unique 6-tuple $(m_1, m_2, m_3, m_4, m_5, m_6)$ such that

$$f(x) = (m_1 x_1 + m_2 x_2 + m_3, m_4 x_1 + m_5 x_2 + m_6) \tag{1}$$

and

$$m_1^2 + m_4^2 = 1, \qquad m_2^2 + m_5^2 = 1, \qquad m_1 m_2 + m_4 m_5 = 0. \tag{2}$$

Each 6-tuple that satisfies (2) determines a function f, defined by (1), that is a motion of $\mathscr{E}^2$.

As Theorem 5 shows, there is a one-to-one correspondence between the motions of $\mathscr{E}^2$ and the subspace $\mathscr{M}$ of $\mathscr{E}^6$ defined by

$$\mathscr{M} = \{m \in \mathscr{E}^6 : m_1^2 + m_4^2 = 1, m_2^2 + m_5^2 = 1, m_1 m_2 + m_4 m_5 = 0\},$$

and we call $\mathscr{M}$ "the motion space of $\mathscr{E}^2$." If P is a point of $\mathscr{M}$, we denote the motion corresponding to P by f_p and

$$f_p(x) = (p_1 x_1 + p_2 x_2 + p_3, p_4 x_1 + p_5 x_2 + p_6).$$

Now let $\mathscr{M}_1 = \{m \in \mathscr{E}^6 : m_1^2 + m_4^2 = 1\}$, $\mathscr{M}_2 = \{m \in \mathscr{E}^6 : m_2^2 + m_5^2 = 1\}$, and $\mathscr{M}_3 = \{m \in \mathscr{E}^6 : m_1 m_2 + m_4 m_5 = 0\}$. The mapping g of $\mathscr{E}^6$ into $\mathscr{E}^1$ defined by $g(m) = m_1^2 + m_4^2$ is clearly continuous. Because the set $\{1\}$ in the range space $\mathscr{E}^1$ is closed, and the domain $\mathscr{E}^6$ is closed, then, by Corollary 4.2, Sec. 1.5, $\mathscr{M}_1 = \{m \in \mathscr{E}^6 : g(m) = 1\}$ is a closed set. By the same reasoning, $\mathscr{M}_2$ and $\mathscr{M}_3$ are closed sets. Therefore $\mathscr{M} = \mathscr{M}_1 \cap \mathscr{M}_2 \cap \mathscr{M}_3$ is closed. We state this fact for reference.

THEOREM 6. The motion space $\mathscr{M}$ of $\mathscr{E}^2$ is a closed subset of $\mathscr{E}^6$.

We can use the motion space $\mathcal{M}$ to obtain a simple representation of the congruence equivalence class of any set $\mathcal{S}$ in $\mathcal{E}^2$. First, if f is any motion, then $f(\mathcal{S}) \cong \mathcal{S}$, by definition. But also by definition, if $\mathcal{R}$ is any set congruent to $\mathcal{S}$, then there is a motion f such that $f(\mathcal{S}) \cong \mathcal{R}$. Thus the congruence equivalence class of $\mathcal{S}$ is simply the collection of all the images of $\mathcal{S}$ under the motions of $\mathcal{E}^2$. If we denote this equivalence class by $\mathcal{M}(\mathcal{S})$, then

$$\mathcal{M}(\mathcal{S}) = \{f_m(\mathcal{S}) : m \in \mathcal{M}\}. \tag{1}$$

In particular, if $\mathcal{S}$ is a non-empty compact set, then it is a "point" in the space $\mathcal{H}$–$\mathcal{E}^2$ and so is every set in $\mathcal{M}(\mathcal{S})$. If $\mathcal{R}$ is a set in $\mathcal{M}(\mathcal{S})$, it is a "point" in $\mathcal{H}$–$\mathcal{E}^2$, and it is also, in $\mathcal{E}^2$, the f_p-image of $\mathcal{S}$ for some $P \in \mathcal{M}$. Thus we may regard the compact set $\mathcal{S}$ as determining a mapping of $\mathcal{M}$ into $\mathcal{H}$–$\mathcal{E}^2$ via a function $f_{\mathcal{S}}$ defined by

$$f_{\mathcal{S}}(P) = f_p(\mathcal{S}), \qquad P \in \mathcal{M}. \tag{2}$$

Since the range of $f_{\mathcal{S}}$ is just the congruence equivalence class of $\mathcal{S}$, we call $f_{\mathcal{S}}$ "the equivalence class function corresponding to $\mathcal{S}$."

If the euclidean distance between points P and Q in the motion space $\mathcal{M}$ is small, it means that the corresponding numerical coefficients in the mappings f_p and f_q are nearly the same. One would therefore expect the congruent sets $f_p(\mathcal{S})$ and $f_q(\mathcal{S})$ to be close to each other, and hence $\mathrm{D}[f_p(\mathcal{S}), f_q(\mathcal{S})]$ to be small. This continuity expectation is verified in our next theorem.

THEOREM 7. If $\mathcal{S}$ is a non-empty compact set in $\mathcal{E}^2$, then the equivalence class function $f_{\mathcal{S}}$ corresponding to $\mathcal{S}$ is a continuous mapping of the motion space $\mathcal{M}$ of $\mathcal{E}^2$ into $\mathcal{H}$–$\mathcal{E}^2$.

PROOF. First, we observe that since $\mathcal{S}$ is compact it is bounded, so there exists a neighborhood of the origin, say $\mathrm{N}(O, a)$, such that $\mathcal{S} \subset \mathrm{N}(O, a)$. Then, for $X \in \mathcal{S}$,

$$|x_1| + |x_2| + 1 \le \sqrt{x_1^2 + x_2^2} + 1 < a + 1. \tag{1}$$

Now let P be an arbitrary point of $\mathcal{M}$ and let ε be any positive number. The continuity of $f_{\mathcal{S}}$ at P (and hence on $\mathcal{M}$) is established if we can show the existence of $\delta > 0$ such that $Q \in \mathcal{M}$ and $\mathrm{d}(P, Q) < \delta$ imply that $\mathrm{D}[f_{\mathcal{S}}(P), f_{\mathcal{S}}(Q)] = \mathrm{D}[f_p(\mathcal{S}), f_q(\mathcal{S})] < \varepsilon$. Define δ by $\delta = \varepsilon/\sqrt{2}(a + 1)$, and let Q be a point of $\mathcal{M}$ such that $\mathrm{d}(P, Q) < \delta$. Then if

$$b = \max\{|p_1 - q_1|, |p_2 - q_2|, \ldots, |p_6 - q_6|\}, \tag{2}$$

we have

$$b \leq \left[\sum_{i=1}^{6} |p_i - q_i|^2 \right]^{1/2} = \mathrm{d}(P, Q) < \delta. \tag{3}$$

Next, consider $X \in \mathscr{S}$. Then

$$\begin{aligned} f_p(x) &= (p_1 x_1 + p_2 x_2 + p_3, p_4 x_1 + p_5 x_2 + p_6), \\ f_q(x) &= (q_1 x_1 + q_2 x_2 + q_3, q_4 x_1 + q_5 x_2 + q_6) \end{aligned} \tag{4}$$

so

$$\begin{aligned} \mathrm{d}[f_p(x), f_q(x)] = \{&[(p_1 - q_1)x_1 + (p_2 - q_2)x_2 + (p_3 - q_3)]^2 \\ &+ [(p_4 - q_4)x_1 + (p_5 - q_5)x_2 + (p_6 - q_6)]^2\}^{1/2}. \end{aligned} \tag{5}$$

Then (using $\sqrt{c^2 + d^2} \leq \sqrt{2} \max\{|c|, |d|\}$), (5) implies that

$$\begin{aligned} \mathrm{d}[f_p(x), f_q(x)] &\leq \sqrt{2} \max\{|(p_1 - q_1)x_1 + (p_2 - q_2)x_2 + (p_3 - q_3)|, \\ &\qquad |(p_4 - q_4)x_1 + (p_5 - q_5)x_2 + (p_6 - q_6)|\} \\ &\leq \sqrt{2} \max\{|p_1 - q_1||x_1| + |p_2 - q_2||x_2| + |p_3 - q_3|, \\ &\qquad |p_4 - q_4||x_1| + |p_5 - q_5||x_2| + |p_6 - q_6|\} \\ &\leq \sqrt{2} \max\{b|x_1| + b|x_2| + b, b|x_1| + b|x_2| + b\} \\ &\leq \sqrt{2}\, b[|x_1| + |x_2| + 1]. \end{aligned} \tag{6}$$

With (1) and the definition of δ, (6) implies that

$$\mathrm{d}[f_p(x), f_q(x)] \leq \sqrt{2}\, b[|x_1| + |x_2| + 1] < \sqrt{2}\, \delta(a + 1) = \varepsilon. \tag{7}$$

The calculations just made show that if $\mathrm{d}(P, Q) < \delta$ and $X \in \mathscr{S}$, then $\mathrm{d}[f_p(x), f_q(x)] < \varepsilon$. Since X is any point of $\mathscr{S}$, it follows that each point of $f_p(\mathscr{S})$ is at a distance less than ε from a point of $f_q(\mathscr{S})$. Similarly, each point of $f_q(\mathscr{S})$ is at a distance less than ε from some point of $f_p(\mathscr{S})$. This is just another way of saying that $\mathrm{D}[f_p(\mathscr{S}), f_q(\mathscr{S})] < \varepsilon$. □

With Theorem 7, we can establish a rather fundamental fact, namely, that if a non-empty, compact set $\mathscr{K}_o$ is a limit figure, in the Hausdorff sense, to a collection of mutually congruent sets, then $\mathscr{K}_o$ is congruent to these sets. For example, a limit figure to a collection of discs of radius r is itself a disc of radius r.

THEOREM 8. If $\mathscr{S}$ is a non-empty, compact set in $\mathscr{E}^2$, its congruence equivalent class $\mathscr{M}(\mathscr{S})$ is a closed "point" set in $\mathscr{H}$-$\mathscr{E}^2$.

PROOF. Let $\mathscr{K}_o$ be a limit "point" in the Hausdorff sense to the collection $\mathscr{M}(\mathscr{S})$. We want to show that there exists $P_o \in \mathscr{M}$ such that $f_{p_o}(\mathscr{S}) = \mathscr{K}_o$.

Because the set $\mathscr{K}_o$ in $\mathscr{E}^2$ is a "point" of $\mathscr{H}$-$\mathscr{E}^2$, the set $\mathscr{K}_o$ is compact and hence bounded. The set $\mathscr{S}$ is compact, so it too is bounded. Thus there exists some neighborhood of the origin, say $\mathrm{N}(O, r)$, such that

$$\mathscr{S} \subset \mathrm{N}(O, r) \qquad \text{and} \qquad \mathscr{K}_o \subset \mathrm{N}(O, r). \tag{1}$$

Now consider the Hausdorff neighborhood of $\mathscr{K}_o$ with radius 1, that is, the collection $\mathscr{N}$ of compact sets $\mathscr{K}$ in $\mathscr{E}^2$ defined by

$$\mathscr{N} = \{\mathscr{K} : \mathrm{D}(\mathscr{K}_o, \mathscr{K}) < 1\}.$$

Because $\mathscr{K}_o$ is a limit "point" to $\mathscr{M}(\mathscr{S})$, then $\mathscr{N} \cap \mathscr{M}(\mathscr{S}) \neq \varnothing$. Let $P \in \mathscr{M}$ be such that

$$f_p(\mathscr{S}) \in \mathscr{N} \cap \mathscr{M}(\mathscr{S}). \tag{2}$$

To see how the relation (2) restricts P, let Z be an arbitrary point of $\mathscr{S}$. Then

$$f_p(z) = (p_1 z_1 + p_2 z_2 + p_3, p_4 z_1 + p_5 z_2 + p_6). \tag{3}$$

The point $Q: (p_1, p_2, 0, p_4, p_5, 0)$ is also in $\mathscr{M}$, and

$$f_q(z) = (p_1 z_1 + p_2 z_2, p_4 z_1 + p_5 z_2). \tag{4}$$

Because f_q is a motion of $\mathscr{E}^2$ that leaves the origin fixed,

$$|f_q(z)| = |z|. \tag{5}$$

From (3) and (4), $(p_3, p_6) = f_p(z) - f_q(z)$, so

$$|(p_3, p_6)| = |f_p(z) - f_q(z)| \leq |f_p(z)| + |f_q(z)|,$$

which, with (5), gives

$$|(p_3, p_6)| \leq |f_p(z)| + |z|. \tag{6}$$

Because $Z \in \mathscr{S} \subset \mathrm{N}(O, r)$, (6) implies that

$$|(p_3, p_6)| \leq |f_p(z)| + r. \tag{7}$$

Next, because $Z \in \mathscr{S}$, it follows that $f_p(z) \in f_p(\mathscr{S})$ and, from (2), $f_p(\mathscr{S})$ is in $\mathscr{N}$, so $D[f_p(\mathscr{S}), \mathscr{K}_o] < 1$. Thus there is some point Y in $\mathscr{K}_o$ whose distance from $f_p(z)$ is less than 1. By ordinary norm properties,

$$|f_p(z)| = |f_p(z) - y + y| \leq |f_p(z) - y| + |y|. \tag{8}$$

Because $|f_p(z) - y| < 1$, and $Y \in \mathscr{K}_o \subset N(O, r)$, (8) implies that

$$|f_p(z)| < 1 + r. \tag{9}$$

Thus, from (7) and (9) we have

$$\sqrt{p_3^2 + p_6^2} \leq 1 + 2r. \tag{10}$$

Setting $k = 1 + 2r$, it follows from (10) that if $f_p(\mathscr{S})$ in $\mathscr{M}(\mathscr{S})$ has Hausdorff distance less than 1 from $\mathscr{K}_o$, then

$$|p| = \left[\sum_{i=1}^{6} p_i^2\right]^{1/2} = [2 + p_3^2 + p_6^2]^{1/2} \leq \sqrt{2 + k^2}. \tag{11}$$

Now, let $\mathscr{M}_k$ be the subset of $\mathscr{M}$ defined by

$$\mathscr{M}_k = \{P: P \in \mathscr{M}, |p| \leq \sqrt{2 + k^2}\}. \tag{12}$$

The set $\mathscr{M}_k$ is just the ball $B(O, |p|)$ in $\mathscr{E}^6$. Such balls are compact sets (cf Exercise 3, Sec. 1.8), so $\mathscr{M}_k$ is compact. By Theorem 7, the equivalence class function $f_{\mathscr{S}}$ corresponding to $\mathscr{S}$ is a continuous mapping of $\mathscr{M}$ into $\mathscr{H}$–$\mathscr{E}^2$. Therefore its restriction to $\mathscr{M}_k$ is also continuous, and since $\mathscr{M}_k$ is compact, the range of the restriction is a compact set in $\mathscr{H}$–$\mathscr{E}^2$. This range is just $\mathscr{M}_k(\mathscr{S})$, that is, the sets $f_p(\mathscr{S})$ for $P \in \mathscr{M}_k$. Because $\mathscr{M}_k(\mathscr{S})$ is the subset of $\mathscr{M}(\mathscr{S})$ consisting of the sets in $\mathscr{M}(\mathscr{S})$ at distance 1 or less from $\mathscr{K}_o$, then $\mathscr{K}_o$ is also a limit "point" to $\mathscr{M}_k(\mathscr{S})$. Because $\mathscr{M}_k(\mathscr{S})$ is closed, $\mathscr{K}_o \in \mathscr{M}_k(\mathscr{S})$. Thus there exists $P_o \in \mathscr{M}_k$ such that $f_{p_o}(\mathscr{S}) = \mathscr{K}_o$. □

Returning now to the minimum cover problem, we can describe a strategy for a proof that the problem has a solution. Let $\mathscr{F}_h$ be the collection of all closed sets in $\mathscr{E}^2$ with diameter h. We want to find a collection $\mathscr{G}$ of compact, convex equivalence covers of $\mathscr{F}_h$ such that

(i) every cc-equivalence cover not in $\mathscr{G}$ is either congruent to one in $\mathscr{G}$ or has greater area than one in $\mathscr{G}$

(ii) the collection $\mathscr{G}$ is a compact "point" set in $\mathscr{H}$–$\mathscr{E}^2$

Property (i) establishes that if $\mathscr{G}$ has a set of minimal area, it is a solution to the problem. Since the area function A is continuous on $\mathscr{H}$-$\mathscr{E}^2$, by Assumption 2, it is continuous on $\mathscr{G}$. Thus property (ii) and the min–max theorem establish that $\mathscr{G}$ does have a set of minimal area. Our next theorem is a first step in this program.

THEOREM 9. Let $\mathscr{F}_h$ denote the collection of all closed sets in $\mathscr{E}^2$ with diameter h. There exists a cc-equivalence cover of $\mathscr{F}_h$ that has area $4h^2$. Moreover, if $\mathscr{W}$ is the full square

$$\mathscr{W} = \{(x_1, x_2): -16h \leq x_1 \leq 16h, -16h \leq x_2 \leq 16h\},$$

then every cc-equivalence cover of $\mathscr{F}_h$ that has area equal to or less than $4h^2$ has diameter equal to or less than $16h$ and is either a subset of $\mathscr{W}$ or congruent to a subset of $\mathscr{W}$.

PROOF. First consider the full square

$$\mathscr{W}_1 = \{(x_1, x_2): -h \leq x_1 \leq h, -h \leq x_2 \leq h\}.$$

Since $\mathscr{W}_1$ is the intersection of closed half-planes, it is convex and, by Theorem 5, Sec. 1.7, it is compact. By Assumption 4, the area of $\mathscr{W}_1$ is $4h^2$.

Now let $\mathscr{S}$ be any set in $\mathscr{F}_h$. Because $\text{Dm}(\mathscr{S}) = h > 0$, $\mathscr{S} \neq \varnothing$, so there exists $P \in \mathscr{S}$. The translation of $\mathscr{E}^2$ by $-p$ maps $\mathscr{S}$ to a set $\mathscr{S}'$, and $\mathscr{S}' \cong \mathscr{S}$, since translations are motions. If $X \in \mathscr{S}'$, then $\text{d}(O, X) \leq \text{Dm}(\mathscr{S}') = \text{Dm}(\mathscr{S}) = h$ implies that $X \in \text{B}(O, h)$. Thus $\mathscr{S}' \subset \text{B}(O, h)$. Clearly, $\text{B}(O, h) \subset \mathscr{W}_1$, so $\mathscr{S}' \subset \mathscr{W}_1$. Thus, by Theorem 4, $\mathscr{W}_1$ is an equivalence cover of $\mathscr{F}_h$.

Next let $\mathscr{K}$ be a cc-equivalence cover of $\mathscr{F}_h$ such that $\text{A}(\mathscr{K}) \leq 4h^2$, and let $k = \text{Dm}(\mathscr{K})$. Because $\mathscr{K}$ is closed, it contains a diameter segment $\text{Sg}[CD]$ of length k. Let $X' = f(X)$ be a motion of $\mathscr{E}^2$ that maps C to $C' = (-k/2, O)$ and D to $D' = (k/2, O)$. By Theorem 3, $\mathscr{K}' = f(\mathscr{K})$ is also a cc-equivalence cover of $\mathscr{F}_h$, and, since motions preserve areas and diameters, $\text{A}(\mathscr{K}') \leq 4h^2$ and $\text{Dm}(\mathscr{K}') = k$. Because the collection $\mathscr{F}_h$ contains all discs of diameter h (the balls in $\mathscr{E}^2$ of radius $h/2$), it follows from Theorem 4 that $\mathscr{K}'$ contains such a disc. This disc must contain a point (p_1, p_2) such that $|p_2| \geq h/2$, otherwise the disc would lie entirely in the open strip $\{(x_1, x_2): -h/2 < x_2 < h/2\}$ and its width in the direction of the x_2-axis would be less than h. Since $\mathscr{K}'$ contains the points C', D' and P, it contains the full triangle $\mathscr{T}$ with these vertices. Since $\text{A}(\mathscr{T}) = \frac{1}{2}\text{d}(C', D')|p_2| \geq \frac{1}{2}k\, h/2$, we have

$$4h^2 \geq \text{A}(\mathscr{K}') \geq \text{A}(\mathscr{T}) \geq \tfrac{1}{4}kh,$$

which implies that

$$k = \text{Dm}(\mathscr{K}') = \text{Dm}(\mathscr{K}) \leq 16h.$$

The origin O is in $\mathscr{K}'$ and the distance of O from all four sides of $\mathscr{W}$ is $16h$. Thus $X \in \mathrm{Cp}(\mathscr{W})$ and $X \in \mathscr{K}'$ would imply $\mathrm{d}(O, X) > 16h$, contradicting $\mathrm{Dm}(\mathscr{K}') \leq 16h$. Hence $\mathscr{K}' \subset \mathscr{W}$. Thus $\mathscr{K}$ is either a subset of $\mathscr{W}$ or congruent to $\mathscr{K}'$ which is a subset of $\mathscr{W}$. □

COROLLARY 9. If $\mathscr{G}$ is the collection of all cc-equivalence covers of $\mathscr{F}_h$ that are contained in $\mathscr{W}$ and have diameters equal to or less than $16h$, then any set in $\mathscr{G}$ of minimal area is a solution to the minimum cover problem.

PROOF. Assume that there exists $\mathscr{K}_o \in \mathscr{G}$ such that $\mathrm{A}(\mathscr{K}_o) \leq \mathrm{A}(\mathscr{K})$ for all $\mathscr{K} \in \mathscr{G}$. Let $\mathscr{R}$ be a cc-equivalence cover of $\mathscr{F}_h$. If $\mathrm{A}(\mathscr{R}) \leq 4h^2$, then, by Theorem 9, $\mathrm{Dm}(\mathscr{R}) \leq 16h$ and either $\mathscr{R} \subset \mathscr{W}$ or $\mathscr{R} \cong \mathscr{R}'$ and $\mathscr{R}' \subset \mathscr{W}$. In the first of these cases, $\mathscr{R} \in \mathscr{G}$, so $\mathrm{A}(\mathscr{K}_o) \leq \mathrm{A}(\mathscr{R})$. In the second case, $\mathscr{R}' \in \mathscr{G}$ and $\mathrm{A}(\mathscr{K}_o) \leq \mathrm{A}(\mathscr{R}') = \mathrm{A}(\mathscr{R})$. If $\mathrm{A}(\mathscr{R}) > 4h^2$, then $\mathrm{A}(\mathscr{R}) > \mathrm{A}(\mathscr{W}_1) = 4h^2$. But $\mathscr{W}_1 \subset \mathscr{W}$ and $\mathrm{Dm}(\mathscr{W}_1) = \sqrt{8}h < 16h$, so $\mathscr{W}_1 \in \mathscr{G}$. Therefore $\mathrm{A}(\mathscr{K}_o) \leq \mathrm{A}(\mathscr{W}_1) < \mathrm{A}(\mathscr{R})$. □

By the strategy discussed prior to Theorem 9, if we can show that the collection $\mathscr{G}$ in Corollary 9 is compact in $\mathscr{H}$–$\mathscr{E}^2$, we have the proof we seek. In establishing this compactness, we need the following lemmas.

LEMMA 2. The mapping $f: \mathscr{H}\text{–}\mathscr{E}^2 \to \mathscr{E}^1$ defined by $f(\mathscr{K}) = \mathrm{Dm}(\mathscr{K})$, $\mathscr{K} \in \mathscr{H}\text{–}\mathscr{E}^2$, is continuous in the Hausdorff metric. (E)

LEMMA 3. If $\mathscr{S}$ is a closed subset of the full square $\mathscr{W}$ in Theorem 9 and if $\mathscr{H}(\mathscr{S})$ is the collection of all compact, convex subsets of $\mathscr{W}$ that contain $\mathscr{S}$, then $\mathscr{H}(\mathscr{S})$ is a compact "point" set in $\mathscr{H}$–$\mathscr{E}^2$.

PROOF. Let $\mathscr{K}_o$ in $\mathscr{E}^2$ be a Hausdorff limit "point" to the "point" set $\mathscr{H}(\mathscr{S})$ in $\mathscr{H}$–$\mathscr{E}^2$, and let $\mathscr{C}^*$ be the collection of all the compact, convex subsets of $\mathscr{W}$. Since $\mathscr{H}(\mathscr{S}) \subset \mathscr{C}^*$, $\mathscr{K}_o$ is a Hausdorff limit "point" to $\mathscr{C}^*$. By Corollary 6, Sec. 6.3, $\mathscr{C}^*$ is compact in $\mathscr{H}$–$\mathscr{E}^2$, so $\mathscr{K}_o \in \mathscr{C}^*$. Thus, in $\mathscr{E}^2$, $\mathscr{K}_o$ is a compact, convex subset of $\mathscr{W}$.

Now consider $X \in \mathscr{S}$ and an arbitrary neighborhood of X, say $\mathrm{N}(X, \varepsilon)$. Because $\mathscr{K}_o$ is a Hausdorff limit "point" to $\mathscr{H}(\mathscr{S})$, then corresponding to $\varepsilon > 0$ there exists $\mathscr{K} \in \mathscr{H}(\mathscr{S})$ such that $\mathrm{D}(\mathscr{K}_o, \mathscr{K}) < \varepsilon$. Therefore $\mathscr{S} \subset \mathscr{K} \subset \mathscr{K}_o + \mathrm{N}(O, \varepsilon)$, so there exists some point $Y \in \mathscr{K}_o$ such that $X \in \mathrm{N}(Y, \varepsilon)$. Then $Y \in \mathrm{N}(X, \varepsilon)$, so $\mathrm{N}(X, \varepsilon) \cap \mathscr{K}_o \neq \varnothing$. Because every neighborhood of X intersects $\mathscr{K}_o$, then $X \in \mathrm{Cl}(\mathscr{K}_o)$. Since $\mathscr{K}_o$ is closed, $X \in \mathscr{K}_o$. Thus $X \in \mathscr{S}$ implies $X \in \mathscr{K}_o$, so $\mathscr{S} \subset \mathscr{K}_o$. Because $\mathscr{K}_o$ is a compact, convex subset of $\mathscr{W}$ that contains $\mathscr{S}$, $\mathscr{K}_o \in \mathscr{H}(\mathscr{S})$. Therefore $\mathscr{H}(\mathscr{S})$ contains its Hausdorff limit "points" and so is closed in $\mathscr{H}$–$\mathscr{E}^2$. Because $\mathscr{H}(\mathscr{S})$ is a closed subset of $\mathscr{C}^*$, which is compact in $\mathscr{H}$–$\mathscr{E}^2$, then, by Theorem 11, Sec. 1.4, $\mathscr{H}(\mathscr{S})$ is compact in $\mathscr{H}$–$\mathscr{E}^2$. □

We now have all the properties we need to settle the existence question for the minimum cover problem.

THEOREM 10. There exists a solution to the minimum cover problem.

PROOF. Let $\mathscr{F}_h$ be the collection of all closed sets in $\mathscr{E}^2$ with diameter h, $\mathscr{W}$ the full square of Theorem 9, and $\mathscr{G}$ the collection of all cc-equivalence covers of $\mathscr{F}_h$ that have diameters equal to or less than $16h$ and are subsets of $\mathscr{W}$. We want to show that $\mathscr{G}$ is closed in $\mathscr{H}$-$\mathscr{E}^2$, so we consider any set $\mathscr{K}$ that is a Hausdorff limit "point" to $\mathscr{G}$. Because $\mathscr{G} \subset \mathscr{C}^*$, the set of all compact, convex subsets of $\mathscr{W}$, $\mathscr{K}$ is a Hausdorff limit "point" to $\mathscr{C}^*$. By Theorem 6, Sec. 6.3, $\mathscr{C}^*$ is compact in $\mathscr{H}$-$\mathscr{E}^2$, so $\mathscr{K} \in \mathscr{C}^*$. Therefore, in $\mathscr{E}^2$, $\mathscr{K}$ is a compact, convex subset of $\mathscr{W}$. Also, because $\mathscr{K}$ is a Hausdorff limit "point" to $\mathscr{G}$, and all the sets in $\mathscr{G}$ have diameters equal to or less than $16h$, it follows from Lemma 2 that $\mathrm{Dm}(\mathscr{K}) \leq 16h$.

Now, let $\mathscr{R}$ be an arbitrary set in $\mathscr{F}_h$. Then there exists some v-translate of $\mathscr{R}$ that maps $\mathscr{R}$ to a set $\mathscr{S}$ in $\mathscr{F}_h$ such that $\mathscr{S}$ contains the origin O. Clearly, $\mathscr{S} \subset \mathscr{W}$. Let $\mathscr{H}(\mathscr{S})$ be the collection of all compact, convex subsets of $\mathscr{W}$ that contain $\mathscr{S}$ and let $\mathscr{M}(\mathscr{K})$ be the congruence equivalence class of $\mathscr{K}$. We contend that $\mathscr{H}(\mathscr{S}) \cap \mathscr{M}(\mathscr{K}) \neq \varnothing$. For an indirect proof,

$$\text{assume that } \mathscr{H}(\mathscr{S}) \cap \mathscr{M}(\mathscr{K}) = \varnothing. \tag{*}$$

By Lemma 3, $\mathscr{H}(\mathscr{S})$ is compact in $\mathscr{H}$-$\mathscr{E}^2$ and, by Theorem 8, $\mathscr{M}(\mathscr{K})$ is closed in $\mathscr{H}$-$\mathscr{E}^2$. Therefore, by Corollary 7.2, Sec. 1.8, the (*) assumption implies that the Hausdorff nearness of $\mathscr{H}(\mathscr{S})$ and $\mathscr{M}(\mathscr{K})$ is some positive number n, that is,

$$\mathrm{D}(\mathscr{T}_1, \mathscr{T}_2) \geq n \ \text{ for } \mathscr{T}_1 \in \mathscr{H}(\mathscr{S}), \mathscr{T}_2 \in \mathscr{M}(\mathscr{K}). \tag{1}$$

Because $\mathscr{K}$ is a Hausdorff limit "point" to $\mathscr{G}$, the n-Hausdorff neighborhood of $\mathscr{K}$ intersects $\mathscr{G}$, hence there exists $\mathscr{K}' \in \mathscr{G}$ such that

$$\mathrm{D}(\mathscr{K}, \mathscr{K}') < n. \tag{2}$$

Since $\mathscr{K}'$ is in $\mathscr{G}$, it is an equivalence cover of the set $\mathscr{S}$ in $\mathscr{F}_h$. Therefore there exists a motion f of $\mathscr{E}^2$ such that $f(\mathscr{K}')$ contains $\mathscr{S}$. Because $f(\mathscr{K}) \cong \mathscr{K}'$, then $f(\mathscr{K}')$ is a compact, convex set with diameter equal to or less than $16h$, and since $O \in \mathscr{S} \subset f(\mathscr{K}')$, then $f(\mathscr{K}') \subset \mathscr{W}$. Therefore $f(\mathscr{K}') \in \mathscr{H}(\mathscr{S})$. The motion f also maps $\mathscr{K}$ to a set $f(\mathscr{K})$ that is congruent to $\mathscr{K}$, hence $f(\mathscr{K}) \in \mathscr{M}(\mathscr{K})$. Because the motion f of $\mathscr{E}^2$ preserves the Hausdorff distance between sets in $\mathscr{E}^2$, we have

$$\mathrm{D}[f(\mathscr{K}'), f(\mathscr{K})] = \mathrm{D}[\mathscr{K}', \mathscr{K}] < n. \tag{3}$$

But, since $f(\mathscr{K}') \in \mathscr{H}(\mathscr{S})$ and $f(\mathscr{K}) \in \mathscr{M}(\mathscr{K})$, (3) is a contradiction of (1). This contradiction shows that the (*) assumption cannot hold, so there exists a set $\mathscr{K}^*$ such that

$$\mathscr{K}^* \in \mathscr{H}(\mathscr{S}) \cap \mathscr{M}(\mathscr{K}). \tag{4}$$

Since the set $\mathscr{K}^*$ in (4) is in $\mathscr{M}(\mathscr{K})$, it is congruent to $\mathscr{K}$. Because $\mathscr{K}^*$ is in $\mathscr{H}(\mathscr{S})$, $\mathscr{K}^* \supset \mathscr{S}$. Therefore the $-v$ translation of $\mathscr{E}^2$, which maps $\mathscr{S}$ onto $\mathscr{R}$, maps $\mathscr{K}^*$ to a set that contains $\mathscr{R}$ and is congruent to $\mathscr{K}^*$ and therefore to $\mathscr{K}$. Thus $\mathscr{K}$ is an equivalence cover of $\mathscr{R}$ for every $\mathscr{R} \in \mathscr{F}_h$, so $\mathscr{K}$ is an equivalence cover of $\mathscr{F}_h$. Therefore $\mathscr{K} \in \mathscr{G}$, and since $\mathscr{G}$ contains its Hausdorff limit "points," $\mathscr{G}$ is closed in $\mathscr{H}$-$\mathscr{E}^2$. Thus $\mathscr{G}$ is a closed subset of $\mathscr{C}^*$, which is compact in $\mathscr{H}$–$\mathscr{E}^2$, so $\mathscr{G}$ is compact in $\mathscr{H}$–$\mathscr{E}^2$. Because the area function A is continuous on $\mathscr{H}$–$\mathscr{E}^2$, it is continuous on $\mathscr{G}$. Because $\mathscr{G}$ is compact, it follows from the min–max theorem that A attains a minimum value at some set $\mathscr{K}_o \in \mathscr{G}$. By Corollary 9, $\mathscr{K}_o$ is a solution to the minimum cover problem. □

It is interesting to compare the progress toward a complete solution of the isoperimetric problem with that of the minimum cover problem. In the case of the isoperimetric problem, what was first established was that at most one solution was possible and that the disc was the only candidate for this solution. Later it was proved that a solution did exist, so a complete answer to the problem was obtained. In the case of the minimum cover problem, it is known, as Theorem 10 shows, that a solution does exist. However, at present it is not known how many solutions there are nor the nature of any particular solution. It was proved by H. W. E. Jung that among discs that are equivalence covers of $\mathscr{F}_h$ the disc of radius $h/\sqrt{3}$ has minimal area.* Later it was shown that the full regular hexagon (of side length $h/\sqrt{3}$) inscribed in the minimal disc is also an equivalence cover of $\mathscr{F}_h$. How much smaller a cc-equivalence cover of $\mathscr{F}_h$ can be is an open question.

Turning to the last of our three problems, the maximum inclusion problem, let $\mathscr{F}_h$ now denote the collection of all compact convex sets in $\mathscr{E}^2$ whose minimum width is h. The problem asks for the compact convex set of maximum area that will fit inside every set in $\mathscr{F}_h$. To make "will fit" more precise, we again introduce an equivalence concept.

DEFINITION (Equivalence subset). Set $\mathscr{R}$ is an *equivalence subset* of set $\mathscr{S}$ if some set in the congruence equivalence class of $\mathscr{R}$ is a subset of $\mathscr{S}$. Set $\mathscr{R}$ is an equivalence subset of a collection of sets $\mathscr{F}$ if $\mathscr{R}$ is an equivalence subset of each set in $\mathscr{F}$.

We can now restate the problem.

* See Exercise 3, Sec. 5.1.

Maximum inclusion problem. Let $\mathscr{F}_h$ be the collection of all compact, convex sets in $\mathscr{E}^2$ with minimum width h. Among all compact, convex equivalence subsets of $\mathscr{F}_h$, which ones have maximum area?

The present state of affairs regarding the maximum inclusion problem is the same as that for the minimum cover problem. That is, it is known that a solution to the maximum inclusion problem exists, but neither the number of solutions nor the nature of any particular solution has been established. It was proved by W. Blaschke that among discs that are equivalence subsets of $\mathscr{F}_h$ the disc of radius $h/3$ has maximum area,* but just how much larger an equivalence subset can be is an open question.

The pattern used for the proof of Theorem 10 can be followed to establish our last theorem.

THEOREM 11. A solution to the maximum inclusion problem exists. (E)

Exercises – Section 6.4

1. Prove Theorem 1.
2. *Dual isoperimetric problem*: Among all compact, convex sets in $\mathscr{E}^2$ with fixed, positive area a_o, find the sets of minimum perimeter. To establish that this dual problem has a solution (without using Theorem 2), an extra assumption, such as the following, is needed. Assumption 6. If $\mathscr{F}_{a_o}$ is the collection of all compact, convex sets in $\mathscr{E}^2$ with fixed positive area a_o, there exists a full square $\mathscr{W}$ and a positive number p such that if $\mathscr{K} \in \mathscr{F}_{a_o}$ and $L(\mathscr{K}) \leq p$, then there exists $\mathscr{K}' \subset \mathscr{W}$ and $\mathscr{K}' \cong \mathscr{K}$. Using this assumption, prove that the dual isoperimetric problem has a solution. (It is, of course, a disc with radius $\sqrt{a_o/\pi}$.)
3. Prove Theorem 3.
4. Prove Theorem 4.
5. Prove Lemma 2.
6. If $\mathscr{R}$ and $\mathscr{S}$ are non-empty, compact sets in $\mathscr{E}^n$ and if $X' = f(X)$ is a motion of $\mathscr{E}^n$ (an isometry of $\mathscr{E}^n$ onto itself) show that $D[f(\mathscr{R}), f(\mathscr{S})] = D(\mathscr{R}, \mathscr{S})$, hence that euclidean motions preserve Hausdorff distance.
7. The following is an outline of a proof for Theorem 11. Give the details that justify the steps (a)–(i).

   Let $\mathscr{F}_h$ be the collection of all compact, convex sets in $\mathscr{E}^2$ with minimal width h. Let $\mathscr{W}$ be the full square

$$\mathscr{W} = \{x : |x_1| \leq h, |x_2| \leq h\}.$$

* See Exercise 7, Sec. 5.2.

Let $\mathscr{F}$ be the family of all equivalence subsets of $\mathscr{F}_h$ which are ordinary subsets of $\mathscr{W}$.

a. If $\mathscr{F}$ has an element of maximum area, then a solution to the maximum inclusion problem exists.
b. If $\mathscr{F}$ is closed in $\mathscr{H}$–$\mathscr{E}^2$, then $\mathscr{F}$ does have an element of maximum area.

From (a) and (b), it follows that Theorem 11 is proved if we can prove that $\mathscr{F}$ is closed in $\mathscr{H}$–$\mathscr{E}^2$. Toward this objective, let $\mathscr{K}_o$ be a limit point to $\mathscr{F}$ in $\mathscr{H}$–$\mathscr{E}^2$.

c. $\mathscr{K}_o \subset \mathscr{W}$.

It remains to be shown that $\mathscr{K}_o$ is an equivalence subset of $\mathscr{F}_h$, and hence, because of (c), that $\mathscr{K}_o \in \mathscr{F}$. Consider $\mathscr{R} \in \mathscr{F}_h$, and define

$$\mathscr{G} = \{\mathscr{S} : \mathscr{K}_o \subset \mathscr{S}, \mathscr{S} \text{ is a compact, convex set}\},$$

$$\mathscr{M}_o(\mathscr{R}) = \{f(\mathscr{R}) : f \text{ is a motion of } \mathscr{E}^2 \text{ and } f(\mathscr{R}) \cap \mathscr{W} \neq \varnothing\}.$$

d. $\mathscr{G}$ is closed in $\mathscr{H}$–$\mathscr{E}^2$.

Now let $\mathscr{M}(\mathscr{R}) = \{f(\mathscr{R}) : f \text{ is a motion of } \mathscr{E}^2\}$, and let $\mathscr{C} = \{\mathscr{S} : \mathscr{S}$ is compact and convex, and $\mathscr{S} \cap \mathscr{W} \neq \varnothing\}$.

e. $\mathscr{M}(\mathscr{R})$ is closed in $\mathscr{H}$–$\mathscr{E}^2$.
f. $\mathscr{C}$ is closed in $\mathscr{H}$–$\mathscr{E}^2$.
g. $\mathscr{M}_o(\mathscr{R}) = \mathscr{M}(\mathscr{R}) \cap \mathscr{C}$ is compact in $\mathscr{H}$–$\mathscr{E}^2$.
h. Assume that $\mathscr{G} \cap \mathscr{M}_o(\mathscr{R}) = \varnothing$. Then, for some $\varepsilon_o > 0$, $\mathscr{A} \in \mathscr{M}_o(\mathscr{R})$ and $\mathscr{B} \in \mathscr{G}$ imply that $D(\mathscr{A}, \mathscr{B}) \geq \varepsilon_o$.

Because $\mathscr{K}_o \in \mathrm{Lp}(\mathscr{F})$, there exists $\mathscr{K} \in \mathscr{F}$ such that $D(\mathscr{K}, \mathscr{K}_o) < \frac{1}{3}\varepsilon_o$. There also exists a motion f such that $f(\mathscr{R}) \supset \mathscr{K}$. In particular, $f(\mathscr{R}) \cap \mathscr{W} \neq \varnothing$, so $f(\mathscr{R}) \in \mathscr{M}_o(\mathscr{R})$.

Now let $\mathscr{S} = f(\mathscr{R}) + B(O, \frac{2}{3}\varepsilon_o)$. Since $f(\mathscr{R}) \supset \mathscr{K}$, then $\mathscr{S} \supset \mathscr{K} + N(O, \frac{2}{3}\varepsilon_o) \supset \mathscr{K}_o$. Thus $\mathscr{S} \in \mathscr{G}$. However, $\mathscr{S} + N(O, \frac{5}{6}\varepsilon_o) \supset \mathscr{S} \supset f(\mathscr{R})$, and $f(\mathscr{R}) + N(O, \frac{5}{6}\varepsilon_o) \supset f(\mathscr{R}) + B(O, \frac{2}{3}\varepsilon_o) = \mathscr{S}$. Consequently, $D(f(\mathscr{R}), \mathscr{S}) < \frac{5}{6}\varepsilon_o < \varepsilon_o$, which is contradictory. Therefore $\mathscr{G} \cap \mathscr{M}_o(\mathscr{R}) \neq \varnothing$.

i. $\mathscr{K}_o \in \mathscr{F}$.

Thus $\mathscr{F}$ is closed in $\mathscr{H}$–$\mathscr{E}^2$, and the proof is complete.

INDEX

–E–

–F–

–G–

–H–

–I–

–J–

–L–

–M–